VOLUME ONE HUNDRED AND THIRTY EIGHT

CURRENT TOPICS IN DEVELOPMENTAL BIOLOGY

Stem Cell Proliferation and Differentiation

CURRENT TOPICS IN DEVELOPMENTAL BIOLOGY

"A meeting-ground for critical review and discussion of developmental processes"
A.A. Moscona and Alberto Monroy (Volume 1, 1966)

VOLUME ONE HUNDRED AND THIRTY EIGHT

CURRENT TOPICS IN DEVELOPMENTAL BIOLOGY

Stem Cell Proliferation and Differentiation

Edited by

THOMAS G. FAZZIO
Department of Molecular, Cell, and Cancer Biology, University of Massachusetts Medical School, Worcester, MA, United States

ELSEVIER

ACADEMIC PRESS
An imprint of Elsevier

Academic Press is an imprint of Elsevier
50 Hampshire Street, 5th Floor, Cambridge, MA 02139, United States
525 B Street, Suite 1650, San Diego, CA 92101, United States
The Boulevard, Langford Lane, Kidlington, Oxford OX5 1GB, United Kingdom
125 London Wall, London, EC2Y 5AS, United Kingdom

First edition 2020

ISBN: 978-0-12-812890-9
ISSN: 0070-2153

For information on all Academic Press publications
visit our website at https://www.elsevier.com/books-and-journals

Publisher: Zoe Kruze
Editorial Project Manager: Shellie Bryant
Production Project Manager: Denny Mansingh
Cover Designer: Greg Harris

Typeset by SPi Global, India

Contents

Contributors *ix*
Preface *xi*

1. Chromatin regulation and dynamics in stem cells **1**
David C. Klein and Sarah J. Hainer

1. Chromatin compaction, structure, and function 3
2. Chromatin dynamics regulate gene expression 8
3. ATP-dependent nucleosome remodeling complexes establish and maintain chromatin state 9
4. Histone modifications provide an additional layer of gene regulation 18
5. Histone chaperones and histone variants regulate chromatin structure 25
6. Histone H2A variants 29
7. Histone H3 variants 31
8. Histone chaperones 32
9. Chromatin structure is dynamic and highly regulated 36
10. Stem cell chromatin is dynamic and tuned to regulate cell fate 36
11. ES cells carefully regulate their chromatin via specialized transcription factors 37
12. Embryonic stem cell chromatin is poised for action 39
13. Histone modifications are specifically regulated in stem cells to maintain pluripotency and facilitate differentiation 39
14. Chromatin state is precisely regulated by nucleosome remodeling factors in ES cells 42
15. Long-range chromatin interactions are critical for regulation of pluripotency 51
16. ES cells regulate chromatin by common processes to preserve pluripotency 54

Acknowledgments 55
References 55

2. Role of lncRNAs in stem cell maintenance and differentiation **73**
Meghali Aich and Debojyoti Chakraborty

1. Introduction 74
2. Origin of noncoding RNAs 75
3. Core regulatory circuit in ESCs 76
4. LncRNAs: New determinants of ES cell fate 78

5. Long noncoding RNAs (lncRNAs) and their biological function 78
6. Discovery of lncRNAs: From sequences to function 80
7. Long noncoding RNAs and epigenetic regulation 80
8. Dissecting functional lncRNAs from transcriptional noise 81
9. LncRNAs in ESC pluripotency and somatic cell reprogramming 81
10. LncRNAs play a role in the differentiation of pluripotent stem cells 87
11. LncRNAs regulating the epigenome 88
12. The role of lncRNAs in dosage composition 89
13. LncRNAs implicated in imprinting developmentally associated genes 90
14. LncRNAs regulating signaling pathways in ESCs 91
15. LncRNAs regulating organ development 92
16. LncRNAs affecting neural development 93
17. LncRNAs regulating organogenesis 94
18. Cellular localization and maturation of lncRNAs 95
19. LncRNAs regulating the stability and functions of other RNAs 95
20. LncRNAs functioning in protein modification pathways 96
21. Mechanisms of lncRNA:DNA/RNA interaction 96
22. Allosteric regulation of proteins by lncRNAs 98
23. Single cell analysis of lncRNA functions 99
24. LncRNAs in disease progression 100
25. LncRNA knockouts often show lack of phenotype: The importance of context and redundancy 101
26. Conclusions 102
References 102
Further reading 111

3. Regulation of pluripotency and reprogramming by RNA binding proteins 113

Dan Li, Mohamed S. Kishta, and Jianlong Wang

1. Pluripotency and reprogramming 114
2. RNA binding proteins 115
3. RNA helicases and DEAD-box helicase family 124
4. Conclusions 130
Acknowledgments 130
References 130

4. Generating primed pluripotent epiblast stem cells: A methodology chapter 139

Milan Samanta and Sundeep Kalantry

1. Introduction 140
2. Materials 141

3. Methods 149
4. Discussion 166
5. Recipes 166
6. Notes 171
Acknowledgments 173
References 173

5. Differentiation of human pluripotent stem cells toward pharyngeal endoderm derivatives: Current status and potential 175

Margaret E. Magaletta, Richard Siller, and René Maehr

1. Introduction 176
2. Overview of the pharyngeal apparatus formation within the gut tube 178
3. Pharyngeal endoderm development and lineage specification within the pharyngeal pouches 180
4. Pharynx derivative pluripotent stem cell differentiation protocols: Current status 182
5. Applications of hPSCs for studying pharyngeal endoderm development and disease 195
6. Future directions for hPSC differentiation approaches toward pharyngeal derivatives 196
7. Concluding remarks 199

Acknowledgments 199
References 199

6. Epigenetic metabolites license stem cell states 209

Logeshwaran Somasundaram, Shiri Levy, Abdiasis M. Hussein, Devon D. Ehnes, Julie Mathieu, and Hannele Ruohola-Baker

1. Introduction 210
2. Stem cell energetics 210
3. Metabolism of quiescent stem cells 212
4. Metabolism of active stem cells 217
5. HIF, the master regulator of metabolism 222
6. Epigenetic signatures and epigenetic metabolites 224
7. Conclusion 229

Acknowledgments 230
References 230
Further reading 240

Contributors

Meghali Aich
CSIR-Institute of Genomics & Integrative Biology; Academy of Scientific & Innovative Research, New Delhi, India

Debojyoti Chakraborty
CSIR-Institute of Genomics & Integrative Biology; Academy of Scientific & Innovative Research, New Delhi, India

Devon D. Ehnes
Department of Biochemistry; Institute for Stem Cell and Regenerative Medicine, University of Washington, Seattle, WA, United States

Sarah J. Hainer
Department of Biological Sciences, University of Pittsburgh, Pittsburgh, PA, United States

Abdiasis M. Hussein
Department of Biochemistry; Institute for Stem Cell and Regenerative Medicine, University of Washington, Seattle, WA, United States

Sundeep Kalantry
Department of Human Genetics, University of Michigan Medical School, Ann Arbor, MI, United States

Mohamed S. Kishta
Hormones Department, Medical Research Division; Stem Cell Lab., Center of Excellence for Advanced Sciences, National Research Centre, Cairo, Egypt; Department of Medicine, Columbia Center for Human Development, Columbia University Irving Medical Center, New York, NY, United States

David C. Klein
Department of Biological Sciences, University of Pittsburgh, Pittsburgh, PA, United States

Shiri Levy
Department of Biochemistry; Institute for Stem Cell and Regenerative Medicine, University of Washington, Seattle, WA, United States

Dan Li
Department of Cell, Developmental and Regenerative Biology; The Black Family Stem Cell Institute; The Graduate School of Biomedical Sciences, Icahn School of Medicine at Mount Sinai, New York, NY, United States

René Maehr
Program in Molecular Medicine, Diabetes Center of Excellence, University of Massachusetts Medical School, Worcester, MA, United States

Margaret E. Magaletta
Program in Molecular Medicine, Diabetes Center of Excellence, University of Massachusetts Medical School, Worcester, MA, United States

Julie Mathieu
Institute for Stem Cell and Regenerative Medicine; Department of Comparative Medicine, University of Washington, Seattle, WA, United States

Hannele Ruohola-Baker
Department of Biochemistry; Institute for Stem Cell and Regenerative Medicine, University of Washington, Seattle, WA, United States

Milan Samanta
Department of Human Genetics, University of Michigan Medical School, Ann Arbor, MI, United States

Richard Siller
Program in Molecular Medicine, Diabetes Center of Excellence, University of Massachusetts Medical School, Worcester, MA, United States

Logeshwaran Somasundaram
Department of Biochemistry; Institute for Stem Cell and Regenerative Medicine, University of Washington, Seattle, WA, United States

Jianlong Wang
Department of Cell, Developmental and Regenerative Biology; The Black Family Stem Cell Institute; The Graduate School of Biomedical Sciences, Icahn School of Medicine at Mount Sinai; Department of Medicine, Columbia Center for Human Development, Columbia University Irving Medical Center, New York, NY, United States

Preface

Different types of stem cells—broadly defined as cells with the capacity to maintain their own undifferentiated population as well as to produce one or more mature cell types—play critical roles in embryos, adult tissues, and some types of cancer. Consequently, the term "stem cell biology" might best be considered a thread connecting researchers across a wide spectrum of interests rather than a single, cohesive field of study. Indeed, the functions and regulation of stem cells constitute a key focus of both basic and applied biological sciences, including the fields of developmental biology, regenerative medicine, cancer biology, and many other areas.

Stem cells exhibit a range of properties that differ among the many types of stem cells, including their metabolic profiles, physical characteristics, gene expression profiles, and proliferation rates. In spite of these differences, many types of stem cells share common markers, epigenetic features, and phenotypic properties. Two major questions are how developmental potency is maintained in stem cells and how cells remodel their identities to transform into mature cell types. Considerable progress has been made on each of these topics in recent years, as we have come to understand that numerous signaling pathways and metabolic stimuli impinge on an ever-expanding array of transcriptional and posttranscriptional regulatory mechanisms ranging from sequence-specific transcription factors to epigenetic mechanisms to noncoding RNAs. These regulators collectively control cell type-specific gene expression at the transcriptional and posttranscriptional levels.

The emergence of new single-cell technologies for identifying regulatory elements, mapping transcription factor binding, and quantifying gene expression on a genome-wide scale has enabled a surge of progress toward understanding of stem cell properties and differentiation pathways. As many of these technologies are still quite young, we are likely seeing only the first wave of breakthroughs in this domain. In addition, sensitive new methods for quantifying changes in protein abundance or identity as cells differentiate, as well as metabolite profiles and flux add additional layers to our understanding of stem cells and their progeny. A major challenge for future studies will be to better decipher these dense multifactorial datasets to understand when and how cell fate decisions are made, when they become irreversible, and how these processes can be manipulated for therapeutic purposes.

This volume of *Current Topics in Developmental Biology* entitled "Stem Cell Proliferation and Differentiation" discusses numerous advances in the field of stem cells, as well as some of the current challenges in stem cell biology and areas of focus for future research. Owing to the enormous breadth of the stem cell field, not all topics in this area could reasonably be covered in one volume. Instead, we focus on a few specific areas that have seen exciting new developments or considerable progress in recent years.

1. Chromatin regulation and dynamics

Work in multiple model systems has identified conserved mechanisms of epigenetic regulation and the protein machines responsible for remodeling the epigenome. More recently, an explosion of studies has examined how chromatin structure controls stem cell state and how chromatin is remodeled during differentiation to lock in gene expression programs defining mature cells. In Chapter 1, Klein and Hainer discuss the unique features of chromatin structure in some types of stem cells, the numerous forms of chromatin regulation, and how chromatin structure is remodeled to control cell state.

Notably, while the roles of covalent histone modifications in regulation of stem cell state have been studied extensively, the general mechanisms by which histone modifications increase or decrease target gene expression and shape the gene regulatory network of pluripotent stem cells have only recently come into focus. One such mechanism involves the recruitment of proteins that regulate the placement or positions of histones on DNA. Of particular note, nucleosome remodeling complexes containing members of the SWI/SNF family of ATPases have been shown to exhibit diverse roles in regulation of stem cell identity. In addition, key roles for histone variants such as histones H3.3 and H2AZ in pluripotency have been identified in recent years.

With the discoveries of critical roles of chromatin remodeling complexes and histone variants in pluripotency, along with recent developments in protocols for directed differentiation of pluripotent stem cells, we are now in position to understand where and when in the differentiation process chromatin remodeling factors function and how they help direct specification of different cell types. Although considerable focus has been directed at uncovering regulatory nodes impacting developmental regulatory transcription factors, recent studies have revealed that most chromatin regulatory complexes bind and regulate the expression of hundreds to thousands of genes in different cell types, suggesting that a focus on individual targets of any one regulatory complex may be of limited utility in understanding how these factors function during development. In addition, recent studies have revealed important noncatalytic roles for numerous chromatin regulatory factors in pluripotency and development. Therefore, a renewed focus on the mechanisms underlying the gene regulatory functions of chromatin regulatory complexes will likely lead to fundamentally new insights.

2. Long noncoding RNAs and RNA-binding proteins

Important noncoding functions of RNAs in cell type specification have been discovered in recent years. Indeed, critical developmental roles for multiple classes of noncoding RNAs, both small (e.g., microRNAs, piRNAs, tRNA fragments) and large (e.g., lncRNAs, enhancer RNAs) have been described in multiple systems. In Chapters 2 and 3, we explore the functions of lncRNAs and RNA-binding proteins (RBPs) in regulation of the cell state in stem and progenitor cells.

Although individual lncRNAs were identified decades ago, the extent to which lncRNAs are expressed, their differential expression patterns in stem cells and differentiation cell types, and their molecular roles in gene expression and development are only now beginning to be understood. Numerous factors have obscured the general and cell type-specific functions of lncRNAs. First, the discovery that transcription is pervasive throughout the genome—which is 98% noncoding in humans—has revealed that mRNAs corresponding to coding genes constitute only a small fraction of the number of unique transcripts within any individual cell, although coding transcripts are often much more abundant than most lncRNAs. Many of these pervasive transcripts are not stable and are therefore present at very low levels, raising the question of what, if any, function they serve in the cell. However, many characterized lncRNAs are expressed from individual promoters, have gene structures that are similar to coding genes (other than the lack of significant coding potential), have relatively high stability, and (as with mRNAs) can be many kilobases in length. Although sequence conservation of lncRNAs is typically very low, their locations vis-à-vis coding genes and predicted secondary structures are often more highly conserved. A second barrier to our understanding of lncRNA functions is the lack of straightforward outputs (such as proteins) for which we have effective tools to assay their functions.

Despite these obstacles, in Chapter 2, Aich and Chakraborty describe substantial progress in our understanding of where and how lncRNAs contribute to developmental potency and cell fate. The expression of a number of lncRNAs expressed in pluripotent stem cells is regulated by pluripotency transcription factors OCT4, SOX2, and NANOG, and several lncRNAs have been shown to perform functions critical for pluripotency and/or cell viability. In addition, some lncRNAs have been shown to help regulate cell fate upon differentiation of pluripotent stem cells. Besides regulation of pluripotency, numerous lncRNAs have been shown to be critical for later stages of embryonic development. For example, multiple studies have identified lncRNAs with key roles in organogenesis, including both heart and brain development.

How do lncRNAs function in regulation of gene expression and cell fate? Several molecular mechanisms have been described by which lncRNAs contribute to gene regulation. In some cases, lncRNAs appear to impact chromatin structure near promoter regions of target genes. Consistent with these findings, a number of chromatin regulatory factors have been shown to interact broadly with coding and noncoding RNAs. However, the degree to which chromatin remodeling complexes are directed to specific loci by individual lncRNAs remains unclear and is an active area of current research. A second mechanism by which lncRNAs control gene expression is through interactions with other regulatory RNAs, such as miRNAs, or in some cases by regulating the enzymatic activities of proteins. Finally, some lncRNAs appear to alter chromatin structure directly through interactions with DNA, either in the form of R-loops (RNA/DNA hybrids) or through triplex structures.

A rapidly progressing area of stem cell and developmental biology centers on the identification and characterization of RBPs that play critical roles in regulation of cell fate. Along with the transcriptional regulatory functions of noncoding RNAs, including lncRNAs discussed above, Li, Kishta, and Wang describe in Chapter 3 an ever-expanding list of posttranscriptional mechanisms controlling the functions of developmental regulators, as well as an expanding network of RBPs that mediate each type of regulation. The development of efficient technologies for crosslinking proteins to RNA and proteomics methods for identification of RNA-binding proteins has uncovered numerous RNA-binding proteins in pluripotent stem cells. Several recently identified RBPs have been shown to play key roles in maintenance of the pluripotent state. While some RBPs control the identities of proteins expressed from RNAs to which they bind (e.g., through alternative splicing), many others control protein levels through numerous posttranscriptional mechanisms, including control of RNA export, stability, and translation.

A major class of regulators of RNA binding and function are DEAD box helicases of the DDX family. DDX helicases play roles in transcription, RNA maturation, ribosome biogenesis, and translation, among other processes. Given their functions in essential cellular processes, it is perhaps not surprising that DDX family proteins play important roles in stem and progenitor cells. However, the distinct effects of mutation or depletion of various DDX proteins on reprogramming to pluripotency, differentiation, and cell fate may be unexpected. Additional studies are needed to understand how these factors contribute to the regulation of cellular identity.

Finally, an exciting new development in this field is the discovery of widespread covalent RNA modifications, and the finding that these modifications play important roles in stem and progenitor cells. For example, methylation at position six within the purine ring of adenine bases on RNA (m6A) has been shown to be critical for transition from the naïve to the primed state of pluripotent stem cells. While many of the targets of RNA modifications and the proteins performing these modifications are now beginning to be characterized, their precise roles in regulation of cellular state are not completely understood.

The onslaught of studies implicating lncRNAs and other noncoding RNAs in regulation of developmental potency has precipitated a recent explosion in the identification and characterization of RBPs in these processes. Additional efforts will be needed to understand the temporal and tissue specificity of factors regulating lncRNA function, as well as how different lncRNAs mediate specific effects within stem cell gene regulatory networks. In addition, a comprehensive examination of how RNA-binding proteins control the activities of developmental regulators, and how these activities are altered during cell state changes (such as stem cell differentiation) will illuminate this aspect of developmental gene regulation.

3. Distinct modes of pluripotency

A large fraction of pluripotent stem cell research focuses on pluripotent cells most similar to early blastocyst stage embryos, specifically embryonic stem cells and induced pluripotent stem cells. During the early stages of formation of the embryonic body plan, one of the first developmental stages is formation of the epiblast, wherein the pluripotent cells of the inner cell mass of the blastocyst transition to a distinct form of pluripotent cells in the epiblast stage embryo. This transition has been described as progression from a "naïve" state to a "primed" state. In this process, naïve cells—a collection of spherical cells that grow in three dimensions *in vitro* and *in vivo*, which express the full repertoire of pluripotency transcription factors—morph into cells that grow as a monolayer that expresses a subset of pluripotency transcription factors, as well as a number of genes not expressed in naïve cells. In addition, these primed pluripotent cells exhibit a number of features more commonly found in mature cells.

Although many studies with stem cells involve directed differentiation from naïve pluripotent stem cells to specific differentiation pathways, *in vivo* pluripotent cells transition through the primed state, which can be modeled in culture in what are known as epiblast stem cells (EpiSCs). Studies with EpiSCs have revealed features observed in cells during gastrulation, where the primary germ layers begin to differentiate *in vivo*, and have informed numerous directed differentiation protocols that utilize cultured pluripotent stem cells. To facilitate the use of these cells as models for differentiation studies, in Chapter 4, Samanta and Kalantry describe the isolation and culture of these cells from embryos, along with their characterization in a special methods chapter.

4. Differentiation of pluripotent stem cells

One area of study that has seen enormous growth in the past several years is the development of new methods and protocols to more precisely differentiate pluripotent stem cells into mature cells or lineage-restricted progenitors that can contribute to functional tissues upon transplant into animal models. Robust methods for differentiation of pluripotent stem cells into distinct neuronal lineages, cardiac muscle, pancreatic beta cells, and many other cell types have been described. In Chapter 5, Magaletta, Siller, and Maehr take a deep dive into pluripotent stem cell-derived pharyngeal endoderm—a field that has seen significant progress in recent years in both human and mouse pluripotent stem cell models.

Pharyngeal endoderm makes up critical components of the thymus, parathyroid, and thyroid, and defects in pharyngeal endoderm development contribute to a number of congenital diseases. Using information accumulated about key developmental regulators and signaling pathways necessary for pharyngeal endoderm development *in vivo*, several groups have developed methods for generation of *in vitro* cell types analogous to those that make up thyroid, parathyroid, and thymus. More detailed characterization of these cells using new genomic and single-cell methods will lead to a more complete understanding of how these tissue types develop. In turn, a better understanding of pharyngeal endoderm differentiation should lead to refinement of differentiation methods to increase their fidelity and robustness, facilitating the use of pluripotent stem cells in therapies.

5. Metabolic control of cell state

Although most efforts in developmental and stem cell biology have focused primarily on the signaling pathways, transcriptional regulators, and (more recently) RNA-based mechanisms that impinge on cell type-specific gene regulatory networks, major roles for metabolites in cell type specification have emerged. Early efforts to understand how different cells differentially employ metabolic pathways used for energy and biomolecule production have focused mainly on tumors. However, as discussed by Ruohola-Baker and colleagues in Chapter 6, it has recently become evident that alterations in metabolism accompany changes in stem cell state, and in some cases help direct cell fate changes.

Several mechanisms govern how metabolic inputs impact cell fate in different types of stem cells. For example, hematopoietic stem cells, which are largely quiescent, utilize anaerobic glycolysis during periods of self-renewal, which is critical for maintenance of their undifferentiated state. In contrast, the transition of rapidly dividing naïve pluripotent stem cells to primed cells is met with a transition from a combined metabolic program of glycolysis and oxidative phosphorylation to one in which glycolysis is used exclusively.

Interestingly, changes in metabolism when cells transition from stem cells to differentiation often impact the epigenome. This is perhaps not surprising when one considers that the cofactors utilized in deposition of multiple epigenetic marks, such as histone methylation, histone acetylation, and DNA methylation, are key metabolites that often reside at the interface of multiple metabolic pathways. Consequently, changes in the concentrations of these metabolites are sometimes met with alterations in the epigenome, leading to changes in gene expression. Undoubtedly, new examples of metabolic regulation of cell fate will be uncovered in coming years as this field continues to progress.

6. Conclusions and perspectives

In recent years, the stem cell community has made tremendous advances in mapping the gene regulatory networks and signaling pathways regulating cell fate, both in pluripotent stem cells and tissue-restricted stem cells. The ability to isolate and characterize stem cell populations *in vitro* and direct their differentiation in cell culture has facilitated these advances. Consequently, one of the ongoing questions is how closely *in vitro* stem cell models reflect the processes of cell type specification that occur *in vivo*. Parallel advances in genomics methodologies have enabled characterization of populations of cells *in vivo* and *in vitro* at the single-cell level, facilitating a more detailed understanding of the similarities and differences among cell populations derived from *in vitro* differentiation of pluripotent stem cells and developing tissues.

Given these increasingly powerful tools for identifying and characterizing the epigenome and gene expression profiles of cells, one major challenge going forward will be to understand how differences between pluripotent stem cell-derived cells and *in vivo* cell populations arise and how they can be eliminated. Two major barriers to the therapeutic use of cells derived from pluripotent stem cell differentiation are incomplete differentiation and differentiated cells that incompletely mimic mature cells *in vivo*. Whether the increasingly detailed maps of the epigenomes and transcriptomes of *in vitro* differentiated cells will lead to better stem cell therapies remains an open question. These concerns aside, stem cell models have uncovered many important new insights into cell type specification that have informed our understanding of mammalian development. There is every indication that stem cell models will continue to represent a powerful conduit for discoveries that enrich our understanding of embryogenesis, as well as models for numerous congenital diseases.

Thomas G. Fazzio
Department of Molecular, Cell, and Cancer Biology, University of Massachusetts Medical School, Worcester, MA, United States

CHAPTER ONE

Chromatin regulation and dynamics in stem cells

David C. Klein, Sarah J. Hainer*
Department of Biological Sciences, University of Pittsburgh, Pittsburgh, PA, United States
*Corresponding author: e-mail address: sarah.hainer@pitt.edu

Contents

1. Chromatin compaction, structure, and function 3
   1.1 Nucleosomes are formed from DNA interacting with an octamer of four histone proteins 3
   1.2 Chromatin structure balances DNA compaction and accessibility 6
2. Chromatin dynamics regulate gene expression 8
3. ATP-dependent nucleosome remodeling complexes establish and maintain chromatin state 9
   3.1 SWI/SNF family nucleosome remodeling factors 11
   3.2 INO80 family nucleosome remodeling factors 13
   3.3 ISWI family nucleosome remodeling factors 15
   3.4 CHD family nucleosome remodeling factors 16
   3.5 Summary 17
4. Histone modifications provide an additional layer of gene regulation 18
   4.1 Histone acetylation and deacetylation 19
   4.2 Histone methylation and demethylation 20
   4.3 Polycomb group proteins mediate H3K27me3 and silencing of developmental genes 24
5. Histone chaperones and histone variants regulate chromatin structure 25
6. Histone H2A variants 29
   6.1 H2A.Z is a histone variant associated with active transcription 29
   6.2 H2A.X is a marker of DNA double-strand breaks 30
   6.3 macroH2A is associated with repression and heterochromatin 30
7. Histone H3 variants 31
   7.1 H3.3 marks regulatory, repetitive, and actively transcribed regions 31
   7.2 CENP-A is a centromere-specific variant of histone H3 32
8. Histone chaperones 32
   8.1 FACT facilitates H2A/H2B dimer exchange to promote nucleosome-templated activities 32
   8.2 CAF1 and ASF1 promote incorporation of H3 and H4 onto newly synthesized DNA 33
   8.3 HIRA deposits H3.3 at actively transcribed and regulatory regions of chromatin 34

Current Topics in Developmental Biology, Volume 138
ISSN 0070-2153
https://doi.org/10.1016/bs.ctdb.2019.11.002

8.4 ATRX/DAXX deposit H3.3 at telomeres and pericentric heterochromatin 35
8.5 The NAP1 family import newly translated histones from the cytoplasm to the nucleus 35
8.6 INO80 family members possess both nucleosome remodeling and histone chaperone activities 35
9. Chromatin structure is dynamic and highly regulated 36
10. Stem cell chromatin is dynamic and tuned to regulate cell fate 36
11. ES cells carefully regulate their chromatin via specialized transcription factors 37
11.1 Master regulators of pluripotency 37
11.2 Pioneer transcription factors 38
12. Embryonic stem cell chromatin is poised for action 39
13. Histone modifications are specifically regulated in stem cells to maintain pluripotency and facilitate differentiation 39
13.1 H3K56ac regulates pluripotency factors and developmental regulators 39
13.2 Bivalent promoters mark lowly expressed but poised genes in ES cells 40
14. Chromatin state is precisely regulated by nucleosome remodeling factors in ES cells 42
14.1 esBAF maintains stem cell pluripotency by preserving chromatin state 42
14.2 CHD proteins are regulators of ES cell pluripotency 43
14.3 MBD3/NuRD generally represses expression of differentiation genes 44
14.4 The ISWI remodeler ATPase SNF2H is essential during development 45
14.5 INO80 remodelers repress transcription of differentiation-associated genes 46
14.6 Polycomb group proteins silence developmental genes in ES cells 48
14.7 Variants of H2A and H3 have specialized roles in pluripotent cells 50
15. Long-range chromatin interactions are critical for regulation of pluripotency 51
16. ES cells regulate chromatin by common processes to preserve pluripotency 54
Acknowledgments 55
References 55

Abstract

In eukaryotes, DNA is highly compacted within the nucleus into a structure known as chromatin. Modulation of chromatin structure allows for precise regulation of gene expression, and thereby controls cell fate decisions. Specific chromatin organization is established and preserved by numerous factors to generate desired cellular outcomes.

In embryonic stem (ES) cells, chromatin is precisely regulated to preserve their two defining characteristics: self-renewal and pluripotent state. This action is accomplished by a litany of nucleosome remodelers, histone variants, epigenetic marks, and other chromatin regulatory factors. These highly dynamic regulatory factors come together to precisely define a chromatin state that is conducive to ES cell maintenance and development, where dysregulation threatens the survival and fitness of the developing organism.

1. Chromatin compaction, structure, and function

To retain the genomic information required to code for a eukaryote, each cell must compact 1.7 m of DNA into a nucleus of between 5 and 20 μm (Lammerding, 2011). To accomplish this compaction, cells package genomic DNA with histone proteins into a structure known as chromatin, the basic repeating unit of which is the nucleosome. Histone proteins are extraordinarily conserved throughout eukaryotes at both the structure and sequence levels, with the shared histone fold domain being especially conserved (Postberg, Forcob, Chang, & Lipps, 2010; Sullivan & Landsman, 2003). In general, the stable association of DNA and histone proteins poses a significant obstacle to cellular processes that rely on protein-DNA interaction, including transcription, DNA replication, and DNA repair (reviewed in Bai & Morozov, 2010; Duina, 2011; Li, Carey, & Workman, 2007; Luger, 2006).

1.1 Nucleosomes are formed from DNA interacting with an octamer of four histone proteins

The nucleosome is composed of ~147 base pairs of double-stranded DNA wrapped around an octamer of histone proteins (Kornberg & Lorch, 1999). The histone octamer is comprised of an inner core tetramer of histones H3 and H4 and two outer histone H2A/H2B dimers (Luger, Mäder, Richmond, Sargent, & Richmond, 1997) (Fig. 1). Histones are small, positively charged proteins and are composed of a folded domain (termed the histone fold) that forms the nucleosome globular core and highly unstructured N- and C-terminal tails extending from the core. The positive charge of histone proteins allows for electrostatic interactions between the histones and the negatively charged DNA. Specifically, the minor groove of DNA interacts with 14 arginine residues on the histone proteins through non-covalent hydrogen bonds. Additionally, salt bridges allow DNA and histone proteins to maintain loose association that allows nucleosome remodeling factors to reposition nucleosomes. Individual histone protein tails are often post-translationally modified (Kouzarides, 2007). These modifications have been shown to be extremely important in a number of biological processes and will be discussed; however, for reviews of histone modifications see Bannister and Kouzarides (2011), Harikumar and Meshorer (2015), Kouzarides (2007). Most of these covalent modifications are found on the

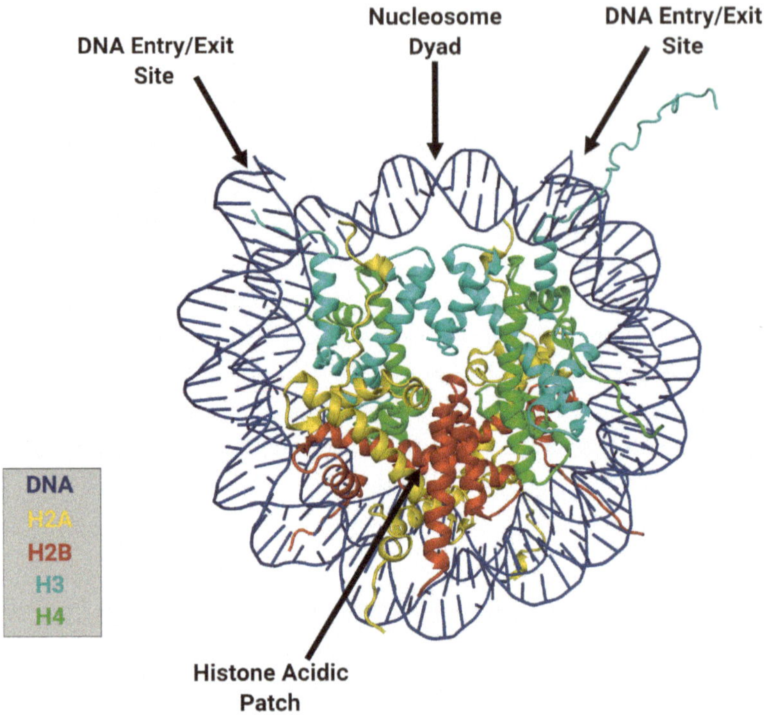

Fig. 1 Model of the crystal structure of the nucleosome. The nucleosome is comprised of 147 base pairs of double-stranded DNA wrapped around an octamer of histone proteins, including two copies each of H2A, H2B, H3, and H4. Nucleosomes contain numerous structural features, including the nucleosome dyad, the DNA entry/exit sites, the histone acidic patch, and the histone fold domain. PDB ID: 1AO1, Luger et al., 1997. Created with Biorender.com.

N-terminal tails of histone proteins, though there are notable exceptions, such as H2BK120, a residue that is monoubiquitinated in histone crosstalk pathways (Tomson & Arndt, 2013).

Each nucleosome contains numerous structural features that can affect DNA packaging—and therefore DNA-templated activities (Fig. 2). These features include the histone fold domain, the nucleosome dyad, the DNA entry and exit sites, and the histone acidic patch (Arents & Moudrianakis, 1995). The histone fold domain is a C-terminal structural motif that is found across all core histones and is important in DNA compaction and histone dimerization (Arents & Moudrianakis, 1995; Hammond, Stromme, Huang, Patel, & Groth, 2017). The nucleosome dyad covers the H3/H3 interaction interface and is largely referred to as the center of the nucleosome (given the position "0" in the DNA superhelix). The dyad is an important

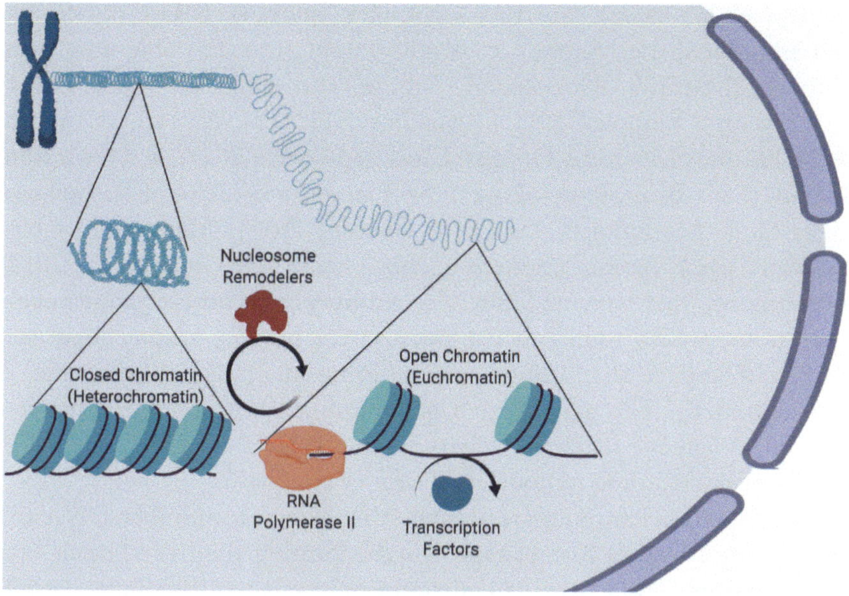

Fig. 2 Chromatin packaging regulates DNA-templated processes. Nucleosome remodeling factors facilitate transitions between accessible chromatin (euchromatin) and closed chromatin (heterochromatin). Open chromatin is permissive of transcription factor binding and recruitment of other factors, in turn promoting DNA-templated processes such as transcription. Created with Biorender.com.

site for interaction of proteins, such as FOXA/HNF3 (Cirillo et al., 1998), AMT1 (White & Luger, 2004), SOX2 (Zhu et al., 2018), and TBX2 (Zhu et al., 2018), as well as the linker histone H1 (Zhou et al., 2013). At the DNA entry/exit sites, the histone-interacting DNA participates in a process known as "nucleosome breathing" in which the DNA base pairs nearest these sites transiently release from and rewrap around the histone proteins (Anderson & Widom, 2000; Osberg, Nuebler, Korber, & Gerland, 2014); this activity can alter contacts with specific regulatory factors, such as the elongation factor and histone chaperone FACT (Hondele & Ladurner, 2013). The histone acidic patch is a region of the nucleosome composed of eight H2A amino acid residues and two H2B residues which together create a highly negatively charged surface on the nucleosome, in contrast to the rest of the nucleosome surface, which is positively charged to promote interaction with DNA (Luger et al., 1997). The acidic patch is a hotspot for interacting factors, including the SAGA coactivator (Morgan et al., 2016), the DOT1L H3K79 histone methyltransferase (Anderson et al., 2019), the Kaposi's sarcoma herpesvirus LANA

(Barbera et al., 2006), the PAF1 complex member RTF1 (Cucinotta, Hildreth, McShane, Shirra, & Arndt, 2019), the E3 ubiquitin ligase RNF168 (Mattiroli, Uckelmann, Sahtoe, van Dijk, & Sixma, 2014), the Polycomb complex PRC1 (McGinty, Henrici, & Tan, 2014), the FACT histone chaperone (Hodges, Gloss, & Wyrick, 2017), and the histone H4 tail (Fan, Rangasamy, Luger, & Tremethick, 2004; Kalashnikova, Porter-Goff, Muthurajan, Luger, & Hansen, 2013). Many factors bind the acidic patch by an "arginine anchor" motif (McGinty et al., 2014). Furthermore, this surface limits the number of possible simultaneous interactions, which results in competition for binding among regulatory proteins (Dann et al., 2017; Gamarra, Johnson, Trnka, Burlingame, & Narlikar, 2018). The acidic patch is additionally intriguing in that it is required to achieve maximal activity of the SWI/SNF, ISWI, and CHD families of nucleosome remodelers (Dann et al., 2017).

Individual nucleosomes are separated by a region of linker DNA that ranges from 20 to 90 base pairs in length, forming the basic "beads on a string" model often used to describe chromatin (Bradbury, 1989). Importantly, nucleosomes do not exist in a vacuum, and are therefore subject to the actions of neighboring nucleosomes, transcribing polymerases, nucleosome remodelers, and a host of other factors that can disrupt this periodic spacing. The sliding action of SWI/SNF family remodelers, for example, is thought to bring individual nucleosomes crashing together, forcibly evicting an H2A/H2B dimer and creating a structure known as an overlapping dinucleosome (Engeholm et al., 2009; Kato et al., 2017). Nucleosomes can also be partially dismantled, involving the loss of an H2A/H2B dimer, leaving a hexasome of histone proteins that is known as a subnucleosome (Rhee, Bataille, Zhang, & Pugh, 2014). A mammalian SWI/SNF remodeler, esBAF, regulates subnucleosome formation in murine embryonic stem (ES) cells (Hainer & Fazzio, 2015). These non-canonical nucleosome structures are depicted in Fig. 3.

1.2 Chromatin structure balances DNA compaction and accessibility

Broadly, chromatin structure can be grouped into two types: heterochromatin, which is compact and transcriptionally silent, and euchromatin, which is more accessible and therefore more permissive of DNA-templated activities. Heterochromatin can be further subdivided into constitutive heterochromatin, which remains compacted throughout the cell cycle, and facultative heterochromatin, which is preferentially, but not exclusively,

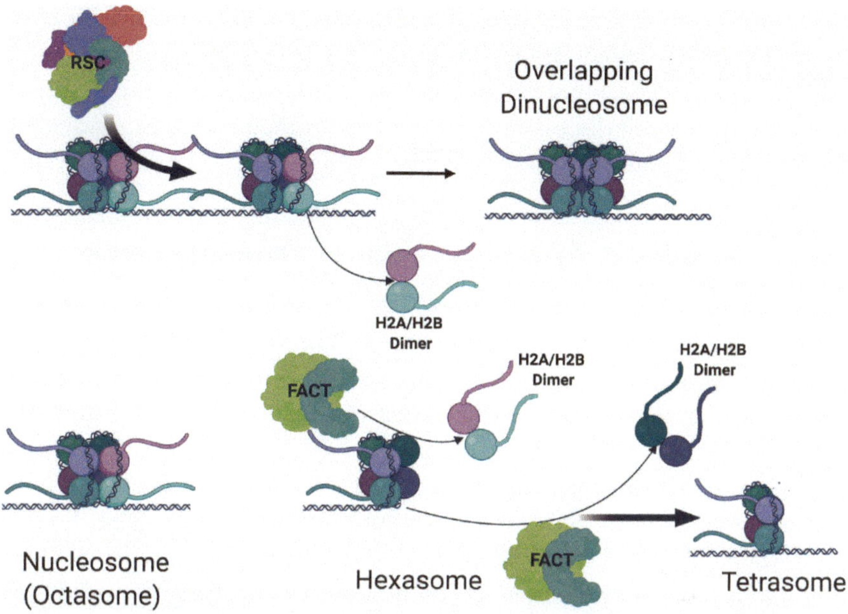

Fig. 3 Non-canonical nucleosome structures are generated through the action of chromatin modifying proteins. Overlapping dinucleosomes are 14-mers of histone proteins generated when the sliding action of a nucleosome remodeler (e.g., yeast RSC) forces a nucleosome to crash into the neighboring nucleosome, thereby displacing an H2A/H2B dimer. Subnucleosomes are generated when H2A/H2B dimers are removed from the octasome (canonical nucleosome), a process often carried out by H2A/H2B chaperones (e.g., the FACT complex). This dimer removal can occur once, leaving behind a hexasome composed of the H3/H4 tetramer and one H2A/H2B dimer, or twice, leaving behind the H3/H4 tetramer alone. Created with Biorender.com.

heterochromatic. Constitutive heterochromatin is enriched for di- and tri-methylation of H3K9, and often marks repeat-heavy regions such as satellite DNA at centromeres (Lehnertz et al., 2003; Nishibuchi & Dejardin, 2017). Facultative heterochromatin, on the other hand, is marked by HP1-dependent H3K27 trimethylation (H3K27me3) (Jamieson et al., 2016) and is closely associated with inactive gene promoters, such as developmental genes that are repressed in ES cells but may require activation at a later time. H3K27me3 is placed largely by EZH2, a member of the PRC2 complex (Kuzmichev, Nishioka, Erdjument-Bromage, Tempst, & Reinberg, 2002), which will be discussed in greater detail later.

In contrast to heterochromatin, euchromatin is a more accessible structure (Noma, Allis, & Grewal, 2001). Euchromatic regions are the main sites of transcription by RNA Polymerase II, regulatory factor binding, and nucleosome remodeling activity. Importantly, euchromatic DNA is still compacted into an array of nucleosomes, and other factors are required to maintain gene accessibility in highly context-dependent manners. In the next section, we will discuss ways in which chromatin structure is dynamic and can be modulated to tightly regulate transcription and gene expression.

2. Chromatin dynamics regulate gene expression

Chromatin structure impacts all DNA-templated activities. The regulation of gene expression is especially important to cell identity and cell fate decisions. Transcription to RNA from a DNA template is carried out by RNA polymerases in three stages: initiation, elongation, and termination (reviewed in Liu, Bushnell, & Kornberg, 2013). In general, most gene promoters are nucleosome-depleted regions (NDRs), which permit binding of transcription factors and successful transcription initiation (Lee, Shibata, Rao, Strahl, & Lieb, 2004; Lee et al., 2007; Varshavsky, Sundin, & Bohn, 1979). Transcription initiation can be hindered, however, when promoter DNA is wrapped into a nucleosome, making the DNA sequence more difficult for DNA-binding factors to recognize (Kornberg & Lorch, 1999). Transcription elongation can also be physically hindered by nucleosome occupancy in that transcription rates of RNA Polymerase II are slowed due to increased pausing and backtracking (Churchman & Weissman, 2011; Izban & Luse, 1991; Lee, Teyssier, Strahl, & Stallcup, 2005). Transcription termination is coordinated by numerous factors affecting polymerase function and chromatin state, including the CHD1, ISW1, and ISW2 nucleosome remodelers (Alén et al., 2002; Morillon et al., 2003; Simic et al., 2003; Xu et al., 1986). Therefore, mechanisms by which eukaryotes control chromatin dynamics are utilized during all stages of transcription to regulate gene expression.

Changes in gene transcription are tightly correlated with changes in chromatin structure (Field et al., 2008; Radman-Livaja & Rando, 2010; Schwabish & Struhl, 2004; Shivaswamy & Iyer, 2008; Weiner, Hughes, Yassour, Rando, & Friedman, 2010; Zawadzki, Morozov, & Broach, 2009). When genes are highly transcribed, the nucleosome preceding the transcription start site is evicted, generating an open and accessible promoter (Shivaswamy & Iyer, 2008; Zawadzki et al., 2009). Conversely, genes that are not expressed, or lowly expressed, have existing promoter nucleosome structure that is not altered until the gene's expression is upregulated. A highly regulated and well-positioned nucleosome, known as the

+1 nucleosome, typically resides at a canonical distance downstream (in the direction of genic transcription) of each major transcriptional start site and forms the downstream border of the nucleosome-depleted region. A well-positioned −1 nucleosome forms the upstream border of the promoter. Nucleosome positioning within gene bodies is initially defined by the +1 nucleosome—the most well-positioned nucleosome in the array—with subsequent nucleosomes arranged as a result of the +1 nucleosome position (Kornberg, 1981; Kornberg & Stryer, 1988). Additionally, over the coding regions of highly transcribed genes, nucleosome occupancy decreases with high rates of transcription, whereas over lowly transcribed genes, nucleosome occupancy is not significantly disrupted (Lee et al., 2004; Shivaswamy et al., 2008). Changes to nucleosome structure, such as those mentioned above, occur through the activity of many factors, including chromatin remodeling factors, histone modifying enzymes, and histone chaperones; however, RNA Polymerase II itself is also responsible for some of the alterations which occur to nucleosome architecture. *In vitro*, RNA Polymerase II is able to transcribe DNA compacted into chromatin without evicting the nucleosome, or by creating subspecies of nucleosomes, such as hexasomes (Kulaeva, Hsieh, & Studitsky, 2010; Studitsky, Clark, & Felsenfeld, 1994; Studitsky, Kassavetis, Geiduschek, & Felsenfeld, 1997). Additionally, deactivation of RNA Polymerase II in yeast has been shown to increase −1 nucleosome occupancy, lending more support to the role of RNA Polymerase II in regulating the chromatin environment of genes (Weiner et al., 2010). Furthermore, inappropriate reassembly of nucleosomes in the wake of RNA Polymerase II transcription can lead to aberrant transcription initiation from within gene bodies (Cheung et al., 2008; Kaplan, Laprade, & Winston, 2003). Nucleosome occupancy and spacing are tightly regulated through numerous mechanisms, the most prominent of which is direct nucleosome remodeling by ATP-dependent complexes.

3. ATP-dependent nucleosome remodeling complexes establish and maintain chromatin state

Eukaryotic cells must actively balance chromatin compaction with DNA accessibility for appropriate gene expression. To maintain a balance between compaction and accessibility, cells make use of a wide array of nucleosome remodeling factors that can alter nucleosome composition and positioning genome-wide. Nucleosome remodeling factors are protein complexes that use the energy from ATP hydrolysis to reposition

(Fazzio & Tsukiyama, 2003; Lomvardas & Thanos, 2001) or remove nucleosomes (Boeger, Griesenbeck, Strattan, & Kornberg, 2004; Cairns, 2005) by altering histone-DNA contacts. Broadly, these factors use ATP to slide DNA around histone proteins, remove histone dimers or octamers, and to alter histone variant composition. Nucleosome remodeling factors are highly variable, allow other proteins to bind regulatory regions, and can permit or repress DNA-templated activities on chromatin. There are ~30 nucleosome remodeling factor ATPases in ES cells, each of which fulfills a distinct niche within the cell—often including both activating and repressing roles (Becker & Workman, 2013; Clapier & Cairns, 2009; Clapier, Iwasa, Cairns, & Peterson, 2017). By modulating histone octamer positioning, nucleosome remodelers can open or close binding sites for regulatory factors, further expanding the possible outcomes of nucleosome repositioning, even by the same ATPase. Nucleosome remodeling factors carry out their function through DNA translocation, which results in repositioning of histones along the DNA, thereby allowing them to facilitate or impede transcription and other DNA-templated processes at target loci.

ATP-dependent nucleosome remodelers are members of RNA/DNA helicase superfamily 2, also referred to as the DEAD/H superfamily (Clapier et al., 2017). There are four main families of ATP-dependent nucleosome remodelers: SWI/SNF, ISWI, INO80, and CHD (Clapier et al., 2017). Each of these families share distinct domains on their ATPase subunits that catalyze nucleosome remodeling by the complex to fulfill extraordinarily dynamic and powerful roles in regulating chromatin structure and gene expression (Fig. 4). A list of mammalian nucleosome remodeling ATPases can be found in Table 1. Broadly, SWI/SNF remodelers destabilize nucleosomes, ISWI nucleosome remodelers function to slide nucleosomes laterally, and other remodeling complexes, such as INO80, exchange H2A/H2B dimers (Yen, Vinayachandran, Batta, Koerber, & Pugh, 2012). Additionally, certain chromatin remodelers have been associated with transcription activation or repression, based on the mechanism of altering accessibility of nucleosomal DNA to other regulatory proteins, such as transcription factors. SWI/SNF family members, for example, tend to disorganize nucleosomes via sliding and ejection, and are therefore thought to promote transcription (reviewed in Rando & Winston, 2012). Alternatively, members of the ISWI family have been shown to remodel and organize nucleosomes over transcriptionally silent regions (Tirosh, Sigal, & Barkai, 2010). Across (and even within) the four families, nucleosome remodeling factors perform highly distinct

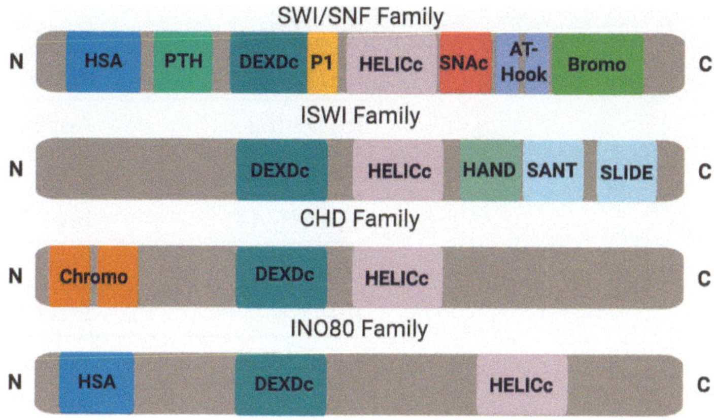

Fig. 4 ATPase domains of the four nucleosome remodeling factor families. ATPases in the SWI/SNF, ISWI, CHD, and INO80 nucleosome remodeling factor families all share related DEXDc (RecA-like Lobe 1) and HELICc (RecA-like Lobe 2) domains. SWI/SNF family ATPases are defined by an N-terminal HSA domain and a C-terminal bromodomain. ISWI family ATPases are defined by a C-terminal module consisting of HAND, SANT, and SLIDE domains. CHD family ATPases are defined by an N-terminal dual tandem chromodomain. INO80 family ATPases feature an N-terminal HSA domain like SWI/SNF remodelers, but not a C-terminal bromodomain. Created with Biorender.com.

activities to regulate chromatin structure and subsequent gene expression. These factors all use the shared activity of DNA translocation around histone proteins to fulfill both activating and repressing roles with extreme precision. For comprehensive reviews of ATP-dependent nucleosome remodeling factors, see Clapier et al. (2017) and Becker and Workman (2013).

3.1 SWI/SNF family nucleosome remodeling factors

The first characterized nucleosome remodelers were identified from two convergent screens in *Saccharomyces cerevisiae* (Neigeborn & Carlson, 1984; Stern, Jensen, & Herskowitz, 1984). Over the next decade, it became clear that these genes belonged to the same complex, and this complex became known as SWI/SNF in yeast. Although the SWI/SNF complex is known as BAF (Brahma-associated factors) in higher eukaryotes, the family of nucleosome remodeling factors retains the SWI/SNF name. The SWI/SNF family is defined by the presence of a bromodomain and a helicase/SANT ATPase (HSA) domain that facilitates binding of actin and/or actin-related proteins (Szerlong et al., 2008). SWI/SNF family remodelers are typically associated with increased chromatin accessibility via nucleosome sliding or eviction, which can have downstream activating

Table 1 Families of nucleosome remodeling factor ATPases.

Nucleosome remodeler family	Gene (ATPase)	Protein (ATPase)
SWI/SNF	*ATRX*	Atrx
	RAD54B	Rad54B
	RAD54L	Rad54
	RAD54L2	Arip4
	SMARCA2	Brm
	SMARCA4	Brg1
	HELLS	Hells
	ERCC6	Ercc6
	ERCC6L	Ercc6L
	ZRANB3	Zranb3
	SMARCAL	Harp
	SMARCAD	Hel1
	BTAF1	Btaf1
	HLTF	Hltf
	TTF2	Ttf2
	SHPRH	Shprh
ISWI	*SMARCA1*	Snf2h
	SMARCA5	Snf2l
INO80	*1NO80*	Ino80
	SRCAP	Srcap
	EP400	p400
CHD	*CHD1*	Chd1
	CHD1L	Chd1L
	CHD2	Chd2
	CHD3	Chd3
	CHD4	Chd4
	CHD5	Chd5
	CHD6	Chd6
	CHD7	Chd7
	CHD8	Chd8
	CHD9	Chd9

Mammals express 31 nucleosome remodeling factor ATPases, each of which belongs to one of the following families (with number of ATPases listed in parentheses): SWI/SNF (16), ISWI (2), INO80 (3), and CHD (10). Familial classifications are primarily based on structural characteristics; as such, nucleosome remodeling factors within the same family can have wildly different mechanisms of action and functional outcomes.

or repressive effects (Kasten, Clapier, & Cairns, 2011). The SWI/SNF-defining bromodomain binds acetylated histone lysines to promote nucleosome remodeling at these sites (Lee, Park, Kim, Kwon, & Kwon, 2010). In addition to complex targeting, the SWI/SNF family member Sth1's bromodomain is necessary for enhanced remodeling activity, suggesting a regulatory role (Chatterjee et al., 2011).

In part because of their extensive variety in mammalian cells, nucleosome remodelers can have many interchangeable subunits that are often differently utilized by cell type. At least six cell type-specific BAF complexes have been identified in mammalian cells, including those specific to cardiac cells (cBAF, containing BAF45C and BAF60C) (Lickert et al., 2004; Takeuchi & Bruneau, 2009), neural progenitors (npBAF, containing BAF45A and BAF53A) (Lamba, Hayes, Karl, & Reh, 2008; Lessard et al., 2007; Wu et al., 2007), neural cells (nBAF, containing BAF45B and BAF53B) (Olave, Wang, Xue, Kuo, & Crabtree, 2002; Wu et al., 2007), hematopoietic stem cells (hscBAF) (Buscarlet et al., 2014), and ES cells (esBAF, containing BRG1 and BAF155 but not BAF170) (Ho, Rohan, et al., 2009). Other non-cell type-specific BAF complexes have been defined by subunit composition, including GLTSCR1-containing BAF (GBAF or non-canonical/ncBAF, containing BRD9 and GLTSCR1/1L instead of an ARID subunit) (Alpsoy & Dykhuizen, 2018; Middeljans et al., 2012) and polybromo-containing BAF (PBAF, containing ARID2, PBRM1, and PHF10) (Wang et al., 1996). Ectopic expression of specific BAF subunits can hijack embryonic development; for example, expression of BAF60C in a developing embryo, rather than BAF60A, leads to beating cardiomyocytes outside of the heart precursor region (Takeuchi & Bruneau, 2009). BAF complexes are combinatorially assembled in a stepwise manner: first, a dimer of BAF155 and/or BAF170 subunits is formed; followed by loading of BAF60 subunits to form the initial core complex; then BAF47 and BAF57 to form the BAF core complex (Fig. 5) (Mashtalir et al., 2018). The immense variety of BAF complexes allows for specialization and target specificity, including roles in development that are too extensive to outline in this chapter; for a targeted review of BAF complexes in development, see Alfert, Moreno, and Kerl (2019) and Ho and Crabtree (2010).

3.2 INO80 family nucleosome remodeling factors

The Inositol-requiring 80 (INO80) family's namesake was first identified in *S. cerevisiae* as a mutant causing inositol auxotrophy (Ebbert, Birkmann, & Schuller, 1999); since then, roles for INO80 have been uncovered in

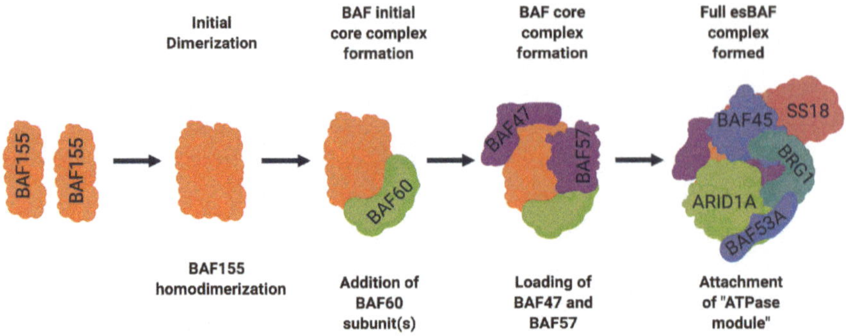

Fig. 5 Nucleosome remodeling factor complexes are assembled in a stepwise manner. The assembly process of the SWI/SNF family nucleosome remodeler esBAF is depicted. An initial homodimerization event between BAF155 dimers is necessary before any other esBAF subunits can be loaded. Following this dimerization, a BAF60 subunit is loaded onto the complex to form the "BAF initial core complex." Core subunits BAF47 and BAF57 are subsequently assembled to form the BAF core complex, at which point an ATPase module containing BRG1, BAF53A, ARID1A, a BAF45 subunit, and SS18 is added to form the full esBAF complex. Created with Biorender.com.

DNA repair, replication, and transcription (Poli, Gasser, & Papamichos-Chronakis, 2017; Poli et al., 2016; Xue et al., 2015). The INO80 family is defined by the presence of an HSA domain like that of SWI/SNF family, but without an ATPase C-terminal bromodomain. There are two general groups of INO80 members, those belonging to the INO80 class and the SWR1 class. Mammals have two SWR1-class members—SRCAP and p400—as well as an INO80-class member (INO80). Although INO80 is the lone member of the INO80 class, the complex may exist in different compositions to act at both euchromatic and heterochromatic regions of chromatin (Runge, Raab, & Magnuson, 2018).

INO80 mediates nucleosome spacing, histone eviction, and replication fork progression, including a role for eviction of RNA Polymerase II and the PAF1 complex at replication-transcription fork collisions (Lafon et al., 2015; Poli et al., 2017, 2016). Specifically, INO80 translocates DNA proximal to H2A/H2B dimers to destabilize nucleosomes and promote both nucleosome sliding and histone variant exchange (Brahma et al., 2017). INO80 can also be found at upward of 90% of nucleosome-depleted regions in *S. cerevisiae* (Yen, Vinayachandran, & Pugh, 2013) and its presence at promoters correlates positively with gene expression (Klopf, Schmidt, Clauder-Munster, Steinmetz, & Schuller, 2017).

The 17-subunit Tat-Interactive Protein 60 kDa (Tip60)/p400 complex is one of two members of the SWRI class of INO80 remodelers, the primary functions of which are to perform histone acetylation and histone H2A.Z/H2B dimer exchange (Doyon, Selleck, Lane, Tan, & Cote, 2004; Kobor et al., 2004). Tip60 is the mammalian homolog of the histone acetyltransferase NuA4, while p400 is the homolog of the nucleosome remodeler SWR1 (Doyon & Cote, 2004; Doyon et al., 2004; Latrick et al., 2016). While the acetylation and incorporation of H2A.Z are generally associated with transcriptional activation, Tip60-p400 instead acts as a repressor of differentiation-associated genes in murine ES cells (Fazzio, Huff, & Panning, 2008a). The complex does so by localizing to gene promoters marked by H3K4me3 and acetylating histones in that region (Fazzio et al., 2008a). The presence of active transcription marks combined with the repressive role for Tip60-p400 in ES cells suggests a heightened importance for the state of chromatin decorated by canonical active and repressive marks, rather than a simple on/off program of gene expression.

Along with p400, the SNF2-related CREBBP activating protein (SRCAP) is a homolog of SWR1 that also promotes exchange of H2A.Z, including deposition at promoters (Wong, Cox, & Chrivia, 2007). SRCAP shares the YL1 subunit with Tip60-p400 complex (Cai et al., 2005) and makes use of YL1 to facilitate transfer of H2A.Z from the remodeler to the nucleosome (Liang et al., 2016). SRCAP interacts with INO80 and has roles in chromosome double-strand break recognition and repair, as well as in regulating transcription (Gerhold & Gasser, 2014).

3.3 ISWI family nucleosome remodeling factors

The Imitation Switch (ISWI) family of nucleosome remodelers was first identified in *D. melanogaster* as a homolog of the yeast SWI/SNF complex (Elfring, Deuring, Mccallum, Peterson, & Tamkun, 1994). Members of the ISWI family of nucleosome remodelers are defined by the presence of HAND, SANT, and SLIDE domains that mediate nucleosome interaction (Clapier & Cairns, 2012). Broadly, ISWI family remodelers are important for nucleosome assembly and spacing, as well as higher-order chromatin compaction (Clapier & Cairns, 2009). ISWI remodeling complexes are substantially smaller than SWI/SNF complexes—composed of 2–4 subunits—and are canonically associated with repression of transcription (Clapier & Cairns, 2009). Mammalian ISWI complexes include WICH, NORC, WCRF/ACF, CHRAC, and RSF, all of which exclusively use the ATPase SNF2H, while NURF and CERF use the SNF2L ATPase (Längst & Becker, 2001).

ISWI remodelers are dependent on the nucleosome acidic patch for maximum enzymatic activity (Dann et al., 2017). Furthermore, ISWI remodelers display enhanced remodeling activity at H2A.*Z*-containing nucleosomes *in vitro* (Goldman, Garlick, & Kingston, 2010). Beyond repression of transcription, ISWI family remodelers function in three DNA damage repair pathways: homologous recombination, non-homologous end-joining, and nucleotide excision repair (Aydin, Vermeulen, & Lans, 2014; Lan et al., 2010; Nakamura et al., 2011; Yadon & Tsukiyama, 2011). In ES cells, ISWI family remodelers have been implicated in the control of cell fate decisions, neural tube formation, and neurite outgrowth (Andersen, Lu, & Horvitz, 2006; Yadon & Tsukiyama, 2011).

3.4 CHD family nucleosome remodeling factors

The Chromodomain helicase DNA-binding (CHD) superfamily of nucleosome remodeling factors contains nine proteins grouped into three subfamilies: those containing PHD domains (CHD3–5), SANT domains and Brahma and Kismet (BRK) domains (CHD6–9), and neither (CHD1 and CHD2) (Micucci, Sperry, & Martin, 2015). The CHD family is defined by the presence of two N-terminal tandem chromodomains that mediate chromatin interaction by binding to methylated lysine residues in the histone tails, and two SNF2-like ATP-dependent helicase domains (Clapier et al., 2017; Micucci et al., 2015). The first CHD family member to be identified was CHD1 (Delmas, Stokes, & Perry, 1993); however, the chromodomain itself had been discovered two years earlier as a shared region between the repressive proteins HP1 and Polycomb in *D. melanogaster* (Paro & Hogness, 1991). Functionally, CHD1 maintains open chromatin by binding to H3K4 trimethylated regions of chromatin and excluding H3K27me3, a repressive mark associated with heterochromatin (Gaspar-Maia et al., 2009). Although CHD family nucleosome remodelers are grouped into subfamilies based on shared domains, CHD family members can exhibit specialized nucleosome binding preferences and activities even within the same subfamily. CHD7 and CHD8, for example, slide nucleosomes along DNA, while CHD6 does not slide nucleosomes to remodel chromatin (Manning & Yusufzai, 2017).

The NuRD complex (originally known as the Mi-2 complex) was isolated and characterized by four separate groups in the same year (Tong, Hassig, Schnitzler, Kingston, & Schreiber, 1998; Wade, Jones, Vermaak, & Wolffe, 1998; Xue et al., 1998; Zhang, LeRoy, Seelig, Lane, & Reinberg, 1998).

Each of these studies described a complex containing parallel ATP-dependent nucleosome remodeling activity and histone deacetylase activity, a combination unique to NuRD. The enzymatic subunits of the canonical NuRD complex are an ATP-dependent remodeler (CHD3 or CHD4, though CHD3 is not expressed in ES cells), and two histone deacetylases (HDAC1 and HDAC2). The MBD3 and MBD2 subunits are mutually exclusive within the NuRD complex and confer complex naming (MBD2/NuRD or MBD3/NuRD) (Le Guezennec et al., 2006). There are three MBD3 isoforms, termed MBD3A, MBD3B, and MBD3C. MBD3A and MBD3B are found in all cell types examined, but MBD3C is unique to stem cells (Ee et al., 2017). MBD2 contains a domain (MBD) that binds methylated CpG dinucleotides on DNA (Hendrich & Bird, 1998) and a C-terminal transcription repression domain. MBD3 is the only MBD protein that does not bind methylated DNA, although it does bind hydroxymethylated cytosines (Yildirim et al., 2011). Despite mutually exclusive association within NuRD complexes, MBD2 and MBD3 localize to many of the same genomic regions (Gunther et al., 2013; Hainer et al., 2016). MBD3/NuRD is a largely repressive complex and functions in opposition to gene activating complexes like esBAF and STAT3 (Yildirim et al., 2011). In addition to transcriptional control, NuRD activity is also associated with higher-order chromatin assembly, maintenance of genome stability, hematopoietic stem cell differentiation, various human cancers, and aging (Lai & Wade, 2011; Pegoraro et al., 2009; Yoshida et al., 2008).

3.5 Summary

Nucleosome remodeling enzymes harness the power of a single biochemical activity—ATP hydrolysis—to alter DNA-histone interactions and regulate chromatin structure with a seemingly endless number of potential outcomes. Making use of unique domains and complex assemblies, nucleosome remodeling factors establish and preserve a diverse suite of chromatin states as a means by which to control DNA accessibility, and therefore all DNA-templated processes. In ES cells, nucleosome remodeling factors are generally important for maintenance of self-renewal and pluripotency, as well as specification into developmental lineages. The dynamic local chromatin states maintained by ATP-dependent nucleosome remodeling factors provide important context for the various modifications associated with active and repressed gene expression; indeed, local chromatin dynamics can even lead to opposite gene expression profiles from those canonically associated with the modifications decorating these regions, as is the case with Tip60-p400-mediated regions (Fazzio, Huff, & Panning, 2008b). In the next section, we will discuss these histone modifications in greater detail, examining some of the best-characterized modifications, their roles, and the techniques used to examine them.

4. Histone modifications provide an additional layer of gene regulation

Histone proteins are composed of globular alpha-helix domains and N- and C-terminal tail extensions, which are intrinsically disordered (Luger et al., 1997). Both the globular domains and tails of histone proteins are subject to a vast array of post-translational modifications. Histone modifying enzymes attach covalent moieties, including methyl groups, acetyl groups, phosphate groups, ubiquitin, and Small Ubiquitin-like Modifiers (SUMOs) to specific amino acid residues on histone proteins (reviewed in Fuchs, Laribee, & Strahl, 2009; Smith & Shilatifard, 2010). These modifications regulate gene expression and other DNA-templated activities by affecting chromatin structure through altering DNA-histone interactions and providing a platform for recruitment of additional regulatory proteins that can further influence chromatin function and dynamics. Histone post-translational modifications (both acetylation and methylation of histone tails) were first identified by Alfred Mirsky's group (Allfrey, Faulkner, & Mirsky, 1964). Lysine residues are the most commonly modified amino acid in histone proteins, and modified residues are most often found on the N-terminal histone tails, though some modifications have been identified within the globular domains (Kouzarides, 2007). All four core histone proteins can be modified, and a summary of the best-characterized modifications can be found in Fig. 6.

Among the most prominent technical advancements regarding investigating the role and localization of histone modifications is the development of chromatin immunoprecipitation (ChIP) (Gilmour & Lis, 1984; Solomon, Larsen, & Varshavsky, 1988). This technology was coupled with deep sequencing, or ChIP-seq (Albert et al., 2007), to provide previously unparalleled resolution with which to examine protein localization on chromatin. ChIP-seq and its various modifications allow one to profile the localization of a protein on chromatin, including histone modification marks. Recently, an analogous technique to ChIP-seq, CUT&RUN (Skene & Henikoff, 2017), has successfully been used to profile localization of pluripotency

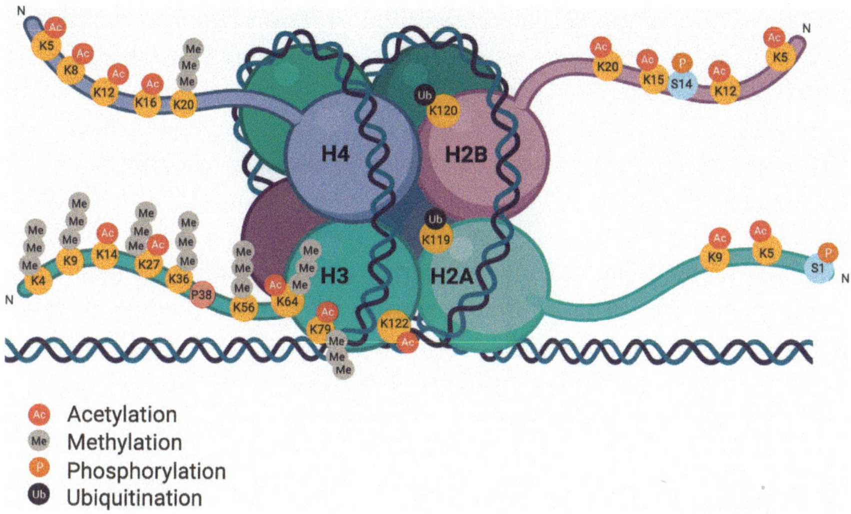

Fig. 6 Histones are post-translationally modified as a mechanism of epigenetic regulation. Histone proteins have N- and C-terminal tails that are highly modified, although residues within the globular domains can also be modified. The most common histone post-translational modifications are methylation and acetylation of lysine residues, although numerous modifications not listed in this figure have been identified. Created with Biorender.com.

factors in individual ES cells (Hainer, Boskovic, McCannell, Rando, & Fazzio, 2019). With the refinement of genomic profiling techniques, it is likely that future experiments will bring additional clarity to the state and roles of histone modifications and the enzymes that deposit and remove them.

4.1 Histone acetylation and deacetylation

The addition of an acetyl group to residues on histone proteins is a modification typically associated with active transcription. Acetyl groups are covalently attached to histone residues post-translationally by histone acetyltransferases (HATs), while histone deacetylases (HDACs) remove acetyl marks (Carrozza et al., 2005; Close et al., 2006; Gilbert, Gore, Herman, & Carducci, 2004; Govind, Zhang, Qiu, Hofmeyer, & Hinnebusch, 2007; Keogh et al., 2005). There are six HAT families, including GCN5/PCAF, MYST, $TAF_{II}250$, CBP/p300, SRC1, HAT1, and ATF-2 (Marmorstein & Zhou, 2014). Individual HAT families are well conserved at the sequence level within families, but not between families (Kuo & Allis, 1998) and include a histone acetyltransferase domain (Marmorstein, 2001).

Histone acetylation facilitates activation of transcription by neutralizing the basic charge of the lysine residue, thereby loosening DNA-histone contacts (Pokholok et al., 2005). In addition, acetylation can serve as a recruitment mark for chromatin modifying factors. For example, HATs enhance SWI/SNF binding to promoter nucleosomes through recognition of acetylated histone residues by the ATPase bromodomain (Hassan, Neely, & Workman, 2001; Lee et al., 2010). Both histone acetylation and ubiquitylation have been shown to interfere with formation of higher-order chromatin structure by disrupting chromatin interactions (Fierz et al., 2011). Commonly acetylated histone residues can be found in Fig. 6. For a more complete review on HATs, see Marmorstein and Zhou (2014).

Histone deacetylases, commonly referred to as HDACs, remove acetyl groups from modified proteins. The 18 human HDACs are sorted into four classes, deemed Class I, Class II, Class III, and Class IV (Seto & Yoshida, 2014). Broadly, Class I, II, and IV HDACs are grouped into the histone deacetylase or classical HDAC family, and are dependent on metals (specifically zinc) to hydrolyze acetylated substrates via a set of shared active site residues, while Class III HDACs use NAD+ to deacetylate modified lysines (Seto & Yoshida, 2014). For a comprehensive review on HDAC structure, function, and classification, see Seto and Yoshida (2014).

4.2 Histone methylation and demethylation

Histone methylation comes in varying states, as residues can be mono-, di-, or tri-methylated in a stepwise manner (Zee et al., 2010), and these marks are associated with different functions depending on the residue modified and the extent of methylation. At H3K4, for example, monomethylation (H3K4me1) is a mark of open chromatin and enhancers, while H3K4 trimethylation (H3K4me3) tends to decorate promoters and actively transcribed regions (Bannister & Kouzarides, 2011). These methyl groups are covalently attached to histone amino acids by a family of enzymes known as histone methyltransferases. Histone methyltransferases can be general or highly specialized, and can include somewhat redundant complexes, like the human H3K4 methyltransferases MLL1–6. Most histone methyltransferases are marked by a SET (Su-39, Enhancer of zeste, Trithorax) domain, with the exception of the DOT1L H3K79 methyltransferase (Dillon, Zhang, Trievel, & Cheng, 2005). Histone methyltransferases catalyze the transfer of methyl groups from the donor molecule S-Adenosyl Methionine (SAM) to target proteins, many—though not all—of which are histones (Dillon et al., 2005). Histone methylation is associated with either transcriptional activation (e.g., H3K4) or, more often, transcriptional repression (e.g., H3K9, H3K27, H4K20, and H1K26), but the function of these marks is often highly context- and location-specific. Commonly methylated histone residues can be found in Fig. 6.

Histone methyl groups are also regulated by enzymes that remove these modifications, known as histone demethylases. There are two main families of histone demethylases: the Lysine-Specific Demethylase (LSD) family, and the Jumonji (JMJ) family. Of the two, JMJ is the larger family, comprised of over 30 demethylases, while LSD is comprised of only two demethylases, LSD1 and LSD2 (Kooistra & Helin, 2012). Both LSDs share a SWI3P, RSC8P, and MOIRA (SWIRM) domain and an amine oxidase domain, along with either a spacer region (LSD1) or a CW-type zinc finger domain (LSD2) (Kooistra & Helin, 2012). LSD1 utilizes the substrates H3K4 (mono- or di-methylated), H3K9 (mono- or di-methylated), plus three non-histone substrates in p53, E2F1, and DNMT1. LSD2, meanwhile, is specific to mono- and di-methylated H3K4 alone (Ciccone et al., 2009; Karytinos et al., 2009). JMJ demethylases exhibit extraordinary diversity of targets. Importantly, only JMJ-family demethylases are known to act upon trimethylated residues (Shi, 2007; Whetstine et al., 2006).

Histone modifications have extensive downstream effects depending on the residue and modification. Marks associated with active transcription include acetylation of H3K27 (H3K27ac), H3K56me3, H3K4me3, and H3K36me, while repression-associated modifications include deacetylation and some methylation marks (H3K9me3, H3K27me3, H3K56me3, and H4K20me). H3K4 methylation is currently the most extensively studied histone modification, likely due to its role in decorating actively transcribed promoters (H3K4me3) and enhancers (H3K4me1). H3K4 methylation is placed by the SET1/COMPASS complex in yeast and by the MLL1–6, SET1D1A, and SET1D1B complexes in humans (Dou et al., 2006; Takahashi et al., 2011). Similar to H3K4me1, H3K27ac decorates regions of active transcription, and is used as the canonical mark by which to identify an active enhancer. H3K27 can also be trimethylated, a repressive mark placed by the PcG proteins in a coordinated positive feedback loop with H2AK119 ubiquitylation, by the PRC2 and PRC1 complexes, respectively (Cao, Tsukada, & Zhang, 2005). Although H3K27me3 is a canonical repressive mark, it also marks "poised" genes in stem cells, along with H3K4me3, leading to what is known as a bivalent gene promoter (Fig. 7) (Azuara et al., 2006; Bernstein et al., 2006). H3K36 methylation is placed

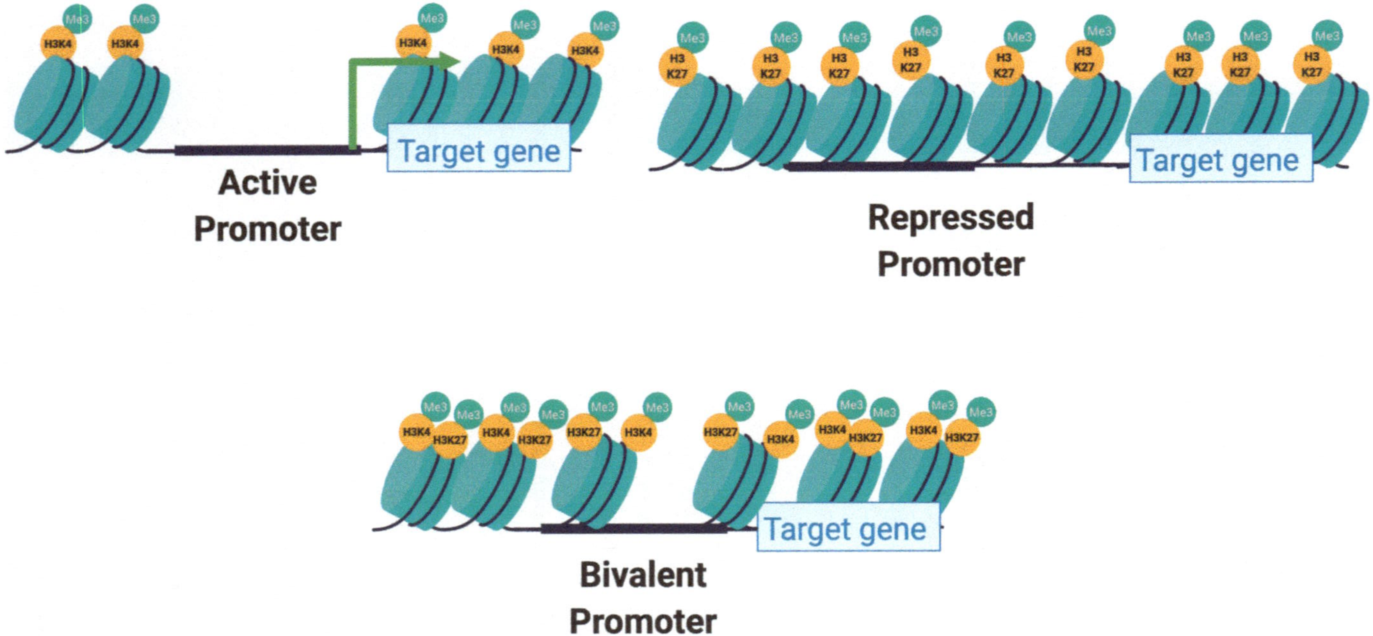

Fig. 7 Bivalent promoters mark lowly expressed but poised genes in stem cells. Between 3000 and 4000 promoters in embryonic stem cells are marked by both an activation-associated modification (H3K4me3) and a repression-associated modification (H3K27me3), and are thus known as bivalent promoters. Bivalent genes remain accessible to transcription factors, but lowly expressed until the H3K27me3 mark is removed (often upon a cellular differentiation event). Created with Biorender.com.

by the SET2 methyltransferase and marks genes that are in the process of being transcribed; as RNA Polymerase II passes along the gene, SET2 is recruited by the phosphorylated C-terminal domain of RNA Polymerase II subunit RPB1 and the methyl mark is left on the gene in its wake (Kizer et al., 2005; Youdell et al., 2008). This mark prevents cryptic initiation of transcripts from within the transcribed gene (Carrozza et al., 2005). Finally, H3K56ac is catalyzed by the RTT109 histone acetyltransferase and ASF1 chaperone in yeast (Driscoll, Hudson, & Jackson, 2007; Schneider, Bajwa, Johnson, Bhaumik, & Shilatifard, 2006; Tsubota et al., 2007) and by p300/CBP and GCN5 in *Drosophila* and mammals (Das, Lucia, Hansen, & Tyler, 2009; Tjeertes, Miller, & Jackson, 2009). Acetylation of H3K56 enhances nucleosome unwrapping and "breathing," allowing for increased DNA accessibility and remodeling by nucleosome remodeling complexes (Neumann et al., 2009). H3K56ac and H2A.Z both mark promoter-proximal nucleosomes with high turnover rate (Kaplan et al., 2008; Raisner et al., 2005; Rufiange, Jacques, Bhat, Robert, & Nourani, 2007) and are thus associated with transcriptional activation.

Importantly, the majority of these marks themselves do not appear to directly activate or silence genes, with acetylation being a notable exception; instead, these marks are correlated with genes activated and silenced through other mechanisms (Bannister & Kouzarides, 2011). Even canonical "activating" marks—like histone acetylation and trimethylation of H3K4—can be present at repressed genes, as is the case with Tip60-p400-repressed genes during stem cell differentiation (Fazzio et al., 2008a). In addition to correlation with activity state, histone modifications can regulate higher-order chromatin structure; H4K16ac, for example, is known to inhibit higher-order chromatin folding and compaction through decreased nucleosome-nucleosome stacking, perhaps due to a decreased interaction of the H4 tail with the histone acidic patch (Shogren-Knaak, Ishii, Pazin, Davie, & Peterson, 2006; Zhang, Erler, & Langowski, 2017).

Histone modifications can also create or eliminate protein binding sites. Upon methylation of H3K9 by a SUV39H methyltransferase, a binding site for HP1 proteins is created that is dependent on a functional HP1 chromodomain and a dimethylated H3K9 (Lachner, O'Carroll, Rea, Mechtler, & Jenuwein, 2001). Similarly, H3K36me3 placed by the SET2 methyltransferase serves as a recruitment cue for the RPD3S deacetylase to coding regions to repress cryptic intragenic transcription (Carrozza et al., 2005). Conversely, complex binding to chromatin can be disrupted by histone modifications, as is the case with H3K4 methylation and

NuRD complex binding (Kouzarides, 2007). Chromatin modifications are highly dependent on the chromatin context in which they are found. For example, the traditionally repression-associated mark H3K27me3 is associated with genes that are inactive, but poised for activation in ES cells. These so-called bivalent genes will be examined in greater detail later in this chapter.

While the above sections have detailed some of the better characterized roles associated with various histone modifications, there are numerous other roles that are less well-studied. These roles include gene bookmarking as a means through which epigenetic information is passed between parent and daughter cells through mitosis (H4K5ac and H4K8ac) (Zhao, Nakamura, Fu, Lazar, & Spector, 2011) and histone "crosstalk" to regulate downstream epigenetic marks (such as H2BK120ub and proline isomerization at H3P38 to regulate H3K36me) (Kouzarides, 2007; Tomson & Arndt, 2013). With new approaches to examine epigenetic marks, powerful single-cell genomic techniques, and decreasing costs of next-generation sequencing, it is likely that new roles for established marks and new epigenetic marks and modifications themselves will be discovered in the near future.

4.3 Polycomb group proteins mediate H3K27me3 and silencing of developmental genes

Polycomb Group (PcG) proteins have been associated with developmental processes since their identification (Lewis, 1947, 1978). PcG includes two complexes, Polycomb-Related Complex 1 (PRC1) and PRC2. PRC1 was purified in 1999 from *D. melanogaster* and found to contain four proteins: Polycomb, Polyhomeotic, dRING, and Posterior Sex Combs (PSC) (Shao et al., 1999). The Sex Combs on Midleg (SCM) protein was later found to copurify with the complex and to be incorporated *in vitro* (Kang et al., 2015; Peterson et al., 2004), and plays a major recruitment role in the PRC1 complex (Kassis & Kennison, 2010), as is the case for the chromobox (CBX) family of proteins that can recognize and bind H3K27me3 (Vincenz & Kerppola, 2008). When added to arrayed nucleosomes, PRC1 was able to block remodeling by BAF, but not to disrupt the arrays by cutting (Shao et al., 1999). PRC1's RING1 and RING2 subunits catalyze the placement of ubiquitin on H2AK119, a mark that serves as a recruitment cue for the PRC2 complex (Cao et al., 2005). In addition to placing the H2AK119Ub mark, PRC1 reads the PRC2-catalyzed H3K27me3 mark as a recruitment cue (Blackledge et al., 2014).

In addition to facilitating gene repression through H3K27me3, Enhancer of Zeste Homolog 2 (EZH2) serves as a scaffold for DNA methyltransferases that methylate cytosines to repress gene promoters (Vire et al., 2006). Functionally, PRC2 binds to chromatin and carries out H3K27me3 via EZH2, an action that works to silence target genes. The exact mechanism by which PRC2 is recruited to target sites is unknown; however, the JARID2 (Pasini et al., 2010) and PCL (Hunkapiller et al., 2012) proteins appear to be linked to this targeting.

5. Histone chaperones and histone variants regulate chromatin structure

Individual histone proteins within the nucleosome octamer can be substituted by a number of variants that provide an additional level of complexity. The canonical histone proteins H2A, H2B, H3, and H4 are regulated in a cell-cycle-dependent manner (Marzluff, Gongidi, Woods, Jin, & Maltais, 2002). In contrast to the canonical histones, genes encoding histone variants are expressed and the variants deposited onto chromatin in a replication-independent manner (Ahmad & Henikoff, 2002). Replacement of canonical histones with histone variants, such as the replacement of H2A with H2A.Z over promoters (Santisteban, Hang, & Smith, 2011; Wan et al., 2009; Zhang, Roberts, & Cairns, 2005), plays an important role in chromatin dynamics during transcription and other DNA-templated activities. Many different variant forms of each histone exist throughout various organisms (Tables 2 and 3) and reviewed in Buschbeck and Hake (2017). Sequence differences between histone variants and canonical histones are found in many regions of the histone proteins: either in the terminal tails, the globular fold domains, or in specific amino acid residues (Doyen, An, et al., 2006; Doyen, Montel, et al., 2006; Henikoff & Ahmad, 2005). While incorporation of histone variants in place of canonical histones impacts chromatin structure in various ways, many post-translational modification sites are conserved between variants and canonical histones (McKittrick, Gafken, Ahmad, & Henikoff, 2004). Therefore, exchanging the canonical histones with these variants may not alter nucleosome recognition by some chromatin regulatory proteins.

Histone exchange is carried out by an extensive class of proteins called histone chaperones. Histone chaperones typically do not require the use of ATP to carry out this exchange; rather, histone chaperones use spontaneous DNA movement around the dyad axis to destabilize the nucleosome

Table 2 Common histone proteins, chaperones that exchange them, and associated localization and processes.

Histone dimer	Exchanging chaperones	Localization	Associated processes
H2A/H2B	FACT, NPM1/2, Nap1, APLF, SET/TAF1b, CINAP, Nucleolin, ISWI, HSP90A/B, HSC70, IPO9, NPM2	Genome-wide	Core histones
H2A.X/H2B	FACT; high sequence similarity so likely many canonical H2A chaperones	DNA damage sites	DNA damage repair
H2A.Z/H2B	FACT, Tip60/p400, SRCAP, IPO9, Nap1, Anp32e	Sites of active transcription, heterochromatin/euchromatin boundaries	Active transcription, regulatory regions
macroH2A/H2B macroH2A.2/H2B	ATRX; unknown but maybe APLF	Deposited genome-wide, but condensed to repressed regions	Heterochromatin compaction
H3/H4	MCM2–7, HAT, IPO4, CAF1, HDAC, NURF, NURD, PRC2, HIRA, ATRX, DAXX, EP400, ASF1a/b, Cabin1, DEK	Genome-wide	Core histones
H3.3/H4	ATRX, DAXX, HIRA	Sites of active transcription, regulatory regions, telomeres, pericentric heterochromatin	Heterochromatin compaction, active transcription, differentiation
CENP-A/H4	HJURP, RBBP4, DAXX	Centromeres	Core histones
H3/H4G	Unknown	Nucleolus, rDNA	Breast cancer progression
H1	HSP90A/B, HAT, IPO9	Heterochromatin	Higher-order chromatin compaction

Core histone proteins conserved throughout eukaryotes are listed, as well as selected histone variants that are highly conserved and/or have important functions in gene regulation. Generally, core histones are synthesized in a replication-dependent manner, while variant histones are synthesized and incorporated throughout the cell cycle.

Table 3 Uncommon histone variants and their associated processes.

Histone variant	Implicated processes	Associated references
H1.1-H1.5	Replication-dependent Histone H1 variants found in somatic cells	Izzo et al. (2017)
H1.0	Replication-independent linker histone found in somatic cells	Izzo et al. (2017)
H1.10	Replication-independent linker histone found in somatic cells	Izzo et al. (2017)
HILS1	Spermatid-specific; important in chromatin remodeling during spermatogenesis	Yan, Ma, Burns, and Matzuk (2003)
H1T2	Spermiogenesis, DNA compaction; exclusive to male haploid germ cells	Martianov et al. (2005)
H1x	Enriched in closed chromatin, accumulates in nucleolus during G1	Happel, Schulze, and Doenecke (2005)
H1oo	Oocyte-specific; maintains expression of pluripotency genes like Nanog, MYC, and KLF9; essential for mouse meiosis	Hayakawa, Ohgane, Tanaka, Yagi, and Shiota (2012) and Furuya et al. (2007)
H3T/H3.4	Development; essential for spermatogenesis	Ueda et al. (2017)
H3.5	Spermatogenesis; transcription initiation; forms an unstable nucleosome	Urahama et al. (2016) and Schenk, Jenke, Zilbauer, Wirth, and Postberg (2011)
H3.X	Primate-specific; transcribed, but not detectable at protein level *in vivo*	Wiedemann et al. (2010)
H3.Y	Primate-specific, differentiates between HIRA and DAXX chaperones and features many novel amino acids near DNA entry/exit sites	Kujirai et al. (2017), Wiedemann et al. (2010), and Zink et al. (2017)
H2B.W	Spermiogenesis, testis-specific	Siuti, Roth, Mizzen, Kelleher, and Pesavento (2006) and Zalensky, Tomilin, Zalenskaya, Teplitz, and Bradbury (1997)

Continued

Table 3 Uncommon histone variants and their associated processes.—cont'd

Histone variant	Implicated processes	Associated references
H2B.Z	H2AZ interaction; apicomplexan specific	Hoeijmakers et al. (2013) and Petter et al. (2013)
sperm H2B	Development; Echinoidea-specific	Oliver et al. (2002)
subH2B	Higher eukaryotes; spermiogenesis-specific	Aul and Oko (2001)
H2B.E	Murine olfactory neurons	Santoro and Dulac (2012)
TH2A	Testis, oocyte, and zygote-specific	Shinagawa et al. (2014)
H2A.B	Associated with transcription upregulation, active gene TSSs in testes, arginine-rich	Arimura et al. (2015) and Molaro, Young, and Malik (2018)
H2A.Bbd	Human-specific (mouse H2A.Lap1); expressed in testis and brain	Arimura et al. (2015)
H2A.L	Pericentric chromatin organization in spermatids and spermiogenesis	Molaro et al. (2018)
H2A.P	Placental mammal-specific; putative variant similar to H2A.B and H2A.L	Molaro et al. (2018)
H2A.Q	Testis-specific, X-chromosome localized; may be involved in sex chromosome-related genetic conflicts	Molaro et al. (2018)
H2A.W	Plant-specific, SPKK motif-containing	Yelagandula et al. (2014)

Histone variants not listed in Table 2 are compiled here; most uncommon histone variants are highly cell-type-specific, and the overwhelming majority are specific to oocytes or testes. Uncommon histone variants have not been extensively studied; seminal papers and comprehensive reviews have been listed here for further information.

and promote histone exchange (Hondele et al., 2013). Histone exchange can be targeted to swap a core histone protein with a histone variant (Venkatesh & Workman, 2015) to transiently promote transcription elongation through the nucleosome (Hsieh et al., 2013; Kulaeva, Hsieh, Chang, Luse, & Studitsky, 2013; Kulaeva et al., 2010; Li et al., 2007) and to replace histone proteins that have been dissociated in the wake of the transcribing polymerase (Hsieh et al., 2013; Li et al., 2007), among other roles. Histone chaperones can be specific to individual histone variants (e.g., Holliday Junction Recognition Protein (HJURP) for CENP-A) or more general (α-thalassemia X-linked mental retardation protein (ATRX) is associated with deposition of both H3.3 and macroH2A) (Skene & Henikoff, 2013).

Among the best-characterized histone variants are those found at *cis*-regulatory regions, H2A.Z and H3.3; however, there are numerous histone variants for H2A and H3, and their incorporation can be regulated by genomic location, depositing histone chaperone, higher-order chromatin state, and numerous other factors (Tables 2 and 3). Although mammalian variants of H2B exist, they are often highly cell type-specific and commonly involved in gametogenesis (Draizen et al., 2016). A variant of histone H4, H4G, has very recently been discovered (Long et al., 2019), but as little is known about it at present, this section will focus on variants of H2A and H3, as they are highly conserved general regulators with numerous well-defined roles.

6. Histone H2A variants

6.1 H2A.Z is a histone variant associated with active transcription

H2A.Z is arguably the best-characterized histone variant to date. Originally identified with H2A.X in *S. cerevisiae* (West & Bonner, 1980), H2A.Z is highly conserved across species and is expressed in every eukaryote examined to date. H2A.Z is highly incorporated at regions of active transcription, especially at promoters and enhancers. Studies suggest that H2A.Z can be deposited into a nucleosome either through ATP-dependent histone exchange (Mizuguchi et al., 2004) or with the help of replication-independent histone chaperones (Park, Chodaparambil, Bao, McBryant, & Luger, 2005). When an H2A.Z/H2B dimer is incorporated into a nucleosome, steric hinderances are induced between the newly incorporated dimer and the H3/H4 tetramer, effectively destabilizing the nucleosome (Giaimo, Ferrante, Herchenrother, Hake, & Borggrefe, 2019). This change may promote active transcription, DNA repair processes, and chromatin domain segregation (Suto, Clarkson, Tremethick, & Luger, 2000). H2A.Z can localize to the 5′ ends of active and inactive genes (the +1 and −1 nucleosomes), though it is most often associated with the +1 and −1 nucleosomes flanking the NDR region of active genes (Guillemette et al., 2005).

H2A.Z also localizes to the borders of euchromatic domains, and it is thought to prevent heterochromatic invasion into euchromatic space, with INO80 complexes helping to accomplish this role (Papamichos-Chronakis, Watanabe, Rando, & Peterson, 2011; Raisner et al., 2005). In addition to INO80, ISWI complexes display enhanced activity at H2A.Z-containing nucleosomes; H2A.Z may therefore regulate ISWI complexes, serving as a stimulatory cue for nucleosome remodeling activity (Goldman et al., 2010). In eukaryotes, SWI/SNF family remodelers (RSC and esBAF) act on H2A.Z-containing nucleosomes (Cakiroglu et al., 2019; Hainer & Fazzio, 2015).

6.2 H2A.X is a marker of DNA double-strand breaks

First characterized in 1980 (West & Bonner, 1980), H2A.X is now best known as a marker of DNA damage. Upon recognition of a DNA double-strand break, S139 on H2A.X becomes phosphorylated, then referenced as γH2A.X (Rogakou, Pilch, Orr, Ivanova, & Bonner, 1998). This phosphorylation, primarily carried out by the ATM, ATR, and DNA-PK kinases (Sharma, Singh, & Almasan, 2012), recruits the INO80 nucleosome remodeling complex (Van Attikum, Fritsch, Hohn, & Gasser, 2004). γH2A.X is found in high levels in ES cells relative to differentiated cells (Turinetto et al., 2012); furthermore, γH2A.X is an important regulator of ES cell proliferative rate, and its deposition at nucleolar ribosomal DNA promoters assembles the nucleolar remodeling complex, thereby repressing rRNA transcription and controlling ES cells' self-renewal rate (Eleuteri, Aranda, & Ernfors, 2018; Turinetto et al., 2012).

6.3 macroH2A is associated with repression and heterochromatin

macroH2A is defined by its C-terminal "macro" domain, which is approximately twice the size of the corresponding histone fold region in core H2A and may allow the variant to resist nucleosome remodeling activity (Buschbeck & Hake, 2017; Chakravarthy et al., 2005). macroH2A associates with heterochromatin but is not preferentially deposited in repressed regions of chromatin; rather, macroH2A is deposited throughout the genome (by the action of multiple histone chaperones) and subsequently pruned away from regions of active transcription by the FACT histone chaperone (Sun et al., 2018). macroH2A variants inhibit transcription initiation by interfering with p300-dependent histone acetylation (Doyen, An, et al., 2006).

Additionally, macroH2A variants interfere with nucleosome remodeling and sliding by SWI/SNF and ACF remodelers through the variant's non-histone domain (Angelov et al., 2003; Doyen, An, et al., 2006; Sun et al., 2018; Timinszky, 2009). Importantly, macroH2A also impedes reactivation of pluripotent genes during reprogramming to iPSCs (Gaspar-Maia et al., 2013).

7. Histone H3 variants

7.1 H3.3 marks regulatory, repetitive, and actively transcribed regions

H3.3 is deposited at regions of active transcription, a process highly regulated by multiple factors including Histone Cell Cycle Regulation Defective Homolog A (HIRA), ATRX, Death domain associated protein (DAXX) and p400 (Ahmad & Henikoff, 2002; Chen, Zhao, et al., 2013; Elsaesser, Goldberg, & Allis, 2010; Hake et al., 2006; Jin & Felsenfeld, 2007; Jin et al., 2009; Lewis, Elsaesser, Noh, Stadler, & Allis, 2010; McKittrick et al., 2004; Pradhan et al., 2016; Wirbelauer, Bell, & Schubeler, 2005). The structure of H3.3 is highly similar to that of H3.1 and H3.2, with the exception of five amino acids found on the surface of the H3/H4 tetramer (Tachiwana, Osakabe, et al., 2011), which suggests that factor binding to H3.3-containing nucleosomes could be different from core H3-containing nucleosomes. Additionally, murine ES cell H3.3 S31 can be phosphorylated, a mark that promotes activity of the p300 HAT and acetylation of enhancers (Martire et al., 2019).

H3.3 can be deposited in both replication-dependent and replication-independent manners (Ahmad & Henikoff, 2002; Tagami, Ray-Gallet, Almouzni, & Nakatani, 2004). Interestingly, H3.3 exchange occurs at both active and, with a lower turnover rate, inactive promoters (Kraushaar et al., 2013). Venkatesh and Workman have proposed that H3.3 exchange at inactive promoters may maintain a poised state, wherein the promoter undergoes rapid histone turnover in preparation for gene activation and subsequent transcription (Venkatesh & Workman, 2015). At many promoters and enhancers, H3.3 co-localizes with H2A.Z; these nucleosomes are believed to be less stable than nucleosomes containing core histones and thus more conducive to open chromatin structure and transcriptional activation (Chen, Zhao, et al., 2013; Jin & Felsenfeld, 2007; Jin et al., 2009).

In addition to actively transcribed and regulatory regions of the genome, H3.3 is specifically deposited at telomeres and other repeat regions.

While HIRA deposits H3.3 at actively transcribed regions, ATRX and DAXX regulate H3.3 deposition at telomeres, pericentric heterochromatin, and other regions of repetitive chromatin (Dyer, Qadeer, Valle-Garcia, & Bernstein, 2017; Lewis et al., 2010). HIRA-mediated H3.3 deposition facilitates PRC2 binding at developmental loci in ES cells and subsequent H3K27me3 of these promoters (Banaszynski et al., 2013), underscoring a specialized role for H3.3 in ES cells that will be expanded upon later in the chapter.

7.2 CENP-A is a centromere-specific variant of histone H3

CENP-A is an H3 variant that is found at centromeres and is conserved from *S. cerevisiae* (in which it is called CENH3) through humans. It bears relatively low sequence identity to canonical H3 (approximately 60% in the histone fold domain and less in the N-terminus) (Chittori et al., 2018; Tachiwana, Kagawa, et al., 2011; Yoda et al., 2000). Regulation of CENP-A synthesis and deposition are key to proper progression through the cell cycle and mitosis (Earnshaw & Rothfield, 1985; Hori et al., 2014; Nardi, Zasadzinska, Stellfox, Knippler, & Foltz, 2016; Niikura et al., 2015; Okada, Okawa, Isobe, & Fukagawa, 2009; Palmer, O'Day, Wener, Andrews, & Margolis, 1987; Shuaib, Ouararhni, Dimitrov, & Hamiche, 2010).

8. Histone chaperones

8.1 FACT facilitates H2A/H2B dimer exchange to promote nucleosome-templated activities

Facilitates Chromatin Transactions (FACT) is a highly conserved complex that plays important roles in several nuclear processes including DNA replication, DNA repair, and transcription initiation and elongation (reviewed in Duina, 2011; Formosa, 2008, 2012; Gurova, Chang, Valieva, Sandlesh, & Studitsky, 2018). FACT is a heterodimer composed of suppressor of Ty 16 (SPT16) and Structure-Specific Recognition Protein 1 (SSRP1) (Orphanides, LeRoy, Chang, Luse, & Reinberg, 1998; Orphanides, Wu, Lane, Hampsey, & Reinberg, 1999). In human cells, FACT was initially identified by its ability to allow RNA Polymerase II to transcribe through nucleosomal DNA (Orphanides et al., 1998). Additionally, human FACT binds histone H2A/H2B dimers while human SSRP1 and *S. pombe* SPT16 can both bind H3/H4 dimers (Belotserkovskaya et al., 2003; Ransom, Dennehey, & Tyler, 2010; Stuwe et al., 2008).

Therefore, FACT can act as both an H2A/H2B and H3/H4 histone chaperone. FACT activity is also required for proper regulation of transcription initiation. SPT16 nucleosome reassembly occurs over certain gene promoters and this activity is required for proper transcriptional repression of these genes (Adkins & Tyler, 2006). In addition to contributing to transcription initiation and elongation, FACT participates in mRNA nuclear export (Hautbergue et al., 2009; Herold, Teixeira, & Izaurralde, 2003). These studies demonstrate a multifaceted role for the FACT complex in regulating multiple stages of transcription and mRNA processing.

Recently, FACT has been implicated in maintenance of pluripotency in stem cells. While FACT had long been thought of as a marker of actively proliferating cells (Hertel et al., 1999), FACT has more recently been associated with pluripotent cells than with actively proliferating cells (marked by Ki-67 and SPT16 intestinal immunohistochemistry) (Garcia et al., 2011). Recently, it was demonstrated that FACT is required for maintenance of pluripotency (Mylonas & Tessarz, 2018), a role consistent with the previously mentioned work as well as FACT's known interactions with the master regulator of pluripotency OCT4 (Ding, Xu, Faiola, Ma'ayan, & Wang, 2012; Pardo et al., 2010). Subsequent studies have suggested that FACT inhibition can both facilitate and impede establishment of pluripotency but that the complex may not be necessary for maintenance of pluripotency (Kolundzic et al., 2018; Shen, Formosa, & Tantin, 2018). Further studies will be necessary to determine the true extent of FACT's role(s) in stem cells and pluripotency.

8.2 CAF1 and ASF1 promote incorporation of H3 and H4 onto newly synthesized DNA

Chromatin Assembly Factor 1 (CAF1) is a histone chaperone responsible for loading histones H3 and H4 onto newly synthesized DNA (Smith & Stillman, 1989; Stillman, 1986). CAF1 binds multiple H3/H4 dimers at once, thereby promoting tetramerization and nucleosome formation on DNA (Liu, Roemer, Port, & Churchill, 2012). CAF1 does so in parallel with Anti-Silencing Factor 1 (ASF1). Like CAF1, ASF1 binds H3/H4 dimers to promote loading onto newly synthesized DNA; unlike CAF1, however, monomeric ASF1 binds only one H3/H4 dimer at a time, and ASF1 is excluded from a CAF1-H3/H4 complex via a tight interaction between CAF1 and the histone dimers (Liu et al., 2012). CAF1 and ASF1 are most active during S-phase as loading of H3/H4 occurs on newly synthesized

DNA. Beyond post-replication H3/H4 incorporation, CAF1 and ASF1 are also active in passing through the DNA damage checkpoint upon successful completion of DNA repair (Kim & Haber, 2009). In adult stem cells, CAF1 prevents expression of differentiation-associated genes (Clemot, Molla-Herman, Mathieu, Huynh, & Dostatni, 2018).

ASF1 has key epigenetic roles via its interaction with the RTT109 histone acetyltransferase; without the ASF1-H3/H4 interaction, RTT109 cannot perform H3K56 acetylation (Recht et al., 2006). ASF1 can also recruit the HAT1 acetyltransferase to modify H4K5 and H4K12 (Fillingham et al., 2008), two modifications that have been implicated in memory and learning (Park, Rehrauer, & Mansuy, 2013; Peleg et al., 2010). H4K5ac also prevents mitotic compaction of some cell type-specific genes, allowing for quick resumption of transcription following cytokinesis (Zhao et al., 2011), a role that may be particularly important in stem cells for a poised differentiation state.

8.3 HIRA deposits H3.3 at actively transcribed and regulatory regions of chromatin

Histone Cell Cycle Regulation Defective Homolog A (HIRA) performs replication-independent incorporation of histone variants, including H3.3 that is associated with active transcription and regulatory regions (Lamour et al., 1995; Lorain et al., 1996). ASF1 seems to assist HIRA in H3.3 deposition, as sites that are HIRA-bound, but not ASF1-bound, are not enriched for H3.3 (Adam, Polo, & Almouzni, 2013; Pchelintsev et al., 2013; Tang et al., 2006). Like CAF1, HIRA has a role in overcoming DNA damage to prepare chromatin for resumption of transcription after genotoxic events (Adam et al., 2013). While HIRA is essential for H3.3 enrichment at active and repressed genes, it is not essential for H3.3 enrichment at telomeres (Goldberg et al., 2010); rather, the histone chaperones ATRX and DAXX that fill this role (Lewis et al., 2010). Interestingly, upon depletion of HIRA (and/or H3.3 itself), ES cells exhibit diminished PRC2 occupancy and decreased H3K27me3 at bivalent genes, though this H3K27me3 loss does not correlate with a global increase in gene expression (Banaszynski et al., 2013). Furthermore, ES cells depleted of H3.3 display a low nucleosome turnover rate, improper lineage specification despite functional pluripotency, increased expression of trophectoderm markers, and a failure of HIRA to localize to chromatin (Banaszynski et al., 2013).

8.4 ATRX/DAXX deposit H3.3 at telomeres and pericentric heterochromatin

ATRX and DAXX form an H3.3-depositing complex, wherein DAXX acts as the primary H3.3 chaperone (Drane, Ouararhni, Depaux, Shuaib, & Hamiche, 2010; Lewis et al., 2010) and ATRX appears to recognize H3K9me3 (Dyer et al., 2017). ATRX is a transcriptional repressor (Gibbons et al., 2000; Gibbons, Picketts, Villard, & Higgs, 1995) that also directly interacts with PRC2; this interaction may be important to recruit PRC2 to genomic loci (Dyer et al., 2017; Sarma et al., 2014). DAXX is essential for H3.3 deposition at telomeres in murine ES cells (Lewis et al., 2010) and at pericentric heterochromatin regions in murine embryonic fibroblasts (Drane et al., 2010).

8.5 The NAP1 family import newly translated histones from the cytoplasm to the nucleus

The Nucleosome Assembly Protein 1 (NAP1) chaperone was first identified based on its ability to facilitate nucleosome assembly *in vitro* (Ishimi, Yasuda, Hirosumi, Hanaoka, & Yamada, 1983). NAP1 is primarily involved in nuclear histone import and regulation of transcription (Avvakumov, Nourani, & Cote, 2011). NAP1 binds newly translated H2A and H2B proteins in the cytosol and shuttles them into the nucleus via a direct interaction with Importin 9 (Mosammaparast, Ewart, & Pemberton, 2002). NAP1 also has H3/H4 tetramer binding activity, including the ability to load an entire H3/H4 tetramer onto DNA *in vitro* (Bowman et al., 2011; McBryant et al., 2003).

8.6 INO80 family members possess both nucleosome remodeling and histone chaperone activities

INO80 family members, such as Tip60-p400, act as histone chaperones as well as ATP-dependent nucleosome remodeling complexes. p400 facilitates incorporation of H3.3 at promoters and enhancers *in vitro* (Pradhan et al., 2016). Additionally, *Drosophila* Tip60 acetylates γH2Av—the *Drosophila* homolog of γH2A.X—and exchanges it with an unphosphorylated H2Av (Kusch et al., 2004). Along with p400, the SRCAP complex is able to incorporate H2A.Z/H2B dimers into mononucleosomes (Ruhl et al., 2006; Wong et al., 2007). Finally, INO80 globally regulates H2A.Z incorporation via direction of H2A.Z/H2B localization and through a histone exchange function that allows the complex to switch H2A.Z/H2B and H2A/H2B dimers (Papamichos-Chronakis et al., 2011). Specifically, INO80 performs this exchange by translocating along DNA at the H2A/H2B dimer and displacing DNA from the nucleosome surface, thereby creating torsional strain and permitting exchange of H2A.Z/H2B dimers without additional chaperones (Brahma et al., 2017).

9. Chromatin structure is dynamic and highly regulated

Chromatin structure is the result of dynamic interplay between histone positioning, variants, modifications, and localization, with the product being DNA compaction and regulated gene expression. Each of these processes must be individually regulated and function synergistically to maintain the integrity of the genome. In spite of often contradictory roles and complex regulatory pathways that require fidelity from each member, cells are able to maintain a balance between gene expression and repression through precise regulation by chromatin modifiers. In the remainder of this chapter, we will discuss in more detail how stem cells make use of unique nucleosome remodeling factors, histone methylation patterns, trends in histone variant incorporation, and long-range chromatin interactions. In doing so, stem cells carefully regulate their chromatin state to preserve stem cell state, as well as to facilitate proper differentiation when required by the organism.

10. Stem cell chromatin is dynamic and tuned to regulate cell fate

ES cells carefully regulate their two defining properties—self-renewal and pluripotency—through a suite of coordinated changes in gene expression modulated by chromatin structure. In ES cells, careful regulation of chromatin state is crucial to establish and preserve the qualities that both prevent and drive differentiation of ES cells into adult cell types. While there are many factors that drive an ES cell's decision to differentiate into a progenitor cell among one of the three germ layers—including cell signaling (Reya & Clevers, 2005), physical strain (Miller & Davidson, 2013), and metabolic changes (Romito & Cobellis, 2016)—the key to ES cell pluripotency lies within their regulated gene expression; chromatin regulation therefore plays an enormous role in ES cell fate and function. Between a suite of transcription factors called master regulators of pluripotency, a largely euchromatic state of compaction, and carefully curated epigenetic marks, ES cells establish and preserve this pluripotent state by diverse mechanisms that harness chromatin's dynamic nature. As one may expect based on the diverse range of specialized eukaryotic cells, this regulation can be carried out through multiple mechanisms. Depending on differentiation signals received, pluripotent ES cells are able to upregulate specific subsets of genes corresponding to any lineage. Thus, the structure and features of ES cell chromatin largely differ from that of most somatic cell types. In this section, we will discuss the ways in which ES cells regulate their chromatin state to maintain proper cellular function and fate.

ES cell chromatin is largely euchromatic (Young, 2011), and thereby permissive of gene transcription. This heightened accessibility does not always directly correspond with increased gene transcription, however. ES cell chromatin structure is preserved by a diverse suite of nucleosome remodeling factors, including the ES cell-specific SWI/SNF family complex esBAF. While the remainder of this chapter will focus on nucleosome remodeling factors, histone-based chromatin organization, and higher-order chromatin structures, it is important to note that pluripotency is affected and regulated by numerous processes that are not direct results of chromatin state changes.

11. ES cells carefully regulate their chromatin via specialized transcription factors

11.1 Master regulators of pluripotency

Among ES cell transcription factors, OCT3/4 (henceforth referred to as OCT4), SOX2, and NANOG are commonly known as core factors that maintain ES cell pluripotency in conjunction with a large network of proteins that includes both transcription factors and chromatin modifiers (Morey, Santanach, & Di Croce, 2015; Orkin & Hochedlinger, 2011). OCT4, SOX2, and NANOG cooperate in a positive feedback loop to maintain their own expression in pluripotent cells as well as the expression of known ES cell regulators such as the LIF signaling pathway or microRNA genes (Chen et al., 2008; Marson et al., 2008; Okumura-Nakanishi, Saito, Niwa, & Ishikawa, 2005; Tomioka et al., 2002; Young, 2011). Chromatin regulatory complexes in both ATP-dependent nucleosome remodeler and histone modifying families have been shown to interact with OCT4, SOX2, or NANOG (Ang et al., 2011; Liang et al., 2008), and the pluripotency network also includes long non-coding RNAs (Guttman et al., 2011).

Differentiated or somatic cells can be reprogrammed to the pluripotent state—termed induced pluripotent stem cells—through ectopic expression of the core pluripotency factors OCT4, SOX2, and KLF4 (Takahashi & Yamanaka, 2006). These cells reactivate their endogenous pluripotency factors, deactivate somatic genes, and re-establish activating histone marks and other hallmarks of ES cell chromatin structure (reviewed in Apostolou & Hochedlinger, 2013).

11.2 Pioneer transcription factors

A subset of transcription factors are able to bind to compacted chromatin and modulate local nucleosome architecture to facilitate downstream events; these transcription factors are known as pioneer transcription factors (reviewed in Iwafuchi-Doi, 2019; Mayran & Drouin, 2018; Zaret & Carroll, 2011). Recently, many traditionally described transcription factors have been identified as pioneer transcription factors, but the mechanism of function for many of these factors remains incompletely understood. Transcription factors with pioneering activity have been shown to move through the nucleus more slowly than other transcription factors in fluorescence recovery after photobleaching experiments, suggesting that these factors may bind to closed chromatin nonspecifically and scan for a binding site at which to remain (Iwafuchi-Doi, 2019; Zaret, Lerner, & Iwafuchi-Doi, 2016). With respect to stem cells, it has been proposed that OCT4 can act as a pioneer transcription factor (Zaret & Carroll, 2011) through recruitment of the esBAF complex to inaccessible regulatory sites. OCT4 interacts with major nucleosome remodeling factors, including esBAF, lending credence to this hypothesis (Ding et al., 2012; Pardo et al., 2010; van den Berg et al., 2010). Nucleosome remodeling by esBAF at these sites subsequently allows for binding and regulation by the remaining core transcription factors (Hainer & Fazzio, 2015; King & Klose, 2017). In addition to OCT4, SOX2 has pioneering activity, and both pluripotency factors (as well as the pioneer factor FOXA) bind to cell-type-restricted enhancers to premark them for activation in differentiated cells (Iwafuchi-Doi et al., 2016; Kim et al., 2018). This premarking function is required for future enhancer activation in differentiated cells, and premarked enhancers are in an open configuration but are not decorated with canonical enhancer histone modifications (H3K27ac, H3K4me1, and H3K4me2) (Kim et al., 2018). Along with OCT4 and SOX2, additional pioneer factors, such as p63, GATA, and AP-1 have been reported to recruit nucleosome remodeling factors to their target sites (Bao et al., 2015; Engelen et al., 2011; Hu et al., 2011; Takaku et al., 2016; Vierbuchen et al., 2017). Recent studies have shown that esBAF is likely required to remodel nucleosome structure prior to factor binding during embryogenesis (Hainer et al., 2019). Pioneer transcription factors are clearly key regulators of pluripotency and cell fate; however, their regulation of, by, and/or with nucleosome remodeling factors (such as OCT4 and esBAF) remains incompletely understood.

12. Embryonic stem cell chromatin is poised for action

ES cell chromatin is uniquely packaged to promote expression of genes required for self-renewal and to prevent differentiation into progenitor cell populations unless acted upon by the proper regulatory pathway. Therefore, ES cells maintain themselves in a poised state, whereby the cells can exit this self-renewal cycle to differentiate into the progenitors for any of the 200+ cell types found in the adult organism. ES cell chromatin is packaged and regulated to accomplish this activity using numerous methods, including a high euchromatin-to-heterochromatin ratio, bivalent gene promoters, and specialized nucleosome remodelers and transcription factors (Azuara et al., 2006; de Dieuleveult et al., 2016; Harikumar & Meshorer, 2015; Orkin & Hochedlinger, 2011). In the following sections, we will discuss the features that define chromatin state in ES cells and the ways in which these features are established and maintained.

13. Histone modifications are specifically regulated in stem cells to maintain pluripotency and facilitate differentiation

13.1 H3K56ac regulates pluripotency factors and developmental regulators

In ES cells, H3K56ac is thought to be important for both pluripotency and differentiation through its interactions with OCT4, SOX2, and NANOG at pluripotency genes, and through its enrichment at developmental regulators such as the *HOX* genes during differentiation (Tan, Xue, Song, & Grunstein, 2013; Xie et al., 2009). Of the core pluripotency factors, only OCT4 directly interacts with H3K56ac (Tan et al., 2013); however, H3K56ac can be found at many of the OCT4/SOX2/NANOG binding sites (Xie et al., 2009). Deletion of *Sirt6,* a NAD-dependent deacetylase that targets H3K56, leads to upregulation of *OCT4*, *SOX2*, and *NANOG* during differentiation, and mis-expression of markers from all three germ layers (Etchegaray & Mostoslavsky, 2015). H3K56 deacetylation by SIRT6 functions to maintain genome integrity, while hyperacetylation of H3K56 is associated with genome instability and sensitivity to genotoxins, such as MMS (Yang, Zwaans, Eckersdorff, & Lombard, 2009).

13.2 Bivalent promoters mark lowly expressed but poised genes in ES cells

Overall, ES cell chromatin is enriched for active histone modifications such as H3K4me3 and H3K9ac (reviewed in Meshorer & Misteli, 2006). Pluripotent chromatin is, however, distinguishable by a subset of gene promoters (between 3000 and 4000) (Li, Lian, Dai, Xiang, & Dai, 2013) marked with both active (H3K4me3) and repressive (H3K27me3) histone modifications, placed, respectively, by complexes in the Trithorax and Polycomb (PRC2) groups (Fig. 7). The first evidence of bivalent chromatin marks in murine ES cells was identified using ChIP and tiling oligonucleotide arrays where a large region of H3K27me3 was accompanied by smaller patches of H3K4me3 (Azuara et al., 2006; Bernstein et al., 2006). These bivalent promoters mark lowly expressed genes in undifferentiated cells, and are therefore poised for activation in response to developmental signaling (Azuara et al., 2006; Bernstein et al., 2006; Voigt, Tee, & Reinberg, 2013). Transcription of these genes is repressed, but not blocked: bivalent genes remain open to transcription factors and are therefore poised for upregulation upon differentiation as required by the organism (Voigt et al., 2013). During differentiation, one mark is typically lost from the promoter while the other becomes enriched depending on whether the gene is expressed or silenced. In ES cells, bivalent chromatin is resolved by SWI/SNF-mediated eviction of PcG proteins (Stanton et al., 2017) and by ASF1A-driven clearing of H3K27me3 (Gao, Gan, Lou, & Zhang, 2018). Eviction of the repressive H3K27me3 mark and the PcG proteins responsible for its placement allows derepression of bivalent lineage specification genes (Stanton et al., 2017). Consistent with their poised state, bivalent genes tend to display low levels of DNA methylation, another repressive epigenetic mark, with the CpG islands of germline and Polycomb genes becoming increasingly methylated as ES cells differentiate and commit to specific lineages (Mohn et al., 2008; Vastenhouw & Schier, 2012). Bivalent genes are largely lineage specification genes (such as members of the SOX, FOX, PAX, IRX, and POU/OCT families; Bernstein et al., 2006) and remain accessible but not expressed in ES cells.

Bivalent chromatin marks are largely specific to stem cells (including ES cells, iPSCs, and hematopoietic stem cells Cui et al., 2009; Harikumar & Meshorer, 2015). Bivalent genes have, however, been identified in non-stem cells, including differentiated T cells and some cancer cell types (Bapat et al., 2010; Barski et al., 2007; Lin et al., 2015; McGarvey et al., 2008; Rodriguez et al., 2008; Roh, Cuddapah, Cui, & Zhao, 2006). While bivalent promoters were initially thought to be restricted to developmentally-regulated genes to enable fast transition between active and repressed expression states, it is clear that bivalency is more complex, as bivalent promoters have been identified in different gene families in numerous different cell types. Furthermore, regulation of the bivalent state is carried out by a wide variety of different proteins and regulators that extend well beyond the Trithorax and Polycomb group members that place the epigenetic marks of bivalency. Some important regulators include nucleosome remodeling complexes such as Tip60-p400, esBAF, CHD7, and NuRD (Alajem et al., 2015; Fazzio et al., 2008b; Ho, Jothi, et al., 2009; Lei et al., 2015; Reynolds, Salmon-Divon, et al., 2012; Schnetz et al., 2010). Tip60-p400 significantly co-localizes with H3K4me3, particularly near the TSS, and the complex mostly acts to repress gene expression in ES cells (Fazzio et al., 2008b). Knockdown of Tip60-p400 results in deregulation of 4% of genes, most of which are upregulated; interestingly, many of those found to be upregulated were classical bivalent early-differentiation genes, which are normally silenced in ES cells (Fazzio et al., 2008b). The esBAF subunit BAF60A was mapped by ChIP-seq to reveal a distribution similar to that of H3K27me3 around TSSs, with significant enrichment at promoters of bivalent genes; after BAF60A depletion, these marks were found to be significantly redistributed genome-wide (Alajem et al., 2015). While esBAF tends to function as an activator of gene expression and PRC2 as a repressor, the two function synergistically to properly regulate expression of differentiation-associated genes (Ho et al., 2011). CHD7 associates with a PcG cluster, containing SUZ12, RING1B, and EZH2, suggesting a connection between CHD7 and H3K27me3 (Schnetz et al., 2010). Finally, the NuRD ATPase CHD4 removes acetyl groups from H3K27 in ES cells, thereby allowing the subsequent recruitment of PcG proteins and H3K27me3 (Reynolds, Salmon-Divon, et al., 2012). CHD4 also interacts with the H3K4 demethylase LSD1, which occupies a majority of active genes and approximately two thirds of bivalent genes (Whyte et al., 2012). Nucleosome remodeling factors play critical roles in cell fate decisions and maintenance of stem cell characteristics; while their relationships with bivalency have been discussed, the next section will substantially expand upon these roles as well as other functions of nucleosome remodeling factors in ES cells.

14. Chromatin state is precisely regulated by nucleosome remodeling factors in ES cells

14.1 esBAF maintains stem cell pluripotency by preserving chromatin state

One of the best described ES cell-specific nucleosome remodeling complex is esBAF. esBAF is characterized by the exclusive use of BRG1 as its ATPase, whereas BAF complexes in differentiated cells use a combination of BRG1 and BRM (Ho, Jothi, et al., 2009; Ho, Ronan, et al., 2009). Murine esBAF is also assembled around a BAF155 homodimer, rather than a BAF155/BAF170 heterodimer, and exclusively uses the ARID1A and BAF53A subunits (Mashtalir et al., 2018), rather than other possible subunits (ARID1B, ARID2, BAF53B). Interestingly, human ES cells are reliant on BAF170 for maintenance of pluripotency, as well as BAF155 (Zhang et al., 2014). In addition to the canonical esBAF complex, a non-canonical BRD9-containing esBAF complex has recently been described, as has a BRD7-containing PBAF complex in ES cells (Gatchalian et al., 2018; Kaeser, Aslanian, Dong, Yates, & Emerson, 2008). BAF complexes are crucial for differentiation of ES cell state. The Panning group confirmed the role of esBAF subunits in ES cell pluripotency via an RNAi screen (Fazzio et al., 2008b). The next year, the Paddison group elucidated a mechanism by which BAF is required for ES cell development (Schaniel et al., 2009). During differentiation and upon RNAi-mediated depletion of the core esBAF components BAF57 and BAF155, NANOG was upregulated and heterochromatin formation was stifled, demonstrating the requirement for esBAF in NANOG silencing and appropriate chromatin compaction during differentiation (Schaniel et al., 2009). Functionally, esBAF preserves pluripotency in ES cells largely by facilitating LIF/STAT3 signaling and by coordinating function (both stimulatory and antagonistic) with PcG proteins (Ho et al., 2011). Interestingly, BAF subunits BAF155 and BRG1 are also major players in regulation of X-chromosome inactivation, creating a nucleosome-depleted region at X-located gene promoters that is necessary for gene silencing (Keniry et al., 2019). Upon depletion of esBAF, H2A.Z occupancy is greatly reduced, and subnucleosome formation appears to be more prevalent at sites of H2A.Z occupancy (Hainer & Fazzio, 2015) suggesting that esBAF regulates H2A.Z localization and incorporation in stem cells, though it is unclear precisely how esBAF does so.

Underscoring the importance of esBAF in development, maternal BRG1 is essential for murine zygotic genome activation (in which the maternally contributed RNA and proteins are silenced and the zygote begins transcription of its own genes) by the 2-cell stage (Bouniol, Nguyen, & Debey, 1995); furthermore, zygotic BRG1 is essential for proliferation of both the inner cell mass and the trophectoderm of blastocysts (Bultman et al., 2000). Deletion of the murine homolog of BAF155, SRG3, is lethal peri-implantation, showing severe deficiencies in vascular formation and circulation when inactivated via a transgenic construct (Han et al., 2008). Haploinsufficiency of SNF5/BAF47 predisposes developing mice to malignant rhabdoid tumors, a cancer brought on by biallelic inactivation of BAF47 (Nakayama et al., 2017; Roberts, Galusha, McMenamin, Fletcher, & Orkin, 2000; Wang et al., 2019). Interestingly, *Brm*$^{-/-}$ mice develop normally, highlighting the divergent roles of the two BAF ATPases (Reyes et al., 1998). Similarly to its role with PcG targets, esBAF functions in opposition to MBD3/NuRD complex at shared targets (Yildirim et al., 2011), suggesting that esBAF is a key regulator of balance between activating and silencing pathways in ES cells. Further highlighting its importance, BRG1 depletion in blastocysts results in reduced expression of OCT4 and NANOG—among other pluripotency-associated genes—while differentiation-associated gene expression rises (Kidder, Palmer, & Knott, 2009). Although we have highlighted specific roles for esBAF as an activator of transcription, it is important to note that this is not an exclusive function; indeed, esBAF suppresses non-coding transcription from over 57,000 nucleosome-depleted regions in ES cells (Hainer et al., 2015), demonstrating a widespread role in transcriptional repression as well.

14.2 CHD proteins are regulators of ES cell pluripotency

Among the most studied nucleosome remodeling factors in ES cells is CHD1. CHD1 uses two N-terminal tandem chromodomains to bind methylated lysine residues in the histone tails and promote nucleosome sliding. CHD1 is essential for maintaining pluripotency in naive stem cells via interaction with the Mediator complex (Gaspar-Maia et al., 2009). Mediator is a large (~30 subunit) coactivator complex that is necessary for RNA Polymerase II transcription and has both permissive and repressive roles in differentiation via interactions with various lineage specification regulators (Yin & Wang, 2014), such as NANOG (Tutter et al., 2009) and SOX9 (Zhou et al., 2002). At gene promoters, Mediator facilitates assembly of the pre-initiation complex and subsequently recruits CHD1 to transcription start sites (Lin et al., 2011). CHD1 features an N-terminal serine-rich region that is not essential for ES cell viability, but is required for pluripotency (Piatti et al., 2015). CHD1 is also essential for establishment of pluripotency during reprogramming from fibroblasts to induced pluripotent stem cells (Gaspar-Maia et al., 2009). Specifically, *Chd1*$^{-/-}$ ES cells cannot give rise to primitive endoderm, but rather differentiate along neural lineages (Gaspar-Maia et al., 2009). CHD1 is enriched at active genes, but specifically depleted at bivalent genes, suggesting a role as an activator of transcription (Gaspar-Maia et al., 2009). Beyond roles that specifically promote pluripotency in ES cells, CHD1 is also important for ES cells to prevent spurious heterochromatin formation over accessible regions of chromatin (Lin et al., 2011). *Chd1*$^{-/-}$ mouse embryos display reduced mRNA and intergenic RNA transcription (Guzman-Ayala et al., 2015), suggesting that the remodeler is a key factor in maintaining ES cells' distinct rapid transcription rate, a property known as "hypertranscription" (Efroni et al., 2008). In addition to facilitating transcription genome-wide, CHD1 has recently been shown to facilitate repair of promoter-proximal DNA double-strand breaks by recruitment of DNA repair proteins (Bulut-Karslioglu et al., 2019). CHD7 has also been implicated in maintenance of pluripotency in ES cells, with a role that may to be similar to that of CHD1 in murine ES cells (Zentner, Tesar, & Scacheri, 2011). CHD7, however, associates with three distinct classes of protein clusters: one associated with enhancers (including CHD7, p400, OCT4, SIX2, NANOG, SMAD1, and STAT3), another with c-MYC and n-MYC-regulated genes, and a third associated with PcG (Schnetz et al., 2010). In addition, CHD7 regulates neural crest formation from multipotent cells through association with the PBAF complex (Bajpai et al., 2010). Therefore, CHD7 regulates the expression of some ES cell-specific genes, possibly through interaction with other regulators of ES cell gene expression.

14.3 MBD3/NuRD generally represses expression of differentiation genes

In an RNAi screen to identify chromatin regulators important for murine ES cell self-renewal, the Panning group identified MBD3 as important for maintenance of the ES cell state (Fazzio et al., 2008b). ES cells derived from *Mbd3*$^{-/-}$ mouse embryos are capable of self-renewal in culture in the absence of LIF and are unable to differentiate properly, with differentiation skewed toward the trophectoderm lineage (Kaji et al., 2006; Kaji, Nichols, & Hendrich, 2007; Zhu, Fang, Li, & Zhang, 2009). MBD3 is essential for MBD3/NuRD complex assembly (Kaji et al., 2006). MBD3/NuRD target genes exhibit increased H3K27ac and decreased H3K27me3 in *Mbd3*$^{-/-}$ ES cells (Reynolds, Latos, et al., 2012; Reynolds, Salmon-Divon, et al., 2012), suggesting that MBD3 regulates pluripotency genes through deacetylation and recruitment of PRC2. MBD3/NuRD also regulates nucleosome positioning across enhancer and promoter regions to control factor access to enhancers during the lineage commitment process, as well as prevent coactivators (e.g., Mediator) and RNA Polymerase II from functioning at these regions (Bornelov et al., 2018). Furthermore, depletion of the MBD3/NuRD member MTA proteins leads to improper expression of differentiation-associated genes, resulting in inability to contribute to embryogenesis (Burgold et al., 2019). MBD3/NuRD is necessary for deacetylation of H3K27, although MBD3/NuRD does not appear to act at bivalent genes, but rather is repelled from chromatin binding by H3K27me3 (Harikumar & Meshorer, 2015; Kaji et al., 2006). MBD3 is also essential for pluripotency—though not viability—of ES cells; specifically, MBD3-deficient cells fail to properly silence lineage specification genes (Kaji et al., 2006). In ES cells, depletion of the MBD3/NuRD ATPase CHD4 leads to loss of self-renewal, decreased proliferation, and increased embryoid body differentiation (Zhao, Han, et al., 2017). In addition, the MBD3/NuRD histone deacetylase HDAC1 binds promoters of pluripotency genes; cells treated with an HDAC inhibitor result in a differentiation phenotype (Kidder & Palmer, 2012). Furthermore, MBD3C—a stem cell-specific isoform—interacts with the MLL1 H3K4 methyltransferase complex component WDR5, an interaction that is unique to stem cells (Ee et al., 2017). It remains to be determined whether this interaction or an alternative function for MBD3C contributes to ES cell pluripotency or self-renewal.

14.4 The ISWI remodeler ATPase SNF2H is essential during development

Deletion of the ISWI ATPase SNF2H is embryonic lethal in mice, suggesting high functional importance in ES cells (Saladi & de la Serna, 2010). Specifically, SNF2H deletion is lethal pre-implantation due to cell growth defects of blastocyst embryos (Stopka & Skoultchi, 2003). It is possible, however, to generate ES cells with a SNF2H functional knockout caused by a frameshift in exon 6 (Barisic, Stadler, Iurlaro, & Schübeler, 2019). Although yeast ISW1 has been known to regulate nucleosome spacing, this functional knockout enabled *in vivo* confirmation of this function in mammalian cells (Barisic et al., 2019). Complicating the issue of SNF2H requirement is the extreme variability of SNF2H-containing complexes. Because SNF2H is incorporated into the RSF, WICH, NoRC, CHRAC, and ACF complexes, it is difficult to identify the precise roles for which ISWI complexes are essential in ES cells. While the alternative ISWI ATPase, SNF2L, is not embryonic lethal in mice, depletion of the ISWI family NURF complex member BPTF disrupts expression of lineage specification genes across all three germ layers, particularly those regulated by SMAD, suggesting a link between BPTF and SMAD signaling in regulation of lineage-specific genes (Landry et al., 2008). Interestingly, depletion of H3K4me3 leads to eviction of BPTF from chromatin, defects in recruitment of SNF2L, and compromised spatial regulation of *HOX* gene expression (Wysocka et al., 2006). One function of SNF2H in cells that contributes to its essential nature is recruitment to DNA double-strand break sites. The deacetylase SIRT6 is rapidly recruited to double-strand breaks and subsequently recruits SNF2H, which then facilitates open chromatin at the break sites (Kokavec et al., 2017). Both SIRT6 and SNF2H are required for recruitment of other repair factors, including BRCA1, 53BP1, and RPA, suggesting a role for ISWI complexes in preventing genome instability (Kokavec et al., 2017). In hematopoietic stem and progenitor cells, SNF2H is necessary to promote maturation into erythroid and myeloid lineages, as well as to allow proliferation of committed erythroid cell populations (Kokavec et al., 2017). SNF2H and SNF2L have been shown to position nucleosomes adjacent to CTCF and other transcription factor binding sites (Wiechens et al., 2016); indeed, SNF2H/L seem to regulate a specific group of transcription factors including CTCF, whereas BAF regulates regulates distinct factors such as OCT4, SOX2, NANOG, and REST (Barisic et al., 2019). Selective reliance of CTCF on SNF2H suggests that higher-order chromatin domains and topologically-associated domains (TADs) may depend upon ISWI activity but not on BRG1.

14.5 INO80 remodelers repress transcription of differentiation-associated genes

Despite roles that are often associated with active transcription—like the histone acetyltransferase activity of Tip60-p400 and deposition of H2A.Z and

H3.3—INO80-family complexes fulfill important transcriptional repression roles in ES cells (Cai et al., 2005; Doyon et al., 2004; Pradhan et al., 2016; Ruhl et al., 2006; Squatrito, Gorrini, & Amati, 2006). As previously noted, the INO80 superfamily member Tip60-p400 is essential for normal self-renewal in ES cells, and *Tip60* knockout is lethal at the blastocyst stage (Fazzio et al., 2008a; Hu et al., 2009). Tip60-p400 maintains pluripotency in ES cells by repressing differentiation-associated genes, despite its role as a HAT, which has traditionally been associated with activation of transcription (Fazzio et al., 2008a). This may be due to a unique interaction of Tip60-p400 with the traditionally cytosolic HDAC6, which shows nuclear localization in ES cells and promotes recruitment of Tip60-p400 to target genes, especially differentiation-associated genes that are normally repressed by Tip60-p400 (Chen, Hung, et al., 2013). While HDAC6 requires its deacetylase domains to silence differentiation genes, it does not regulate gene expression by deacetylating histones near Tip60-p400 target promoters; rather, the catalytic domains of HDAC6 are required for its interaction with Tip60-p400 (Chen, Hung, et al., 2013). Interestingly, Tip60 shows acetyltransferase-dependent activity in differentiation but acetyltransferase-independent activity in promoting self-renewal of ES cells (Acharya et al., 2017). Unlike *Tip60*$^{-/-}$ mice, acetyltransferase-deficient mice do not display compromised self-renewal and are able to proceed past the blastocyst stage (Acharya et al., 2017; Hu et al., 2009).

Comparatively little is known about SRCAP, including its role(s) in pluripotent cells. SRCAP remodels nucleosomes primarily through incorporation of the variant H2A.Z (Ruhl et al., 2006; Wong et al., 2007). SRCAP is recruited to sites of H2A.Z deposition along with Tip60-p400 via their shared subunit, GAS41, a reader of acetylated lysines (Hsu et al., 2018). Interestingly, PCI domain-containing protein 2 (PCID2), a highly expressed protein in multipotent hematopoietic progenitor cells, impedes SRCAP remodeling activity by blocking deposition of H2A.Z and recruitment of transcription factor PU.1 to lymphoid fate regulator genes via an interaction with the zinc finger HIT-type containing 1 (ZNHIT1) protein (Ye et al., 2017). Additionally, SRCAP assembly and ATPase activity is promoted by the transcription factor ZBTB3; depletion of ZBTB3 disrupts ES cell self-renewal and viability. Activity of ZBTB3 is itself activated by the lncRNA LncKdm2b by promoting the assembly of the SRCAP complex in *trans* (Ye et al., 2018). SRCAP therefore fulfills the INO80 paradigm of seemingly disparate roles in stem cells through its role in specifying lymphoid fate as well as its role in preserving stem cell pluripotency.

The INO80 complex itself largely fulfills an activating role with respect to pluripotency genes in stem cells—including *Oct4, Nanog, Sox2, Klf4,* and *Esrrb* (Wang et al., 2014)—by promoting recruitment of Mediator and RNA Polymerase II to these genes.

14.6 Polycomb group proteins silence developmental genes in ES cells

Polycomb group (PcG) complexes mediate H3K27 methylation and are typically associated with gene repression (Di Croce & Helin, 2013). There are two broad classes of PcG complexes, PRC1 and PRC2, each of which includes numerous subcomplexes. PRC1 includes six major complexes, each defined by a Polycomb Group Ring Finger (PCGF) subunit, PCGF1–6. In addition to the PCGF subunit, PRC1 complexes include the RING1/2 E3 ubiquitin ligase, RVBP/YAF2 or a chromobox (CBX) protein, and a unique set of associated proteins (Gao et al., 2012). The canonical PRC1 complexes are PRC1.2 (containing PCGF2) and PRC1.4 (containing PCGF4/BMI1); these complexes are recruited to chromatin by the H3K27me3 mark deposited by PRC2 (Gao et al., 2012). In ES cells, canonical PRC1 functions to maintain pluripotency in a CBX7-dependent manner, whereas CBX2 and CBX4 become more abundant in PRC1 complexes upon ES cell differentiation into embryoid bodies (Morey et al., 2012). This switch from CBX7 to CBX2 and CBX4 is regulated by the miR-125 and miR-181 microRNAs (O'Loghlen et al., 2012). The non-canonical PRC1 complexes (PRC1.1, PRC1.3, PRC1.5, and PRC1.6) are recruited to chromatin through H3K27me3-independent mechanisms and catalyze ubiquitylation of H2A at lysine 119 (H2AK119), a mark that is not necessary for PRC1 target binding but is necessary for efficient repression of genes (specifically drivers of cellular differentiation) in ES cells (Endoh et al., 2012; Tavares et al., 2012). PCGF1 is somewhat distinct in that it functions in gene activation during ES cell lineage specification through positive regulation of endoderm- and mesoderm-associated transcription factor expression (Yan et al., 2017). PRC1.3 and PRC1.5 mainly function to activate transcription in ES cells via an interaction with the TEX10 pluripotency factor (Zhao, Huang, et al., 2017). PCGF6 is necessary for proper self-renewal and differentiation, likely through PRC1.6's role in recruiting RING1B and mediating H2AK119ub (Endoh et al., 2017; Zhao, Tong, et al., 2017).

PRC2, meanwhile, includes the PRC2.1 and PRC2.2 complexes, which are recruited by distinct mechanisms and have divergent gene silencing functions (Hauri et al., 2016; Jones & Wang, 2010; van Mierlo, Veenstra, Vermeulen, & Marks, 2019). The accepted model for PcG activity is that PRC2 methylates H3K27—a process that then creates PRC1 binding sites and enables subsequent H2AK119ub, a mark recognized by the CBX proteins of the PRC1 complex (Endoh et al., 2012; Eskeland et al., 2010). PRC1 can also compact chromatin independently of its ubiquitylation activity (Buchwald et al., 2006; Eskeland et al., 2010). Specific to ES cells, PcG proteins are known to directly repress hundreds of development-associated genes (Aloia, Di Stefano, & Di Croce, 2013; Boyer et al., 2006; Lee et al., 2006). H3K27me3 can be disrupted at various cell stages using direct inhibition of PRC2 function via disruption of the EZH2/EED interaction, a novel technique that was used to show that PRC2 is required for self-renewal at all but the earliest stage of human ES cell development (Moody et al., 2017). Upon differentiation, promoters that are marked by H3K27me3 in pluripotent cells often become DNA methylated, and the H3K27-methylated gene landscape changes dramatically—including loss of bivalency (Mohn et al., 2008). In human ES cells, PcG knockouts cause pluripotency loss and subsequent meso-endoderm fate specification, as well as a failure to differentiate into ectoderm lineages upon EZH1 and EZH2 loss (Shan et al., 2017). In murine ES cells, PcG deficiency causes loss of BMP4 but not pluripotency loss; however, when these cells are converted to a primed state, they undergo similar spontaneous differentiation to PcG-deficient human ES cells (Shan et al., 2017). Interestingly, human ES cells appear more dependent on EZH2 than murine ES cells, as EZH2 deficiency confers a more severe self-renewal and proliferation defect in human ES cells than mouse (Collinson et al., 2016). In ES cells, H3K27me3 deposition and subsequent gene silencing is also regulated by H2A.Z and H3.3 deposition; however, the mechanism by which this silencing occurs has yet to be elucidated (Wang et al., 2018).

Increased DNA methylation at H3K27me3 sites suggests some level of crosstalk between PcG-mediated silencing and DNA methylation. PcG binding and subsequent gene silencing also appears to be antagonistic to esBAF-mediated gene activation, particularly at LIF signaling targets that are dependent on BRG1 to facilitate establishment of STAT3 binding sites (Ho, Jothi, et al., 2009; Ho et al., 2011). It is likely that this STAT3 binding site establishment (and perhaps that of other pluripotency factors) is enabled by binding of BRG1 at the two nucleosomes flanking relatively long nucleosome-depleted promoter regions in ES cells (de Dieuleveult et al., 2016). Although STAT3 binding appears more reliant on BRG1 than vice versa, STAT3 and BRG1 recruitment are mutually dependent, and both LIF and BRG1 must be expressed at sufficient levels to enable proper expression and function of transcription factors that regulate self-renewal, such as TBX3, TFCP2l1, ESRRB, SOCS3, and TCL1 (Ho et al., 2011). Importantly, however, PcG and esBAF do not exist in a simple antagonistic relationship; esBAF also promotes PcG-mediated silencing at *Hox* loci, working with PcG to facilitate their repression in ES cells (Ho et al., 2011). In summary, PcG proteins coordinate gene silencing for hundreds of differentiation-associated genes, working with and against other chromatin modifying factors to facilitate maintenance of pluripotency in ES cells and proper lineage differentiation as is required by the maturing organism.

14.7 Variants of H2A and H3 have specialized roles in pluripotent cells

Along with nucleosome remodeling factors like esBAF, the histone variant H2A.Z also functions at PcG target genes; H2A.Z is enriched in ES cells at PcG targets and is necessary for the lineage commitment process (Creyghton et al., 2008). Furthermore, H2A cannot compensate for H2A.Z loss during early development (Creyghton et al., 2008; Faast et al., 2001; Hu et al., 2013). PRC1 and PRC2 are not, however, required for targeting of H2A.Z to developmental genes in ES cells (Illingworth, Botting, Grimes, Bickmore, & Eskeland, 2012). Like the H3 variant H3.3, H2A.Z can be found at regulatory regions of the ES cell genome, such as enhancers (Hu et al., 2013). H2A.Z regulates chromatin to facilitate access of both active and repressive factors, including OCT4, MLL complexes, and PRC2 (Hu et al., 2013). H2A.Z also promotes ES cell differentiation by regulation of nucleosomes themselves (Li et al., 2012) and of epigenetic histone marks, including the bivalent chromatin mark H3K27me3 (Surface et al., 2016; Wang et al., 2018). H2A.Z can be ubiquitylated, and without this mark, murine ES cells undergo faulty lineage commitment (Surface et al., 2016). Prior work has suggested that PRC1-mediated ubiquitylation of H2A is important for silencing bivalent genes in murine ES cells (de Napoles et al., 2004; Endoh et al., 2012). H2A.Z contains an acidic patch domain distinct from that of core H2A that is also necessary for regulation of lineage commitment by interplay between transcription and chromatin dynamics (Subramanian et al., 2013).

The histone H3 variant H3.3 marks active promoter regions and other regulatory regions genome-wide in ES cells, including super-enhancers (Deaton et al., 2016; Goldberg et al., 2010). In addition to H3.3's presence at enhancer loci, H3.3 can be phosphorylated at S31 to promote enhancer acetylation (Martire et al., 2019). H3.3-containing nucleosomes are interaction hotspots for pluripotency factors, nucleosome remodelers, and other factors that may help to establish and preserve pluripotency in stem cells. Binding sites for ES cell-specific pluripotency factors have been identified at regions of rapid H3.3-containing nucleosome turnover, suggesting that H3.3 may facilitate binding of pluripotency factors in ES cells (Ha, Kraushaar, & Zhao, 2014). Furthermore, the −1 nucleosome of expressed genes in ES cells contains the H3.3 variant; upon differentiation, an H3.3-containing nucleosome can be found shifted into the +1 position, possibly to regulate gene expression (Schlesinger et al., 2017). In addition to interactions with pluripotency factors, H3.3 facilitates silencing of developmental promoters via PRC2 recruitment and subsequent H3K27me3 deposition (Banaszynski et al., 2013). While HIRA and H3.3 are required for H3K27me3 establishment at developmental gene promoters in ES cells, H3K4me3 was largely maintained upon H3.3 depletion (Banaszynski et al., 2013). By forcing H3.1 expression and knocking down H3.3 expression in ES cells, however, myogenic differentiation is impaired, and both H3K4me3 and H3K27me3 are diminished (Harada et al., 2015). It therefore appears that H3.3 deposition and/or placement is essential for the presence of both bivalent promoter marks, and that H3.3 may play a greater role in regulation of bivalent promoters than has previously been known. Further, H3.3 is incorporated by ATRX/DAXX to facilitate silencing of repetitive elements via the KAP1 co-repressor, and this incorporation appears to function upstream of H3K9me3 and heterochromatin formation at these regions (Elsässer, Noh, Diaz, Allis, & Banaszynski, 2015).

Several developmental roles for H3.3 in ES cells have been suggested. Without H3.3, mice show near-complete embryonic lethality by E6.5, and those mice that survived displayed early (E8.5) expression of *Brachyury,* a mesoderm marker, and more closely resembled E7.5 embryos (Jang, Shibata, Starmer, Yee, & Magnuson, 2015). Similarly, the Jiao group demonstrated a requirement for H3.3 in neural stem cells to properly proliferate and differentiate—a requirement that could be overcome by overexpression of H3.3, MOF, and GLI1, but not H3.1 or an H3.3 mutant with K36 mutated to arginine (Xia & Jiao, 2017).

15. Long-range chromatin interactions are critical for regulation of pluripotency

While ES cells precisely regulate chromatin structure, accessibility, and transcription at individual gene loci, long-range chromatin interactions include physical contacts made between enhancers, promoters, insulators, gene clusters, and other features that regulate gene expression beyond the sequence, transcription factor, and individual nucleosome levels. These interactions are mappable genome-wide by Hi-C, ChIA-PET, GAM, and OligoPAINT, among more techniques (Beagrie et al., 2017; Beliveau et al., 2012; Dekker, 2016; Li et al., 2010; Lieberman-Aiden et al., 2009). CCCTC-binding factor (CTCF) facilitates looping interactions within chromosomes, and it is canonically referred to as an insulator protein; however, CTCF has also been implicated in transcriptional activation (Vostrov & Quitschke, 1997; Vostrov, Taheny, & Quitschke, 2002) and splicing (Shukla et al., 2011). CTCF works with cohesin and Mediator complexes to mediate loop formation and establish the 3D genomic landscape, consisting of domains of decreasing size known as chromosomal territories, compartments, topologically-associating domains (TADs), and subTADs (Dekker & Misteli, 2015). TADs feature high interaction frequencies within domains, but low interaction frequencies across domains (Cavalli & Misteli, 2013; Dekker & Misteli, 2015; Dixon et al., 2012; Nora et al., 2012) (Fig. 8). The expression of genes within a single TAD are correlated during differentiation (Nora et al., 2012), but TAD formation, regulation, and function are largely unknown processes at present.

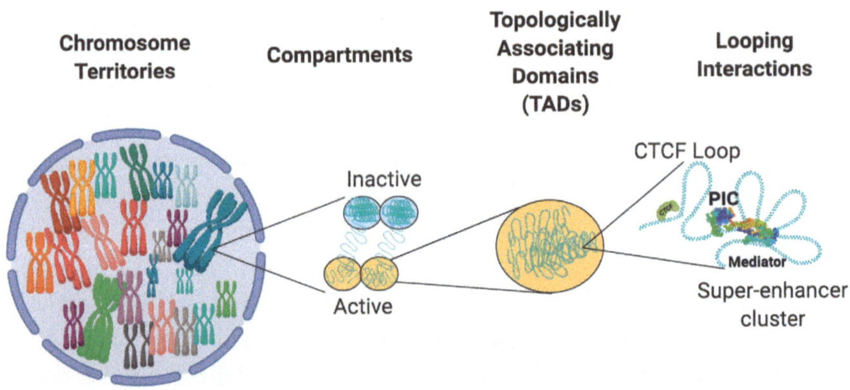

Fig. 8 3D chromatin interactions occur in defined regions of the nucleus. The chromatin that compacts to form each chromosome occupies a defined region of space within the nucleus, known as a chromosome territory. Within these chromosome territories are compartments of active and inactive chromatin that house topologically-associating domains, or TADs. 3D interaction frequency (such as promoter-enhancer looping or CTCF-CTCF looping) is heightened within, but not across, TADs. PDB IDs used: 5N9J, 5SVA. Created with Biorender.com.

Long-range chromatin interactions are key regulatory features of ES cell chromatin (Dixon et al., 2015, 2012). At the *Nanog* gene locus, for example, there are numerous DNase I hypersensitivity sites (DHSs) spread across a 160 kb region between *Nanog* and the next gene, *Dppa3* (Levasseur, Wang, Dorschner, Stamatoyannopoulos, & Orkin, 2008). When the pluripotency factors OCT4, NANOG, ZFP281, and NACI bind at these DHSs, they alter higher-order chromatin structure to loop out the extragenic region and bring the *Nanog*, *Dppa3*, and *Gdf3* promoters into close proximity; furthermore, upon depletion of OCT4, this looping event collapses, and ES cells fail to maintain pluripotency (Keenen & de la Serna, 2009; Levasseur et al., 2008). This long-range interaction highlights another key aspect of pluripotency factors: they are often involved in complex autoregulatory circuits that can interfere with attempts to understand the actions of individual pluripotency factors.

Among the most prominent of long-range chromatin interactions in ES cells are those of clusters of enhancers that function together, known as super-enhancers (also referred to as stretch enhancers) (Parker et al., 2013; Peng & Zhang, 2018). Super-enhancers are prevalent in ES cells, and are reorganized upon exit from a pluripotent state to facilitate cellular differentiation and specification, as shown through promoter capture experiments (Novo et al., 2018). Super-enhancers tend to regulate sets of genes that control cell identity and fate, and they are occupied by between 12% and 36% of enhancer-associated RNA Polymerase II and cofactors, despite making up less than 3% of total identified enhancers in ES cells; this heightened association likely explains super-enhancers high RNA abundance (Hnisz et al., 2013; Whyte et al., 2013). More specifically, super-enhancers in ES cells are highly occupied by the master regulators of pluripotency, OCT4, SOX2, and NANOG, as well as Mediator and ASH2L, an MLL complex subunit that recruits the master regulators to super-enhancer loci and forms a complex with OCT4 to regulate pluripotency genes (Tsai et al., 2019; Whyte et al., 2013). In addition, TEX10 is a pluripotency factor that is enriched at super-enhancers, localizing to them in a SOX2-dependent manner to regulate histone acetylation and DNA demethylation at super-enhancers (Ding et al., 2015). Interestingly, the master regulators of pluripotency are themselves regulated by super-enhancers, highlighting the complex autoregulatory circuits involved in maintaining pluripotency in stem cells (Blinka, Reimer, Pulakanti, & Rao, 2016; Li et al., 2014; Whyte et al., 2013; Zhou et al., 2014). Indeed, deletion of a super-enhancer can alter transcription of genes that are not currently assigned to that super-enhancer (Moorthy et al., 2017), suggesting that super-enhancers may regulate multiple promoters and function as operators within larger regulatory pathways than have currently been identified.

Long-range chromatin interactions are not unique to ES cells; however, they are specific, critical, and highly regulated. Over the course of cellular differentiation, over one third of long-range chromatin interactions change, including changes to both active and inactive compartments throughout the genome (Dixon et al., 2015). These altered long-range chromatin interactions serve to modify expression of developmental genes (both positively and negatively), to silence pluripotency factors, and to generally facilitate differentiation into correct lineages. Although the 3D interactome remains incompletely characterized, it is clear that ES cells undergo drastic changes to long-range chromatin interactions over the course of their differentiation; as such, 3D chromatin interactions represent a diverse and understudied source of gene regulation.

16. ES cells regulate chromatin by common processes to preserve pluripotency

In this chapter, we have discussed the importance of genetic and epigenetic regulators that govern lineage fidelity and permit lineage specification. Specifically, ES cells carefully regulate the expression of developmental and pluripotency genes through unique mechanisms, including bivalent histone modifications over the promoters of developmental genes, highly accessible chromatin structure that is permissive of enhanced transcription, prevalent use of enhancers and super-enhancers largely regulated by pluripotency factors OCT4, SOX2, and NANOG, and unique application of chromatin machinery used in somatic cells (including nucleosome remodelers esBAF, Tip60-p400, and CHD1) and histone variants (including H3.3 and H2A.Z). Despite the vast array of unique and specialized tools utilized to regulate ES cell chromatin, at their core these processes reflect the mechanisms used to regulate chromatin in cells in the adult organism. The effectors and pathways are unique; however, as ES cells must precisely regulate their chromatin architecture to reflect the dynamic nature of this cell type. While chromatin regulatory processes are as critical within all cell types, their function within ES cells is precisely guided to ensure proper development. In recent years, our understanding of mechanisms that govern proliferation of ES cells, and those leading to lineage commitment, has increased. Further characterization of the mechanisms discussed throughout this chapter will continue to help identify effective methods for reprogramming differentiated cells, thereby facilitating the development of stem cell based therapies. Additionally, cells in various disease states take on chromatin characteristics reminiscent of those in ES cells, including hijacking of chromatin machinery, as in MYC-driven "transcription addiction," acquisition of novel super-enhancers to regulate oncogenes, and transition to an un- or less-differentiated state (Hnisz et al., 2013; Jordan, 2007; Loven et al., 2013). In sum, stem cells represent a unique application of fundamental biological processes to maintain a delicate balance and facilitate a rapid shift between gene repression and activation in the pluripotent state and along the pathway of lineage specification.

Acknowledgments

We thank members of the Hainer lab for critical reading of this chapter. This work was supported by a Charles E. Kaufman Foundation New Investigator Award and National Institutes of Health grant 1R35GM133732-01 to S.J.H.

References

Acharya, D., Hainer, S. J., Yoon, Y., Wang, F., Bach, I., Rivera-Perez, J. A., et al. (2017). *Cell Reports, 19*, 671–679.

Adam, S., Polo, S. E., & Almouzni, G. (2013). *Cell, 155*, 94–106.

Adkins, M. W., & Tyler, J. K. (2006). *Molecular Cell, 21*, 405–416.

Ahmad, K., & Henikoff, S. (2002). *Molecular Cell, 9*, 1191–1200.

Alajem, A., Biran, A., Harikumar, A., Sailaja, B. S., Aaronson, Y., Livyatan, I., et al. (2015). *Cell Reports, 10*, 2019–2031.

Albert, I., Mavrich, T. N., Tomsho, L. P., Qi, J., Zanton, S. J., Schuster, S. C., et al. (2007). *Nature, 446*, 572–576.

Alén, C., Kent, N. A., Jones, H. S., O'Sullivan, J., Aranda, A., & Proudfoot, N. J. (2002). *Molecular Cell, 10*, 1441–1452.

Alfert, A., Moreno, N., & Kerl, K. (2019). *Epigenetics & Chromatin, 12*, 19.

Allfrey, V., Faulkner, R., & Mirsky, A. (1964). *Proceedings of the National Academy of Sciences of the United States of America, 51*, 8.

Aloia, L., Di Stefano, B., & Di Croce, L. (2013). *Development, 140*, 2525–2534.

Alpsoy, A., & Dykhuizen, E. C. (2018). *The Journal of Biological Chemistry, 293*, 3892–3903.

Andersen, E. C., Lu, X., & Horvitz, H. R. (2006). *Development, 133*, 2695–2704.

Anderson, C. J., Baird, M. R., Hsu, A., Barbour, E. H., Koyama, Y., Borgnia, M. J., et al. (2019). *Cell Reports, 26*. 1681-1690.e5.

Anderson, J. D., & Widom, J. (2000). *Journal of Molecular Biology, 296*, 979–987.

Ang, Y. S., Tsai, S. Y., Lee, D. F., Monk, J., Su, J., Ratnakumar, K., et al. (2011). *Cell, 145*, 183–197.

Angelov, D., Molla, A., Perche, P.-Y., Hans, F., Côté, J., Khochbin, S., et al. (2003). *Molecular Cell, 11*, 1033–1041.

Apostolou, E., & Hochedlinger, K. (2013). *Nature, 502*, 462–471.

Arents, G., & Moudrianakis, E. N. (1995). *Proceedings of the National Academy of Sciences of the United States of America, 92*, 11170–11174.

Arimura, Y., Kimura, H., Oda, T., Sato, K., Osakabe, A., Tachiwana, H., et al. (2015). *Scientific Reports, 5*, 9628.

Aul, R. B., & Oko, R. J. (2001). *Developmental Biology, 239*, 376–387.

Avvakumov, N., Nourani, A., & Cote, J. (2011). *Molecular Cell, 41*, 502–514.

Aydin, Ö. Z., Vermeulen, W., & Lans, H. (2014). *Cell Cycle (Georgetown, Texas), 13*, 3016–3025.

Azuara, V., Perry, P., Sauer, S., Spivakov, M., Jorgensen, H. F., John, R. M., et al. (2006). *Nature Cell Biology, 8*, 532–538.

Bai, L., & Morozov, A. V. (2010). *Trends in Genetics, 26*, 476–483.

Bajpai, R., Chen, D. A., Rada-Iglesias, A., Zhang, J., Xiong, Y., Helms, J., et al. (2010). *Nature, 463*, 958–962.

Banaszynski, L. A., Wen, D. C., Dewell, S., Whitcomb, S. J., Lin, M. Y., Diaz, N., et al. (2013). *Cell, 155*, 107–120.

Bannister, A. J., & Kouzarides, T. (2011). *Cell Research, 21*, 381–395.

Bao, X., Rubin, A. J., Qu, K., Zhang, J., Giresi, P. G., Chang, H. Y., et al. (2015). *Genome Biology, 16*, 284.

Bapat, S. A., Jin, V., Berry, N., Balch, C., Sharma, N., Kurrey, N., et al. (2010). *Epigenetics, 5*, 716–729.

Barbera, A. J., Chodaparambil, J. V., Kelley-Clarke, B., Joukov, V., Walter, J. C., Luger, K., et al. (2006). *Science, 311*, 856–861.

Barisic, D., Stadler, M. B., Iurlaro, M., & Schübeler, D. (2019). *Nature, 569*, 136–140.

Barski, A., Cuddapah, S., Cui, K., Roh, T. Y., Schones, D. E., Wang, Z., et al. (2007). *Cell, 129*, 823–837.

Beagrie, R. A., Scialdone, A., Schueler, M., Kraemer, D. C., Chotalia, M., Xie, S. Q., et al. (2017). *Nature, 543*, 519–524.

Becker, P. B., & Workman, J. L. (2013). *Cold Spring Harbor Perspectives in Biology, 5*, a017905.

Beliveau, B. J., Joyce, E. F., Apostolopoulos, N., Yilmaz, F., Fonseka, C. Y., McCole, R. B., et al. (2012). *Proceedings of the National Academy of Sciences of the United States of America, 109*, 21301–21306.

Belotserkovskaya, R., Oh, S., Bondarenko, V. A., Orphanides, G., Studitsky, V. M., & Reinberg, D. (2003). *Science, 301*, 1090–1093.

Bernstein, B. E., Mikkelsen, T. S., Xie, X., Kamal, M., Huebert, D. J., Cuff, J., et al. (2006). *Cell, 125*, 315–326.

Blackledge, N. P., Farcas, A. M., Kondo, T., King, H. W., McGouran, J. F., Hanssen, L. L., et al. (2014). *Cell, 157*, 1445–1459.

Blinka, S., Reimer, M. H., Jr., Pulakanti, K., & Rao, S. (2016). *Cell Reports, 17*, 19–28.

Boeger, H., Griesenbeck, J., Strattan, J. S., & Kornberg, R. D. (2004). *Molecular Cell, 14*, 667–673.

Bornelov, S., Reynolds, N., Xenophontos, M., Gharbi, S., Johnstone, E., Floyd, R., et al. (2018). *Molecular Cell, 71*. 56-72.e4.

Bouniol, C., Nguyen, E., & Debey, P. (1995). *Experimental Cell Research, 218*, 57–62.

Bowman, A., Ward, R., Wiechens, N., Singh, V., El-Mkami, H., Norman, D. G., et al. (2011). *Molecular Cell, 41*, 398–408.

Boyer, L. A., Plath, K., Zeitlinger, J., Brambrink, T., Medeiros, L. A., Lee, T. I., et al. (2006). *Nature, 441*, 349–353.

Bradbury, E. M. (1989). *Journal of Molecular Recognition, 2*. i-i.

Brahma, S., Udugama, M. I., Kim, J., Hada, A., Bhardwaj, S. K., Hailu, S. G., et al. (2017). *Nature Communications, 8*, 15616.

Buchwald, G., van der Stoop, P., Weichenrieder, O., Perrakis, A., van Lohuizen, M., & Sixma, T. K. (2006). *The EMBO Journal, 25*, 2465–2474.

Bultman, S., Gebuhr, T., Yee, D., La Mantia, C., Nicholson, J., Gilliam, A., et al. (2000). *Molecular Cell, 6*, 1287–1295.

Bulut-Karslioglu, A., Jin, H., Guzman-Ayala, M., Williamson, A. J. K., Hejna, M., Whetton, A. D., et al. (2019). *BioRxiv*.

Burgold, T., Barber, M., Kloet, S., Cramard, J., Gharbi, S., Floyd, R., et al. (2019). *The EMBO Journal*, *38*, e100788.

Buscarlet, M., Krasteva, V., Ho, L., Simon, C., Hébert, J., Wilhelm, B., et al. (2014). *Blood*, *123*, 1720–1728.

Buschbeck, M., & Hake, S. B. (2017). *Nature Reviews Molecular Cell Biology*, *18*, 299–314.

Cai, Y., Jin, J. J., Florens, L., Swanson, S. K., Kusch, T., Li, B., et al. (2005). *Journal of Biological Chemistry*, *280*, 13665–13670.

Cairns, B. R. (2005). *Current Opinion in Genetics & Development*, *15*, 185–190.

Cakiroglu, A., Clapier, C. R., Ehrensberger, A. H., Darbo, E., Cairns, B. R., Luscombe, N. M., et al. (2019). *Genome Research*, *29*, 988–998.

Cao, R., Tsukada, Y., & Zhang, Y. (2005). *Molecular Cell*, *20*, 845–854.

Carrozza, M. J., Li, B., Florens, L., Suganuma, T., Swanson, S. K., Lee, K. K., et al. (2005). *Cell*, *123*, 581–592.

Cavalli, G., & Misteli, T. (2013). *Nature Structural & Molecular Biology*, *20*, 290–299.

Chakravarthy, S., Gundimella, S. K., Caron, C., Perche, P. Y., Pehrson, J. R., Khochbin, S., et al. (2005). *Molecular and Cellular Biology*, *25*, 7616–7624.

Chatterjee, N., Sinha, D., Lemma-Dechassa, M., Tan, S., Shogren-Knaak, M. A., & Bartholomew, B. (2011). *Nucleic Acids Research*, *39*, 8378–8391.

Chen, P. B., Hung, J. H., Hickman, T. L., Coles, A. H., Carey, J. F., Weng, Z., et al. (2013). *eLife*, *2*, e01557.

Chen, X., Xu, H., Yuan, P., Fang, F., Huss, M., Vega, V. B., et al. (2008). *Cell*, *133*, 1106–1117.

Chen, P., Zhao, J., Wang, Y., Wang, M., Long, H., Liang, D., et al. (2013). *Genes & Development*, *27*, 2109–2124.

Cheung, V., Chua, G., Batada, N. N., Landry, C. R., Michnick, S. W., Hughes, T. R., et al. (2008). *PLoS Biology*, *6*, e277.

Chittori, S., Hong, J., Saunders, H., Feng, H., Ghirlando, R., Kelly, A. E., et al. (2018). *Science*, *359*, 339–343.

Churchman, L., & Weissman, J. (2011). *Nature*, *469*, 368–373.

Ciccone, D. N., Su, H., Hevi, S., Gay, F., Lei, H., Bajko, J., et al. (2009). *Nature*, *461*, 415–418.

Cirillo, L. A., McPherson, C. E., Bossard, P., Stevens, K., Cherian, S., Shim, E. Y., et al. (1998). *The EMBO Journal*, *17*, 244–254.

Clapier, C. R., & Cairns, B. R. (2009). *Annual Review of Biochemistry*, *78*, 273–304.

Clapier, C. R., & Cairns, B. R. (2012). *Nature*, *492*, 280–284.

Clapier, C. R., Iwasa, J., Cairns, B. R., & Peterson, C. L. (2017). *Nature Reviews. Molecular Cell Biology*, *18*, 407–422.

Clemot, M., Molla-Herman, A., Mathieu, J., Huynh, J. R., & Dostatni, N. (2018). *Development*, *145*, 1–13.

Close, P., Hawkes, N., Cornez, I., Creppe, C., Lambert, C. A., Rogister, B., et al. (2006). *Molecular Cell*, *22*, 521–531.

Collinson, A., Collier, A. J., Morgan, N. P., Sienerth, A. R., Chandra, T., Andrews, S., et al. (2016). *Cell Reports*, *17*, 2700–2714.

Creyghton, M. P., Markoulaki, S., Levine, S. S., Hanna, J., Lodato, M. A., Sha, K., et al. (2008). *Cell*, *135*, 649–661.

Cucinotta, C. E., Hildreth, A. E., McShane, B. M., Shirra, M. K., & Arndt, K. M. (2019). *Nucleic Acids Research*, *47*, 8410–8423.

Cui, K., Zang, C., Roh, T.-Y., Schones, D. E., Childs, R. W., Peng, W., et al. (2009). *Cell Stem Cell*, *4*, 80–93.

Dann, G. P., Liszczak, G. P., Bagert, J. D., Muller, M. M., Nguyen, U. T. T., Wojcik, F., et al. (2017). *Nature*, *548*, 607–611.

Das, C., Lucia, M. S., Hansen, K. C., & Tyler, J. K. (2009). *Nature, 459*, 113–117.
de Dieuleveult, M., Yen, K., Hmitou, I., Depaux, A., Boussouar, F., Bou Dargham, D., et al. (2016). *Nature, 530*, 113–116.
de Napoles, M., Mermoud, J. E., Wakao, R., Tang, Y. A., Endoh, M., Appanah, R., et al. (2004). *Developmental Cell, 7*, 663–676.
Deaton, A. M., Gomez-Rodriguez, M., Mieczkowski, J., Tolstorukov, M. Y., Kundu, S., Sadreyev, R. I., et al. (2016). *eLife, 5*, 1–24.
Dekker, J. (2016). *Nature Reviews. Molecular Cell Biology, 17*, 741–742.
Dekker, J., & Misteli, T. (2015). *Cold Spring Harbor Perspectives in Biology, 7*, a019356.
Delmas, V., Stokes, D. G., & Perry, R. P. (1993). *Proceedings of the National Academy of Sciences of the United States of America, 90*, 2414–2418.
Di Croce, L., & Helin, K. (2013). *Nature Structural & Molecular Biology, 20*, 1147.
Dillon, S. C., Zhang, X., Trievel, R. C., & Cheng, X. (2005). *Genome Biology, 6*, 227.
Ding, J., Huang, X., Shao, N., Zhou, H., Lee, D. F., Faiola, F., et al. (2015). *Cell Stem Cell, 16*, 653–668.
Ding, J., Xu, H., Faiola, F., Ma'ayan, A., & Wang, J. (2012). *Cell Research, 22*, 155–167.
Dixon, J. R., Jung, I., Selvaraj, S., Shen, Y., Antosiewicz-Bourget, J. E., Lee, A. Y., et al. (2015). *Nature, 518*, 331–336.
Dixon, J. R., Selvaraj, S., Yue, F., Kim, A., Li, Y., Shen, Y., et al. (2012). *Nature, 485*, 376–380.
Dou, Y., Milne, T. A., Ruthenburg, A. J., Lee, S., Lee, J. W., Verdine, G. L., et al. (2006). *Nature Structural & Molecular Biology, 13*, 713–719.
Doyen, C. M., An, W., Angelov, D., Bondarenko, V., Mietton, F., Studitsky, V. M., et al. (2006). *Molecular Cell Biology, 26*, 1156–1164.
Doyen, C. M., Montel, F., Gautier, T., Menoni, H., Claudet, C., Delacour-Larose, M., et al. (2006). *The EMBO Journal, 25*, 4234–4244.
Doyon, Y., & Cote, J. (2004). *Current Opinion in Genetics & Development, 14*, 147–154.
Doyon, Y., Selleck, W., Lane, W. S., Tan, S., & Cote, J. (2004). *Molecular Cell Biology, 24*, 1884–1896.
Draizen, E. J., Shaytan, A. K., Marino-Ramirez, L., Talbert, P. B., Landsman, D., & Panchenko, A. R.(2016). *Database-the Journal of Biological Databases and Curation, 2016, 1–10.*
Drane, P., Ouararhni, K., Depaux, A., Shuaib, M., & Hamiche, A. (2010). *Genes & Development, 24*, 1253–1265.
Driscoll, R., Hudson, A., & Jackson, S. P. (2007). *Science (New York, N.Y.), 315*, 649–652.
Duina, A. A. (2011). *Genetics Research International, 2011*, 625210.
Dyer, M. A., Qadeer, Z. A., Valle-Garcia, D., & Bernstein, E. (2017). *Cold Spring Harbor Perspectives in Medicine, 7*, a026567.
Earnshaw, W. C., & Rothfield, N. (1985). *Chromosoma, 91*, 313–321.
Ebbert, R., Birkmann, A., & Schuller, H. J. (1999). *Molecular Microbiology, 32*, 741–751.
Ee, L. S., McCannell, K. N., Tang, Y., Fernandes, N., Hardy, W. R., Green, M. R., et al. (2017). *Stem Cell Reports, 8*, 1488–1496.
Efroni, S., Duttagupta, R., Cheng, J., Dehghani, H., Hoeppner, D. J., Dash, C., et al. (2008). *Cell Stem Cell, 2*, 437–447.
Eleuteri, B., Aranda, S., & Ernfors, P. (2018). *Cell Reports, 23*, 1853–1866.
Elfring, L., Deuring, R., Mccallum, C., Peterson, C. L., & Tamkun, J. (1994). *Molecular Cell Biology, 14*, 2225–2234.
Elsaesser, S. J., Goldberg, A. D., & Allis, C. D. (2010). *Current Opinion in Genetics & Development, 20*, 110–117.
Elsässer, S. J., Noh, K.-M., Diaz, N., Allis, C. D., & Banaszynski, L. A. (2015). *Nature, 522*, 240.

Endoh, M., Endo, T. A., Endoh, T., Isono, K., Sharif, J., Ohara, O., et al. (2012). *PLoS Genetics, 8*, 1–10, e1002774.
Endoh, M., Endo, T. A., Shinga, J., Hayashi, K., Farcas, A., Ma, K. W., et al. (2017). *Elife, 6*, 1–26.
Engeholm, M., de Jager, M., Flaus, A., Brenk, R., van Noort, J., & Owen-Hughes, T. (2009). *Nature Structural & Molecular Biology, 16*, 151–158.
Engelen, E., Akinci, U., Bryne, J. C., Hou, J., Gontan, C., Moen, M., et al. (2011). *Nature Genetics, 43*, 607–611.
Eskeland, R., Leeb, M., Grimes, G. R., Kress, C., Boyle, S., Sproul, D., et al. (2010). *Molecular Cell, 38*, 452–464.
Etchegaray, J.-P., & Mostoslavsky, R. (2015). *Cell Cycle (Georgetown, Texas), 14*, 2187–2188.
Faast, R., Thonglairoam, V., Schulz, T. C., Beall, J., Wells, J. R. E., Taylor, H., et al. (2001). *Current Biology, 11*, 1183–1187.
Fan, J. Y., Rangasamy, D., Luger, K., & Tremethick, D. J. (2004). *Molecular Cell, 16*, 655–661.
Fazzio, T. G., Huff, J. T., & Panning, B. (2008a). *Cell Cycle, 7*, 3302–3306.
Fazzio, T. G., Huff, J. T., & Panning, B. (2008b). *Cell, 134*, 162–174.
Fazzio, T. G., & Tsukiyama, T. (2003). *Molecular Cell, 12*, 1333–1340.
Field, Y., Kaplan, N., Fondufe-Mittendorf, Y., Moore, I. K., Sharon, E., Lubling, Y., et al. (2008). *PLoS Computational Biology, 4*, e1000216.
Fierz, B., Chatterjee, C., McGinty, R. K., Bar-Dagan, M., Raleigh, D. P., & Muir, T. W. (2011). *Nature Chemical Biology, 7*, 113–119.
Fillingham, J., Recht, J., Silva, A. C., Suter, B., Emili, A., Stagljar, I., et al. (2008). *Molecular Cell Biology, 28*, 4342–4353.
Formosa, T. (2008). *Molecular BioSystems, 4*, 1085–1093.
Formosa, T. (2012). *Biochimica et Biophysica Acta, 1819*, 247–255.
Fuchs, S., Laribee, R., & Strahl, B. (2009). *Biochimica et biophysica astra, 1789*, 26–36.
Furuya, M., Tanaka, M., Teranishi, T., Matsumoto, K., Hosoi, Y., Saeki, K., et al. (2007). *The Journal of Reproduction and Development, 53*, 895–902.
Gamarra, N., Johnson, S. L., Trnka, M. J., Burlingame, A. L., & Narlikar, G. J. (2018). *eLife, 7*, 1–25.
Gao, Y., Gan, H., Lou, Z., & Zhang, Z. (2018). *Proceedings of the National Academy of Sciences of the United States of America, 115*, E6162–E6171.
Gao, Z., Zhang, J., Bonasio, R., Strino, F., Sawai, A., Parisi, F., et al. (2012). *Molecular Cell, 45*, 344–356.
Garcia, H., Fleyshman, D., Kolesnikova, K., Safina, A., Commane, M., Paszkiewicz, G., et al. (2011). *Oncotarget, 2*, 783–796.
Gaspar-Maia, A., Alajem, A., Polesso, F., Sridharan, R., Mason, M. J., Heidersbach, A., et al. (2009). *Nature, 460*, 863–868.
Gaspar-Maia, A., Qadeer, Z. A., Hasson, D., Ratnakumar, K., Adrian Leu, N., Leroy, G., et al. (2013). *Nature Communications, 4*, 1565.
Gatchalian, J., Malik, S., Ho, J., Lee, D. S., Kelso, T. W. R., Shokhirev, M. N., et al. (2018). *Nature Communications, 9*, 5139.
Gerhold, C., & Gasser, S. (2014). *Trends in Cell Biology, 24*, 12.
Giaimo, B. D., Ferrante, F., Herchenrother, H., Hake, S. B., & Borggrefe, T. (2019). *Epigenetics & Chromatin, 12*, 37.
Gibbons, R. J., McDowell, T. L., Raman, S., O'Rourke, D. M., Garrick, D., Ayyub, H., et al. (2000). *Nature Genetics, 24*, 368–371.
Gibbons, R. J., Picketts, D. J., Villard, L., & Higgs, D. R. (1995). *Cell, 80*, 837–845.
Gilbert, J., Gore, S., Herman, J., & Carducci, M. (2004). *Clinical Cancer Research, 10*, 4589–4596.

Gilmour, D., & Lis, J. (1984). *Proceedings of the National Academy of Sciences of the United States of America*, *81*, 4.
Goldberg, A. D., Banaszynski, L. A., Noh, K. M., Lewis, P. W., Elsaesser, S. J., Stadler, S., et al. (2010). *Cell*, *140*, 678–691.
Goldman, J. A., Garlick, J. D., & Kingston, R. E. (2010). *The Journal of Biological Chemistry*, *285*, 4645–4651.
Govind, C. K., Zhang, F., Qiu, H., Hofmeyer, K., & Hinnebusch, A. G. (2007). *Molecular Cell*, *25*, 31–42.
Guillemette, B., Bataille, A. R., Gevry, N., Adam, M., Blanchette, M., Robert, F., et al. (2005). *PLoS Biology*, *3*, e384.
Gunther, K., Rust, M., Leers, J., Boettger, T., Scharfe, M., Jarek, M., et al. (2013). *Nucleic Acids Research*, *41*, 3010–3021.
Gurova, K., Chang, H. W., Valieva, M. E., Sandlesh, P., & Studitsky, V. M. (2018). *Biochimica et biophysica acta. Gene regulatory mechanisms*, *1861*, 892–904.
Guttman, M., Donaghey, J., Carey, B. W., Garber, M., Grenier, J. K., Munson, G., et al. (2011). *Nature*, *477*, 295–300.
Guzman-Ayala, M., Sachs, M., Koh, F. M., Onodera, C., Bulut-Karslioglu, A., Lin, C.-J., et al. (2015). *Development (Cambridge, England)*, *142*, 118–127.
Ha, M., Kraushaar, D. C., & Zhao, K. (2014). *Epigenetics & Chromatin*, *7*, 38.
Hainer, S. J., Boskovic, A., McCannell, K. N., Rando, O. J., & Fazzio, T. G. (2019). *Cell*, *177*, 1319–1329.e11.
Hainer, S. J., & Fazzio, T. G. (2015). *Cell Reports*, *13*, 61–69.
Hainer, S. J., Gu, W., Carone, B. R., Landry, B. D., Rando, O. J., Mello, C. C., et al. (2015). *Genes & Development*, *29*, 362–378.
Hainer, S. J., McCannell, K. N., Yu, J., Ee, L. S., Zhu, L. J., Rando, O. J., et al. (2016). *Elife*, *5*, 1–24.
Hake, S. B., Garcia, B. A., Duncan, E. M., Kauer, M., Dellaire, G., Shabanowitz, J., et al. (2006). *The Journal of Biological Chemistry*, *281*, 559–568.
Hammond, C. M., Stromme, C. B., Huang, H., Patel, D. J., & Groth, A. (2017). *Nature Reviews. Molecular Cell Biology*, *18*, 141–158.
Han, D., Jeon, S., Sohn, D. H., Lee, C., Ahn, S., Kim, W. K., et al. (2008). *Developmental Biology*, *315*, 136–146.
Happel, N., Schulze, E., & Doenecke, D. (2005). *Biological Chemistry*, *386*, 541–551.
Harada, A., Maehara, K., Sato, Y., Konno, D., Tachibana, T., Kimura, H., et al. (2015). *Nucleic Acids Research*, *43*, 775–786.
Harikumar, A., & Meshorer, E. (2015). *EMBO Reports*, *16*, 1609–1619.
Hassan, A. H., Neely, K. E., & Workman, J. L. (2001). *Cell*, *104*, 817–827.
Hauri, S., Comoglio, F., Seimiya, M., Gerstung, M., Glatter, T., Hansen, K., et al. (2016). *Cell Reports*, *17*, 583–595.
Hautbergue, G. M., Hung, M. L., Walsh, M. J., Snijders, A. P., Chang, C. T., Jones, R., et al. (2009). *Current Biology*, *19*, 1918–1924.
Hayakawa, K., Ohgane, J., Tanaka, S., Yagi, S., & Shiota, K. (2012). *Epigenetics*, *7*, 1029–1036.
Hendrich, B., & Bird, A. (1998). *Molecular Cell Biology*, *18*, 6538–6547.
Henikoff, S., & Ahmad, K. (2005). *Annual Review of Cell and Developmental Biology*, *21*, 133–153.
Herold, A., Teixeira, L., & Izaurralde, E. (2003). *The EMBO Journal*, *22*, 2472–2483.
Hertel, L., De Andrea, M., Bellomo, G., Santoro, P., Landolfo, S., & Gariglio, M. (1999). *Experimental Cell Research*, *250*, 313–328.
Hnisz, D., Abraham, B. J., Lee, T. I., Lau, A., Saint-Andre, V., Sigova, A. A., et al. (2013). *Cell*, *155*, 934–947.
Ho, L., & Crabtree, G. R. (2010). *Nature*, *463*, 474–484.

Ho, L., Jothi, R., Ronan, J. L., Cui, K., Zhao, K., & Crabtree, G. R. (2009). *Proceedings of the National Academy of Sciences of the United States of America, 106*, 5187–5191.

Ho, L., Miller, E. L., Ronan, J. L., Ho, W. Q., Jothi, R., & Crabtree, G. R. (2011). *Nature Cell Biology, 13*, 903.

Ho, L., Ronan, J. L., Wu, J., Staahl, B. T., Chen, L., Kuo, A., et al. (2009). *Proceedings of the National Academy of Sciences of the United States of America, 106*, 5181–5186.

Hodges, A. J., Gloss, L. M., & Wyrick, J. J. (2017). *Genetics, 206*, 1339–1348.

Hoeijmakers, W. A., Salcedo-Amaya, A. M., Smits, A. H., Francoijs, K. J., Treeck, M., Gilberger, T. W., et al. (2013). *Molecular Microbiology, 87*, 1061–1073.

Hondele, M., & Ladurner, A. G. (2013). *Nucleus, 4*, 443–449.

Hondele, M., Stuwe, T., Hassler, M., Halbach, F., Bowman, A., Zhang, E. T., et al. (2013). *Nature, 499*, 111.

Hori, T., Shang, W. H., Toyoda, A., Misu, S., Monma, N., Ikeo, K., et al. (2014). *Developmental Cell, 29*, 740–749.

Hsieh, F. K., Kulaeva, O. I., Patel, S. S., Dyer, P. N., Luger, K., Reinberg, D., et al. (2013). *Proceedings of the National Academy of Sciences of the United States of America, 110*, 7654–7659.

Hsu, C.-C., Zhao, D., Shi, J., Peng, D., Guan, H., Li, Y., et al. (2018). *Cell Discovery, 4*, 28.

Hu, G., Cui, K., Northrup, D., Liu, C., Wang, C., Tang, Q., et al. (2013). *Cell Stem Cell, 12*, 180–192.

Hu, Y., Fisher, J. B., Koprowski, S., McAllister, D., Kim, M. S., & Lough, J. (2009). *Developmental Dynamics, 238*, 2912–2921.

Hu, G., Schones, D. E., Cui, K., Ybarra, R., Northrup, D., Tang, Q., et al. (2011). *Genome Research, 21*, 1650–1658.

Hunkapiller, J., Shen, Y., Diaz, A., Cagney, G., McCleary, D., Ramalho-Santos, M., et al. (2012). *PLoS Genetics, 8*, e1002576.

Illingworth, R. S., Botting, C. H., Grimes, G. R., Bickmore, W. A., & Eskeland, R. (2012). *PLoS One, 7*, e34848.

Ishimi, Y., Yasuda, H., Hirosumi, J., Hanaoka, F., & Yamada, M. (1983). *Journal of Biochemistry, 94*, 735–744.

Iwafuchi-Doi, M. (2019). *Wiley Interdisciplinary Reviews Systems Biology and Medicine, 11* e1427.

Iwafuchi-Doi, M., Donahue, G., Kakumanu, A., Watts, J. A., Mahony, S., Pugh, B. F., et al. (2016). *Molecular Cell, 62*, 79–91.

Izban, M., & Luse, D. (1991). *Genes & Development, 5*, 683–696.

Izzo, A., Ziegler-Birling, C., Hill, P. W. S., Brondani, L., Hajkova, P., Torres-Padilla, M. E., et al. (2017). *The Journal of Cell Biology, 216*, 3017–3028.

Jamieson, K., Wiles, E. T., McNaught, K. J., Sidoli, S., Leggett, N., Shao, Y., et al. (2016). *Genome Research, 26*, 97–107.

Jang, C. W., Shibata, Y., Starmer, J., Yee, D., & Magnuson, T. (2015). *Genes & Development, 29*, 1377–1392.

Jin, C., & Felsenfeld, G. (2007). *Genes & Development, 21*, 1519–1529.

Jin, C., Zang, C., Wei, G., Cui, K., Peng, W., Zhao, K., et al. (2009). *Nature Genetics, 41*, 941–945.

Jones, A., & Wang, H. (2010). *Protein & Cell, 1*, 1056–1062.

Jordan, C. T. (2007). *Best Practice & Research. Clinical Haematology, 20*, 13–18.

Kaeser, M. D., Aslanian, A., Dong, M.-Q., Yates, J. R., 3rd, & Emerson, B. M. (2008). *The Journal of Biological Chemistry, 283*, 32254–32263.

Kaji, K., Caballero, I. M., MacLeod, R., Nichols, J., Wilson, V. A., & Hendrich, B. (2006). *Nature Cell Biology, 8*, 285–292.

Kaji, K., Nichols, J., & Hendrich, B. (2007). *Development, 134*, 1123–1132.

Kalashnikova, A. A., Porter-Goff, M. E., Muthurajan, U. M., Luger, K., & Hansen, J. C. (2013). *Journal of the Royal Society, Interface*, *10*, 20121022.
Kang, H., McElroy, K. A., Jung, Y. L., Alekseyenko, A. A., Zee, B. M., Park, P. J., et al. (2015). *Genes & Development*, *29*, 1136–1150.
Kaplan, C. D., Laprade, L., & Winston, F. (2003). *Science*, *301*, 1096–1099.
Kaplan, T., Liu, C. L., Erkmann, J. A., Holik, J., Grunstein, M., Kaufman, P. D., et al. (2008). *PLoS Genetics*, *4*, e1000270.
Karytinos, A., Forneris, F., Profumo, A., Ciossani, G., Battaglioli, E., Binda, C., et al. (2009). *The Journal of Biological Chemistry*, *284*, 17775–17782.
Kassis, J. A., & Kennison, J. A. (2010). *Molecular Cell Biology*, *30*, 2581–2583.
Kasten, M. M., Clapier, C. R., & Cairns, B. R. (2011). *Cell*, *144*. 310.e1.
Kato, D., Osakabe, A., Arimura, Y., Mizukami, Y., Horikoshi, N., Saikusa, K., et al. (2017). *Science*, *356*, 205–208.
Keenen, B., & de la Serna, I. L. (2009). *Journal of Cellular Physiology*, *219*, 1–7.
Keniry, A., Jansz, N., Gearing, L. J., Wanigasuriya, I., Chen, J., Nefzger, C. M., et al. (2019). *BioRxiv*.
Keogh, M. C., Kurdistani, S. K., Morris, S. A., Ahn, S. H., Podolny, V., Collins, S. R., et al. (2005). *Cell*, *123*, 593–605.
Kidder, B. L., & Palmer, S. (2012). *Nucleic Acids Research*, *40*, 2925–2939.
Kidder, B. L., Palmer, S., & Knott, J. G. (2009). *Stem Cells*, *27*, 317–328.
Kim, J. A., & Haber, J. E. (2009). *Proceedings of the National Academy of Sciences of the United States of America*, *106*, 1151–1156.
Kim, H. S., Tan, Y., Ma, W., Merkurjev, D., Destici, E., Ma, Q., et al. (2018). *Nature*, *556*, 510–514.
King, H. W., & Klose, R. J. (2017). *eLife*, *6*, 1–24.
Kizer, K. O., Phatnani, H. P., Shibata, Y., Hall, H., Greenleaf, A. L., & Strahl, B. D. (2005). *Molecular Cell Biology*, *25*, 3305–3316.
Klopf, E., Schmidt, H. A., Clauder-Munster, S., Steinmetz, L. M., & Schuller, C. (2017). *Nucleic Acids Research*, *45*, 3752–3766.
Kobor, M. S., Venkatasubrahmanyam, S., Meneghini, M. D., Gin, J. W., Jennings, J. L., Link, A. J., et al. (2004). *PLoS Biology*, *2*, E131.
Kokavec, J., Zikmund, T., Savvulidi, F., Kulvait, V., Edelmann, W., Skoultchi, A. I., et al. (2017). *Stem Cells*, *35*, 1614–1623.
Kolundzic, E., Ofenbauer, A., Bulut, S. I., Uyar, B., Baytek, G., Sommermeier, A., et al. (2018). *Development Cell*, *46*, 611–626.
Kooistra, S. M., & Helin, K. (2012). *Nature Reviews. Molecular Cell Biology*, *13*, 297–311.
Kornberg, R. (1981). *Nature*, *292*, 579–580.
Kornberg, R., & Lorch, Y. (1999). *Cell*, *98*, 285–294.
Kornberg, R. D., & Stryer, L. (1988). *Nucleic Acids Research*, *16*, 6677–6690.
Kouzarides, T. (2007). *Cell*, *128*, 693–705.
Kraushaar, D. C., Jin, W., Maunakea, A., Abraham, B., Ha, M., & Zhao, K. (2013). *Genome Biology*, *14*, R121.
Kujirai, T., Horikoshi, N., Sato, K., Maehara, K., Machida, S., Osakabe, A., et al. (2017). *Nucleic Acids Research*, *45*, 3612.
Kulaeva, O. I., Hsieh, F. K., Chang, H. W., Luse, D. S., & Studitsky, V. M. (2013). *Biochimica et Biophysica Acta*, *1829*, 76–83.
Kulaeva, O. I., Hsieh, F., & Studitsky, V. (2010). *Proceedings of the National Academy of Sciences of the United States of America*, *107*, 11325–11330.
Kuo, M.-H., & Allis, C. D. (1998). *BioEssays*, *20*, 615–626.
Kusch, T., Florens, L., Macdonald, W. H., Swanson, S. K., Glaser, R. L., Yates, J. R., 3rd, et al. (2004). *Science*, *306*, 2084–2087.

Kuzmichev, A., Nishioka, K., Erdjument-Bromage, H., Tempst, P., & Reinberg, D. (2002). *Genes & Development, 16*, 2893–2905.
Lachner, M., O'Carroll, D., Rea, S., Mechtler, K., & Jenuwein, T. (2001). *Nature, 410*, 116–120.
Lafon, A., Taranum, S., Pietrocola, F., Dingli, F., Loew, D., Brahma, S., et al. (2015). *Molecular Cell, 60*, 784–796.
Lai, A. Y., & Wade, P. A. (2011). *Nature Reviews. Cancer, 11*, 588–596.
Lamba, D. A., Hayes, S., Karl, M. O., & Reh, T. (2008). *Developmental Dynamics, 237*, 3016–3023.
Lammerding, J. (2011). *Comprehensive Physiology, 1*, 783–807.
Lamour, V., Lecluse, Y., Desmaze, C., Spector, M., Bodescot, M., Aurias, A., et al. (1995). *Human Molecular Genetics, 4*, 791–799.
Lan, L., Ui, A., Nakajima, S., Hatakeyama, K., Hoshi, M., Watanabe, R., et al. (2010). *Molecular Cell, 40*, 976–987.
Landry, J., Sharov, A. A., Piao, Y., Sharova, L. V., Xiao, H., Southon, E., et al. (2008). *PLoS Genetics, 4*, e1000241.
Längst, G., & Becker, P. (2001). *Journal of Cell Science, 114*, 2561–2568.
Latrick, C. M., Marek, M., Ouararhni, K., Papin, C., Stoll, I., Ignatyeva, M., et al. (2016). *Nature Structural & Molecular Biology, 23*, 309–316.
Le Guezennec, X., Vermeulen, M., Brinkman, A. B., Hoeijmakers, W. A., Cohen, A., Lasonder, E., et al. (2006). *Molecular Cell Biology, 26*, 843–851.
Lee, T. I., Jenner, R. G., Boyer, L. A., Guenther, M. G., Levine, S. S., Kumar, R. M., et al. (2006). *Cell, 125*, 301–313.
Lee, H. S., Park, J. H., Kim, S. J., Kwon, S. J., & Kwon, J. (2010). *The EMBO Journal, 29*, 1434–1445.
Lee, C.-K., Shibata, Y., Rao, B., Strahl, B. D., & Lieb, J. D. (2004). *Nature Genetics, 36*, 900–905.
Lee, D. Y., Teyssier, C., Strahl, B. D., & Stallcup, M. R. (2005). *Endocrine Reviews, 26*, 147–170.
Lee, W., Tillo, D., Bray, N., Morse, R., Davis, R., Hughes, T., et al. (2007). *Nature Genetics, 39*, 1235–1244.
Lehnertz, B., Ueda, Y., Derijck, A. A. H. A., Braunschweig, U., Perez-Burgos, L., Kubicek, S., et al. (2003). *Current Biology, 13*, 1192–1200.
Lei, I., West, J., Yan, Z., Gao, X., Fang, P., Dennis, J. H., et al. (2015). *The Journal of Biological Chemistry, 290*, 19343–19352.
Lessard, J., Wu, J. I., Ranish, J. A., Wan, M., Winslow, M. M., Staahl, B. T., et al. (2007). *Neuron, 55*, 201–215.
Levasseur, D. N., Wang, J., Dorschner, M. O., Stamatoyannopoulos, J. A., & Orkin, S. H. (2008). *Genes & Development, 22*, 575–580.
Lewis, P. H. (1947). *Drosophila Information Service, 21*, 69.
Lewis, E. B. (1978). *Nature, 276*, 565–570.
Lewis, P. W., Elsaesser, S. J., Noh, K. M., Stadler, S. C., & Allis, C. D. (2010). *Proceedings of the National Academy of Sciences of the United States of America, 107*, 14075–14080.
Li, B., Carey, M., & Workman, J. L. (2007). *Cell, 128*, 707–719.
Li, G., Fullwood, M. J., Xu, H., Mulawadi, F. H., Velkov, S., Vega, V., et al. (2010). *Genome Biology, 11*, R22.
Li, Z., Gadue, P., Chen, K., Jiao, Y., Tuteja, G., Schug, J., et al. (2012). *Cell, 151*, 1608–1616.
Li, Q., Lian, S., Dai, Z., Xiang, Q., & Dai, X. (2013). *Database: The Journal of Biological Databases and Curation, 2013*, bat057.
Li, Y., Rivera, C. M., Ishii, H., Jin, F., Selvaraj, S., Lee, A. Y., et al. (2014). *PLoS One, 9*, e114485.

Liang, X., Shan, S., Pan, L., Zhao, J., Ranjan, A., Wang, F., et al. (2016). *Nature Structural & Molecular Biology, 23*, 317–323.
Liang, J., Wan, M., Zhang, Y., Gu, P., Xin, H., Jung, S. Y., et al. (2008). *Nature Cell Biology, 10*, 731–739.
Lickert, H., Takeuchi, J. K., von Both, I., Walls, J. R., McAuliffe, F., Lee Adamson, S., et al. (2004). *Nature, 432*, 107–112.
Lieberman-Aiden, E., van Berkum, N. L., Williams, L., Imakaev, M., Ragoczy, T., Telling, A., et al. (2009). *Science, 326*, 289–293.
Lin, B., Lee, H., Yoon, J. G., Madan, A., Wayner, E., Tonning, S., et al. (2015). *Oncotarget, 6*, 5369–5381.
Lin, J. J., Lehmann, L. W., Bonora, G., Sridharan, R., Vashisht, A. A., Tran, N., et al. (2011). *Genes & Development, 25*, 2198–2209.
Liu, X., Bushnell, D. A., & Kornberg, R. D. (2013). *Biochimica et Biophysica Acta, 1829*, 2–8.
Liu, W. H., Roemer, S. C., Port, A. M., & Churchill, M. E. (2012). *Nucleic Acids Research, 40*, 11229–11239.
Lomvardas, S., & Thanos, D. (2001). *Cell, 106*, 685–696.
Long, M., Sun, X., Shi, W., Yanru, A., Leung, S. T. C., Ding, D., et al. (2019). *Nucleic Acids Research, 47*, 8399–8409.
Lorain, S., Demczuk, S., Lamour, V., Toth, S., Aurias, A., Roe, B. A., et al. (1996). *Genome Research, 6*, 43–50.
Loven, J., Hoke, H. A., Lin, C. Y., Lau, A., Orlando, D. A., Vakoc, C. R., et al. (2013). *Cell, 153*, 320–334.
Luger, K. (2006). *Chromosome Research, 14*, 5–16.
Luger, K., Mäder, A. W., Richmond, R. K., Sargent, D. F., & Richmond, T. J. (1997). *Nature*, 389.
Manning, B. J., & Yusufzai, T. (2017). *The Journal of Biological Chemistry, 292*, 11927–11936.
Marmorstein, R. (2001). *Journal of Molecular Biology, 311*, 433–444.
Marmorstein, R., & Zhou, M. M. (2014). *Cold Spring Harbor Perspectives in Biology, 6*, a018762.
Marson, A., Levine, S. S., Cole, M. F., Frampton, G. M., Brambrink, T., Johnstone, S., et al. (2008). *Cell, 134*, 521–533.
Martianov, I., Brancorsini, S., Catena, R., Gansmuller, A., Kotaja, N., Parvinen, M., et al. (2005). *Proceedings of the National Academy of Sciences of the United States of America, 102*, 2808–2813.
Martire, S., Gogate, A. A., Whitmill, A., Tafessu, A., Nguyen, J., Teng, Y. C., et al. (2019). *Nature Genetics, 51*, 941–946.
Marzluff, W. F., Gongidi, P., Woods, K. R., Jin, J., & Maltais, L. J. (2002). *Genomics, 80*, 487–498.
Mashtalir, N., D'Avino, A. R., Michel, B. C., Luo, J., Pan, J., Otto, J. E., et al. (2018). *Cell, 175*. 1272-1288.e20.
Mattiroli, F., Uckelmann, M., Sahtoe, D. D., van Dijk, W. J., & Sixma, T. K. (2014). *Nature Communications, 5*, 3291.
Mayran, A., & Drouin, J. (2018). *The Journal of Biological Chemistry, 293*, 13795–13804.
McBryant, S. J., Park, Y. J., Abernathy, S. M., Laybourn, P. J., Nyborg, J. K., & Luger, K. (2003). *The Journal of Biological Chemistry, 278*, 44574–44583.
McGarvey, K. M., Van Neste, L., Cope, L., Ohm, J. E., Herman, J. G., Van Criekinge, W., et al. (2008). *Cancer Research, 68*, 5753–5759.
McGinty, R. K., Henrici, R. C., & Tan, S. (2014). *Nature, 514*, 591.
McKittrick, E., Gafken, P., Ahmad, K., & Henikoff, S. (2004). *Proceedings of the National Academy of Sciences of the United States of America, 101*, 1525–1530.
Meshorer, E., & Misteli, T. (2006). *Nature Reviews. Molecular Cell Biology*, 7, 540–546.
Micucci, J., Sperry, E., & Martin, D. (2015). *Stem Cells and Development, 24*, 9.

Middeljans, E., Wan, X., Jansen, P. W., Sharma, V., Stunnenberg, H. G., & Logie, C. (2012). *PLoS One, 7*, e33834.
Miller, C. J., & Davidson, L. A. (2013). *Nature Reviews Genetics, 14*, 733–744.
Mizuguchi, G., Shen, X., Landry, J., Wu, W., Sen, S., & Wu, C. (2004). *Science, 303*, 343–348.
Mohn, F., Weber, M., Rebhan, M., Roloff, T. C., Richter, J., Stadler, M. B., et al. (2008). *Molecular Cell, 30*, 755–766.
Molaro, A., Young, J. M., & Malik, H. S. (2018). *Genome Research, 28*, 460–473.
Moody, J. D., Levy, S., Mathieu, J., Xing, Y., Kim, W., Dong, C., et al. (2017). *Proceedings of the National Academy of Sciences of the United States of America, 114*, 10125–10130.
Moorthy, S. D., Davidson, S., Shchuka, V. M., Singh, G., Malek-Gilani, N., Langroudi, L., et al. (2017). *Genome Research, 27*, 246–258.
Morey, L., Pascual, G., Cozzuto, L., Roma, G., Wutz, A., Benitah, S. A., et al. (2012). *Cell Stem Cell, 10*, 47–62.
Morey, L., Santanach, A., & Di Croce, L. (2015). *Molecular Cell Biology, 35*, 2716–2728.
Morgan, M. T., Haj-Yahya, M., Ringel, A. E., Bandi, P., Brik, A., & Wolberger, C. (2016). *Science, 351*, 725–728.
Morillon, A., Karabetsou, N., O'Sullivan, J., Kent, N., Proudfoot, N., & Mellor, J. (2003). *Cell, 115*, 425–435.
Mosammaparast, N., Ewart, C. S., & Pemberton, L. F. (2002). *The EMBO Journal, 21*, 6527–6538.
Mylonas, C., & Tessarz, P. (2018). *Life Science Alliance, 1*, e201800085.
Nakamura, K., Kato, A., Kobayashi, J., Yanagihara, H., Sakamoto, S., Oliveira, D. V. N. P., et al. (2011). *Molecular Cell, 41*, 515–528.
Nakayama, R. T., Pulice, J. L., Valencia, A. M., McBride, M. J., McKenzie, Z. M., Gillespie, M. A., et al. (2017). *Nature Genetics, 49*, 1613–1623.
Nardi, I. K., Zasadzinska, E., Stellfox, M. E., Knippler, C. M., & Foltz, D. R. (2016). *Molecular Cell, 61*, 774–787.
Neigeborn, L., & Carlson, M. (1984). *Genetics, 108*, 845–858.
Neumann, H., Hancock, S. M., Buning, R., Routh, A., Chapman, L., Somers, J., et al. (2009). *Molecular Cell, 36*, 153–163.
Niikura, Y., Kitagawa, R., Ogi, H., Abdulle, R., Pagala, V., & Kitagawa, K. (2015). *Developmental Cell, 32*, 589–603.
Nishibuchi, G., & Dejardin, J. (2017). *Chromosome Research, 25*, 77–87.
Noma, K., Allis, C. D., & Grewal, S. I. (2001). *Science, 293*, 1150–1155.
Nora, E. P., Lajoie, B. R., Schulz, E. G., Giorgetti, L., Okamoto, I., Servant, N., et al. (2012). *Nature, 485*, 381–385.
Novo, C. L., Javierre, B. M., Cairns, J., Segonds-Pichon, A., Wingett, S. W., Freire-Pritchett, P., et al. (2018). *Cell Reports, 22*, 2615–2627.
Okada, M., Okawa, K., Isobe, T., & Fukagawa, T. (2009). *Molecular Biology of the Cell, 20*, 3986–3995.
Okumura-Nakanishi, S., Saito, M., Niwa, H., & Ishikawa, F. (2005). *The Journal of Biological Chemistry, 280*, 5307–5317.
Olave, I., Wang, W., Xue, Y., Kuo, A., & Crabtree, G. R. (2002). *Genes & Development, 16*, 2509–2517.
Oliver, M. I., Concha, C., Gutierrez, S., Bustos, A., Montecino, M., Puchi, M., et al. (2002). *Journal of Cellular Biochemistry, 85*, 851–859.
O'Loghlen, A., Munoz-Cabello, A. M., Gaspar-Maia, A., Wu, H. A., Banito, A., Kunowska, N., et al. (2012). *Cell Stem Cell, 10*, 33–46.
Orkin, S. H., & Hochedlinger, K. (2011). *Cell, 145*, 835–850.
Orphanides, G., LeRoy, G., Chang, C. H., Luse, D. S., & Reinberg, D. (1998). *Cell, 92*, 105–116.

Orphanides, G., Wu, W., Lane, W., Hampsey, M., & Reinberg, D. (1999). *Nature, 400,* 284–288.
Osberg, B., Nuebler, J., Korber, P., & Gerland, U. (2014). *Nucleic Acids Research, 42,* 13633–13645.
Palmer, D., O'Day, K., Wener, M., Andrews, B., & Margolis, R. (1987). *Journal of Cell Biology, 104,* 10.
Papamichos-Chronakis, M., Watanabe, S., Rando, O. J., & Peterson, C. L. (2011). *Cell, 144,* 200–213.
Pardo, M., Lang, B., Yu, L., Prosser, H., Bradley, A., Babu, M. M., et al. (2010). *Cell Stem Cell, 6,* 382–395.
Park, Y. J., Chodaparambil, J. V., Bao, Y., McBryant, S. J., & Luger, K. (2005). *The Journal of Biological Chemistry, 280,* 1817–1825.
Park, C. S., Rehrauer, H., & Mansuy, I. M. (2013). *BMC Genomics, 14,* 539.
Parker, S. C., Stitzel, M. L., Taylor, D. L., Orozco, J. M., Erdos, M. R., Akiyama, J. A., et al. (2013). *Proceedings of the National Academy of Sciences of the United States of America, 110,* 17921–17926.
Paro, R., & Hogness, D. S. (1991). *Proceedings of the National Academy of Sciences of the United States of America, 88,* 263–267.
Pasini, D., Cloos, P. A., Walfridsson, J., Olsson, L., Bukowski, J. P., Johansen, J. V., et al. (2010). *Nature, 464,* 306–310.
Pchelintsev, N. A., McBryan, T., Rai, T. S., van Tuyn, J., Ray-Gallet, D., Almouzni, G., et al. (2013). *Cell Reports, 3,* 1012–1019.
Pegoraro, G., Kubben, N., Wickert, U., Gohler, H., Hoffmann, K., & Misteli, T. (2009). *Nature Cell Biology, 11,* 1261–1267.
Peleg, S., Sananbenesi, F., Zovoilis, A., Burkhardt, S., Bahari-Javan, S., Agis-Balboa, R. C., et al. (2010). *Science, 328,* 753–756.
Peng, Y., & Zhang, Y. (2018). *Animal Models and Experimental Medicine, 1,* 169–179.
Peterson, A. J., Mallin, D. R., Francis, N. J., Ketel, C. S., Stamm, J., Voeller, R. K., et al. (2004). *Genetics, 167,* 1225–1239.
Petter, M., Selvarajah, S. A., Lee, C. C., Chin, W. H., Gupta, A. P., Bozdech, Z., et al. (2013). *Molecular Microbiology, 87,* 1167–1182.
Piatti, P., Lim, C. Y., Nat, R., Villunger, A., Geley, S., Shue, Y. T., et al. (2015). *Scientific Reports, 5,* 8007.
Pokholok, D. K., Harbison, C. T., Levine, S., Cole, M., Hannett, N. M., Lee, T. I., et al. (2005). *Cell, 122,* 517–527.
Poli, J., Gasser, S. M., & Papamichos-Chronakis, M. (2017). *Philosophical Transactions of the Royal Society of London. Series B, Biological Sciences, 372,* 20160289.
Poli, J., Gerhold, C. B., Tosi, A., Hustedt, N., Seeber, A., Sack, R., et al. (2016). *Genes & Development, 30,* 337–354.
Postberg, J., Forcob, S., Chang, W. J., & Lipps, H. J. (2010). *BMC Evolutionary Biology, 10,* 259.
Pradhan, S. K., Su, T., Yen, L., Jacquet, K., Huang, C., Cote, J., et al. (2016). *Molecular Cell, 61,* 27–38.
Radman-Livaja, M., & Rando, O. J. (2010). *Developmental Biology, 339,* 258–266.
Raisner, R. M., Hartley, P. D., Meneghini, M. D., Bao, M. Z., Liu, C. L., Schreiber, S. L., et al. (2005). *Cell, 123,* 233–248.
Rando, O. J., & Winston, F. (2012). *Genetics, 190,* 351–387.
Ransom, M., Dennehey, B. K., & Tyler, J. K. (2010). *Cell, 140,* 183–195.
Recht, J., Tsubota, T., Tanny, J. C., Diaz, R. L., Berger, J. M., Zhang, X., et al. (2006). *Proceedings of the National Academy of Sciences of the United States of America, 103,* 6988–6993.
Reya, T., & Clevers, H. (2005). *Nature, 434,* 843–850.

Reyes, J. C., Barra, J., Muchardt, C., Camus, A., Babinet, C., & Yaniv, M. (1998). *The EMBO Journal, 17*, 6979–6991.
Reynolds, N., Latos, P., Hynes-Allen, A., Loos, R., Leaford, D., O'Shaughnessy, A., et al. (2012). *Cell Stem Cell, 10*, 583–594.
Reynolds, N., Salmon-Divon, M., Dvinge, H., Hynes-Allen, A., Balasooriya, G., Leaford, D., et al. (2012). *The EMBO Journal, 31*, 593–605.
Rhee, H. S., Bataille, A. R., Zhang, L., & Pugh, B. F. (2014). *Cell, 159*, 1377–1388.
Roberts, C. W., Galusha, S. A., McMenamin, M. E., Fletcher, C. D., & Orkin, S. H. (2000). *Proceedings of the National Academy of Sciences of the United States of America, 97*, 13796–13800.
Rodriguez, J., Munoz, M., Vives, L., Frangou, C. G., Groudine, M., & Peinado, M. A. (2008). *Proceedings of the National Academy of Sciences of the United States of America, 105*, 19809–19814.
Rogakou, E. P., Pilch, D. R., Orr, A. H., Ivanova, V. S., & Bonner, W. M. (1998). *The Journal of Biological Chemistry, 273*, 5858–5868.
Roh, T. Y., Cuddapah, S., Cui, K., & Zhao, K. (2006). *Proceedings of the National Academy of Sciences of the United States of America, 103*, 15782–15787.
Romito, A., & Cobellis, G. (2016). *Stem Cells International, 2016*, 9451492.
Rufiange, A., Jacques, P. E., Bhat, W., Robert, F., & Nourani, A. (2007). *Molecular Cell, 27*, 393–405.
Ruhl, D. D., Jin, J., Cai, Y., Swanson, S., Florens, L., Washburn, M. P., et al. (2006). *Biochemistry, 45*, 5671–5677.
Runge, J. S., Raab, J. R., & Magnuson, T. (2018). *G3: genes—genomes—genetics, 8*, 1095–1102.
Saladi, S. V., & de la Serna, I. L. (2010). *Stem Cell Reviews and Reports, 6*, 62–73.
Santisteban, M. S., Hang, M., & Smith, M. M. (2011). *Molecular Cell Biology, 31*, 1848–1860.
Santoro, S. W., & Dulac, C. (2012). *eLife, 1*, e00070.
Sarma, K., Cifuentes-Rojas, C., Ergun, A., Del Rosario, A., Jeon, Y., White, F., et al. (2014). *Cell, 159*, 869–883.
Schaniel, C., Ang, Y.-S., Ratnakumar, K., Cormier, C., James, T., Bernstein, E., et al. (2009). *Stem Cells (Dayton, Ohio), 27*, 2979–2991.
Schenk, R., Jenke, A., Zilbauer, M., Wirth, S., & Postberg, J. (2011). *Chromosoma, 120*, 275–285.
Schlesinger, S., Kaffe, B., Melcer, S., Aguilera, J. D., Sivaraman, D. M., Kaplan, T., et al. (2017). *Nucleic Acids Research, 45*, 12181–12194.
Schneider, J., Bajwa, P., Johnson, F. C., Bhaumik, S. R., & Shilatifard, A. (2006). *The Journal of Biological Chemistry, 281*, 37270–37274.
Schnetz, M. P., Handoko, L., Akhtar-Zaidi, B., Bartels, C. F., Pereira, C. F., Fisher, A. G., et al. (2010). *PLoS Genetics, 6*, e1001023.
Schwabish, M., & Struhl, K. (2004). *Molecular Cell Biology, 24*, 10111–10117.
Seto, E., & Yoshida, M. (2014). *Cold Spring Harbor Perspectives in Biology, 6*, a018713.
Shan, Y., Liang, Z., Xing, Q., Zhang, T., Wang, B., Tian, S., et al. (2017). *Nature Communications, 8*, 672.
Shao, Z., Raible, F., Mollaaghababa, R., Guyon, J. R., Wu, C. T., Bender, W., et al. (1999). *Cell, 98*, 37–46.
Sharma, A., Singh, K., & Almasan, A. (2012). *Methods in Molecular Biology, 920*, 613–626.
Shen, Z., Formosa, T., & Tantin, D. (2018). *Stem Cells and Development, 27*, 1693–1701.
Shi, Y. (2007). *Nature Reviews Genetics, 8*, 829.
Shinagawa, T., Takagi, T., Tsukamoto, D., Tomaru, C., Huynh, L. M., Sivaraman, P., et al. (2014). *Cell Stem Cell, 14*, 217–227.
Shivaswamy, S., Bhinge, A., Zhao, Y., Jones, S., Hirst, M., & Iyer, V. R. (2008). *PLoS Biology, 6*, e65.

Shivaswamy, S., & Iyer, V. (2008). *Molecular and Cell Biology*, *28*, 2221–2234.
Shogren-Knaak, M., Ishii, H., Pazin, M., Davie, J., & Peterson, C. L. (2006). *Science*, *311*, 4.
Shuaib, M., Ouararhni, K., Dimitrov, S., & Hamiche, A. (2010). *Proceedings of the National Academy of Sciences of the United States of America*, *107*, 1349–1354.
Shukla, S., Kavak, E., Gregory, M., Imashimizu, M., Shutinoski, B., Kashlev, M., et al. (2011). *Nature*, *479*, 74.
Simic, R., Lindstrom, D. L., Tran, H. G., Roinick, K. L., Costa, P. J., Johnson, A. D., et al. (2003). *The EMBO Journal*, *22*, 1846–1856.
Siuti, N., Roth, M. J., Mizzen, C. A., Kelleher, N. L., & Pesavento, J. J. (2006). *Journal of Proteome Research*, *5*, 233–239.
Skene, P. J., & Henikoff, S. (2013). *Development*, *140*, 2513–2524.
Skene, P. J., & Henikoff, S. (2017). *eLife*, *6*, 1–35.
Smith, E., & Shilatifard, A. (2010). *Molecular Cell*, *40*, 689–701.
Smith, S., & Stillman, B. (1989). *Cell*, *58*, 15–25.
Solomon, M. J., Larsen, P. L., & Varshavsky, A. (1988). *Cell*, *53*, 937–947.
Squatrito, M., Gorrini, C., & Amati, B. (2006). *Trends in Cell Biology*, *16*, 433–442.
Stanton, B. Z., Hodges, C., Calarco, J. P., Braun, S. M. G., Ku, W. L., Kadoch, C., et al. (2017). *Nature Genetics*, *49*, 282–288.
Stern, M., Jensen, R., & Herskowitz, I. (1984). *Journal of Molecular Biology*, *178*, 853–868.
Stillman, B. (1986). *Cell*, *45*, 555–565.
Stopka, T., & Skoultchi, A. I. (2003). *Proceedings of the National Academy of Sciences of the United States of America*, *100*, 14097–14102.
Studitsky, V. M., Clark, D. J., & Felsenfeld, G. (1994). *Cell*, *76*, 371–382.
Studitsky, V. M., Kassavetis, G. A., Geiduschek, E. P., & Felsenfeld, G. (1997). *Science*, *278*, 1960–1963.
Stuwe, T., Hothorn, M., Lejeune, E., Rybin, V., Bortfeld, M., Scheffzek, K., et al. (2008). *Proceedings of the National Academy of Sciences of the United States of America*, *105*, 8884–8889.
Subramanian, V., Mazumder, A., Surface, L. E., Butty, V. L., Fields, P. A., Alwan, A., et al. (2013). *PLoS Genetics*, *9*, e1003725.
Sullivan, S. A., & Landsman, D. (2003). *Proteins*, *52*, 454–465.
Sun, Z., Filipescu, D., Andrade, J., Gaspar-Maia, A., Ueberheide, B., & Bernstein, E. (2018). *Nature Structural & Molecular Biology*, *25*, 958–970.
Surface, L. E., Fields, P. A., Subramanian, V., Behmer, R., Udeshi, N., Peach, S. E., et al. (2016). *Cell Reports*, *14*, 1142–1155.
Suto, R., Clarkson, M., Tremethick, D., & Luger, K. (2000). *Nature Structural Biology*, *7*, 1121–1124.
Szerlong, H., Hinata, K., Viswanathan, R., Erdjument-Bromage, H., Tempst, P., & Cairns, B. R. (2008). *Nature Structural & Molecular Biology*, *15*, 469–476.
Tachiwana, H., Kagawa, W., Shiga, T., Osakabe, A., Miya, Y., Saito, K., et al. (2011). *Nature*, *476*, 232.
Tachiwana, H., Osakabe, A., Shiga, T., Miya, Y., Kimura, H., Kagawa, W., et al. (2011). *Acta Crystallographica. Section D, Biological Crystallography*, *67*, 578–583.
Tagami, H., Ray-Gallet, D., Almouzni, G., & Nakatani, Y. (2004). *Cell*, *116*, 51–61.
Takahashi, Y. H., Westfield, G. H., Oleskie, A. N., Trievel, R. C., Shilatifard, A., & Skiniotis, G. (2011). *Proceedings of the National Academy of Sciences of the United States of America*, *108*, 20526–20531.
Takahashi, K., & Yamanaka, S. (2006). *Cell*, *126*, 663–676.
Takaku, M., Grimm, S. A., Shimbo, T., Perera, L., Menafra, R., Stunnenberg, H. G., et al. (2016). *Genome Biology*, *17*, 36.
Takeuchi, J. K., & Bruneau, B. G. (2009). *Nature*, *459*, 708–711.
Tan, Y., Xue, Y., Song, C., & Grunstein, M. (2013). *Proceedings of the National Academy of Sciences of the United States of America*, *110*, 11493–11498.

Tang, Y., Poustovoitov, M. V., Zhao, K., Garfinkel, M., Canutescu, A., Dunbrack, R., et al. (2006). *Nature Structural & Molecular Biology, 13*, 921–929.
Tavares, L., Dimitrova, E., Oxley, D., Webster, J., Poot, R., Demmers, J., et al. (2012). *Cell, 148*, 664–678.
Timinszky, G. (2009). *Nature Structural & Molecular Biology, 16*, 7.
Tirosh, I., Sigal, N., & Barkai, N. (2010). *Genome Biology, 11*, 1–13 R49.
Tjeertes, J. V., Miller, K. M., & Jackson, S. P. (2009). *The EMBO Journal, 28*, 1878–1889.
Tomioka, M., Nishimoto, M., Miyagi, S., Katayanagi, T., Fukui, N., Niwa, H., et al. (2002). *Nucleic Acids Research, 30*, 3202–3213.
Tomson, B. N., & Arndt, K. M. (2013). *Biochimica et Biophysica Acta, 1829*, 116–126.
Tong, J. K., Hassig, C. A., Schnitzler, G. R., Kingston, R. E., & Schreiber, S. L. (1998). *Nature, 395*, 917–921.
Tsai, P. H., Chien, Y., Wang, M. L., Hsu, C. H., Laurent, B., Chou, S. J., et al. (2019). *Nucleic Acids Research, 47*, 10115–10133.
Tsubota, T., Berndsen, C. E., Erkmann, J. A., Smith, C. L., Yang, L., Freitas, M. A., et al. (2007). *Molecular Cell, 25*, 703–712.
Turinetto, V., Orlando, L., Sanchez-Ripoll, Y., Kumpfmueller, B., Storm, M. P., Porcedda, P., et al. (2012). *Stem Cells, 30*, 1414–1423.
Tutter, A. V., Kowalski, M. P., Baltus, G. A., Iourgenko, V., Labow, M., Li, E., et al. (2009). *The Journal of Biological Chemistry, 284*, 3709–3718.
Ueda, J., Harada, A., Urahama, T., Machida, S., Maehara, K., Hada, M., et al. (2017). *Cell Reports, 18*, 593–600.
Urahama, T., Harada, A., Maehara, K., Horikoshi, N., Sato, K., Sato, Y., et al. (2016). *Epigenetics & Chromatin, 9*, 2.
Van Attikum, H., Fritsch, O., Hohn, B., & Gasser, S. (2004). *Cell, 119*, 11.
van den Berg, D. L., Snoek, T., Mullin, N. P., Yates, A., Bezstarosti, K., Demmers, J., et al. (2010). *Cell Stem Cell, 6*, 369–381.
van Mierlo, G., Veenstra, G. J. C., Vermeulen, M., & Marks, H. (2019). *Trends in Cell Biology, 29*, 660–671.
Varshavsky, A. J., Sundin, O., & Bohn, M. (1979). *Cell, 16*, 453–466.
Vastenhouw, N. L., & Schier, A. F. (2012). *Current Opinion in Cell Biology, 24*, 374–386.
Venkatesh, S., & Workman, J. L. (2015). *Nature Reviews. Molecular Cell Biology, 16*, 178–189.
Vierbuchen, T., Ling, E., Cowley, C. J., Couch, C. H., Wang, X., Harmin, D. A., et al. (2017). *Molecular Cell, 68*. 1067-1082.e12.
Vincenz, C., & Kerppola, T. K. (2008). *Proceedings of the National Academy of Sciences of the United States of America, 105*, 16572–16577.
Vire, E., Brenner, C., Deplus, R., Blanchon, L., Fraga, M., Didelot, C., et al. (2006). *Nature, 439*, 871–874.
Voigt, P., Tee, W. W., & Reinberg, D. (2013). *Genes & Development, 27*, 1318–1338.
Vostrov, A. A., & Quitschke, W. W. (1997). *The Journal of Biological Chemistry, 272*, 33353–33359.
Vostrov, A. A., Taheny, M. J., & Quitschke, W. W. (2002). *Journal of Biological Chemistry, 277*, 1619–1627.
Wade, P. A., Jones, P. L., Vermaak, D., & Wolffe, A. P. (1998). *Current Biology, 8*, 843–846.
Wan, Y., Saleem, R. A., Ratushny, A. V., Roda, O., Smith, J. J., Lin, C. H., et al. (2009). *Molecular Cell Biology, 29*, 2346–2358.
Wang, W., Côté, J., Xue, Y., Zhou, S., Khavari, P. A., Biggar, S. R., et al. (1996). *The EMBO Journal, 15*, 5370–5382.
Wang, L., Du, Y., Ward, J. M., Shimbo, T., Lackford, B., Zheng, X., et al. (2014). *Cell Stem Cell, 14*, 575–591.
Wang, Y., Long, H., Yu, J., Dong, L., Wassef, M., Zhuo, B., et al. (2018). *BMC Biology, 16*, 107.
Wang, X., Wang, S., Troisi, E. C., Howard, T. P., Haswell, J. R., Wolf, B. K., et al. (2019). *Nature Communications, 10*, 1–11 1881.

Weiner, A., Hughes, A., Yassour, M., Rando, O. J., & Friedman, N. (2010). *Genome Research*, *20*, 90–100.

West, M. H., & Bonner, W. M. (1980). *Biochemistry*, *19*, 3238–3245.

Whetstine, J. R., Nottke, A., Lan, F., Huarte, M., Smolikov, S., Chen, Z., et al. (2006). *Cell*, *125*, 467–481.

White, C. L., & Luger, K. (2004). *Journal of Molecular Biology*, *342*, 1391–1402.

Whyte, W. A., Bilodeau, S., Orlando, D. A., Hoke, H. A., Frampton, G. M., Foster, C. T., et al. (2012). *Nature*, *482*, 221–225.

Whyte, W. A., Orlando, D. A., Hnisz, D., Abraham, B. J., Lin, C. Y., Kagey, M. H., et al. (2013). *Cell*, *153*, 307–319.

Wiechens, N., Singh, V., Gkikopoulos, T., Schofield, P., Rocha, S., & Owen-Hughes, T. (2016). *PLoS Genetics*, *12*, e1005940.

Wiedemann, S. M., Mildner, S. N., Bonisch, C., Israel, L., Maiser, A., Matheisl, S., et al. (2010). *The Journal of Cell Biology*, *190*, 777–791.

Wirbelauer, C., Bell, O., & Schubeler, D. (2005). *Genes & Development*, *19*, 1761–1766.

Wong, M. M., Cox, L. K., & Chrivia, J. C. (2007). *The Journal of Biological Chemistry*, *282*, 26132–26139.

Wu, J. I., Lessard, J., Olave, I. A., Qiu, Z., Ghosh, A., Graef, I. A., et al. (2007). *Neuron*, *56*, 94–108.

Wysocka, J., Swigut, T., Xiao, H., Milne, T. A., Kwon, S. Y., Landry, J., et al. (2006). *Nature*, *442*, 86–90.

Xia, W., & Jiao, J. (2017). *Cell Death and Differentiation*, *24*, 1548–1563.

Xie, W., Song, C., Young, N. L., Sperling, A. S., Xu, F., Sridharan, R., et al. (2009). *Molecular Cell*, *33*, 417–427.

Xu, M., Barnard, M. B., Rose, S. M., Cockerill, P. N., Huang, S. Y., & Garrard, W. T. (1986). *The Journal of Biological Chemistry*, *261*, 3838–3845.

Xue, Y., Van, C., Pradhan, S. K., Su, T., Gehrke, J., Kuryan, B. G., et al. (2015). *Genes & Development*, *29*, 350–355.

Xue, Y., Wong, J., Moreno, G. T., Young, M. K., Côté, J., & Wang, W. (1998). *Molecular Cell*, *2*, 851–861.

Yadon, A. N., & Tsukiyama, T. (2011). *Cell*, *144*, 453–453.e1.

Yan, W., Ma, L., Burns, K. H., & Matzuk, M. M. (2003). *Proceedings of the National Academy of Sciences of the United States of America*, *100*, 10546–10551.

Yan, Y., Zhao, W., Huang, Y., Tong, H., Xia, Y., Jiang, Q., et al. (2017). *Scientific Reports*, *7*, 46276.

Yang, B., Zwaans, B. M., Eckersdorff, M., & Lombard, D. B. (2009). *Cell Cycle*, *8*, 2662–2663.

Ye, B., Liu, B., Yang, L., Huang, G., Hao, L., Xia, P., et al. (2017). *Nature Communications*, *8*, 1518.

Ye, B., Liu, B., Yang, L., Zhu, X., Zhang, D., Wu, W., et al. (2018). *The EMBO Journal*, *37*, *1–15* (2018).

Yelagandula, R., Stroud, H., Holec, S., Zhou, K., Feng, S., Zhong, X., et al. (2014). *Cell*, *158*, 98–109.

Yen, K., Vinayachandran, V., Batta, K., Koerber, R. T., & Pugh, B. F. (2012). *Cell*, *149*, 1461–1473.

Yen, K., Vinayachandran, V., & Pugh, B. F. (2013). *Cell*, *154*, 1246–1256.

Yildirim, O., Li, R., Hung, J.-H., Chen, P. B., Dong, X., Ee, L.-S., et al. (2011). *Cell*, *147*, 1498–1510.

Yin, J.-W., & Wang, G. (2014). *Development*, *141*, 977–987.

Yoda, K., Ando, S., Morishita, S., Houmura, K., Hashimoto, K., Takeyasu, K., et al. (2000). *Proceedings of the National Academy of Sciences of the United States of America*, *97*, 7266–7271.

Yoshida, T., Hazan, I., Zhang, J., Ng, S. Y., Naito, T., Snippert, H. J., et al. (2008). *Genes & Development, 22*, 1174–1189.
Youdell, M. L., Kizer, K. O., Kisseleva-Romanova, E., Fuchs, S. M., Duro, E., Strahl, B. D., et al. (2008). *Molecular Cell Biology, 28*, 4915–4926.
Young, R. A. (2011). *Cell, 144*, 940–954.
Zalensky, A. O., Tomilin, N. V., Zalenskaya, I. A., Teplitz, R. L., & Bradbury, E. M. (1997). *Experimental Cell Research, 232*, 29–41.
Zaret, K. S., & Carroll, J. S. (2011). *Genes & Development, 25*, 2227–2241.
Zaret, K. S., Lerner, J., & Iwafuchi-Doi, M. (2016). *Molecular Cell, 62*, 665–667.
Zawadzki, K., Morozov, A., & Broach, J. (2009). *Molecular Cell Biology, 20*, 3451–3617.
Zee, B. M., Levin, R. S., Xu, B., LeRoy, G., Wingreen, N. S., & Garcia, B. A. (2010). *The Journal of Biological Chemistry, 285*, 3341–3350.
Zentner, G., Tesar, P., & Scacheri, P. (2011). *Genome Research, 21*, 10.
Zhang, R., Erler, J., & Langowski, J. (2017). *Biophysical Journal, 112*, 450–459.
Zhang, Y., LeRoy, G., Seelig, H.-P., Lane, W. S., & Reinberg, D. (1998). *Cell, 95*, 279–289.
Zhang, X., Li, B., Li, W., Ma, L., Zheng, D., Li, L., et al. (2014). *Stem Cell Reports, 3*, 460–474.
Zhang, H., Roberts, D., & Cairns, B. R. (2005). *Cell, 123*, 219–231.
Zhao, H., Han, Z., Liu, X., Gu, J., Tang, F., Wei, G., et al. (2017). *The Journal of Biological Chemistry, 292*, 8507–8519.
Zhao, W., Huang, Y., Zhang, J., Liu, M., Ji, H., Wang, C., et al. (2017). *The Journal of Biological Chemistry, 292*, 21527–21537.
Zhao, R., Nakamura, T., Fu, Y., Lazar, Z., & Spector, D. (2011). *Nature Cell Biology, 13*, 9.
Zhao, W., Tong, H., Huang, Y., Yan, Y., Teng, H., Xia, Y., et al. (2017). *The Journal of Biological Chemistry, 292*, 2773–2784.
Zhou, R., Bonneaud, N., Yuan, C. X., de Santa Barbara, P., Boizet, B., Schomber, T., et al. (2002). *Nucleic Acids Research, 30*, 3245–3252.
Zhou, B. R., Feng, H., Kato, H., Dai, L., Yang, Y., Zhou, Y., et al. (2013). *Proceedings of the National Academy of Sciences of the United States of America, 110*, 19390–19395.
Zhou, H. Y., Katsman, Y., Dhaliwal, N. K., Davidson, S., Macpherson, N. N., Sakthidevi, M., et al. (2014). *Genes & Development, 28*, 2699–2711.
Zhu, D., Fang, J., Li, Y., & Zhang, J. (2009). *PLoS One, 4*, e7684.
Zhu, F., Farnung, L., Kaasinen, E., Sahu, B., Yin, Y., Wei, B., et al. (2018). *Nature, 562*, 76–81.
Zink, L. M., Delbarre, E., Eberl, H. C., Keilhauer, E. C., Bonisch, C., Punzeler, S., et al. (2017). *Nucleic Acids Research, 45*, 5691–5706.

CHAPTER TWO

Role of lncRNAs in stem cell maintenance and differentiation

Meghali Aich[a,b], **Debojyoti Chakraborty**[a,b,*]
[a]CSIR-Institute of Genomics & Integrative Biology, New Delhi, India
[b]Academy of Scientific & Innovative Research, New Delhi, India
*Corresponding author: e-mail address: debojyoti.chakraborty@igib.res.in

Contents

1. Introduction 74
2. Origin of noncoding RNAs 75
3. Core regulatory circuit in ESCs 76
4. LncRNAs: New determinants of ES cell fate 78
5. Long noncoding RNAs (lncRNAs) and their biological function 78
6. Discovery of lncRNAs: From sequences to function 80
7. Long noncoding RNAs and epigenetic regulation 80
8. Dissecting functional lncRNAs from transcriptional noise 81
9. LncRNAs in ESC pluripotency and somatic cell reprogramming 81
10. LncRNAs play a role in the differentiation of pluripotent stem cells 87
11. LncRNAs regulating the epigenome 88
12. The role of lncRNAs in dosage composition 89
13. LncRNAs implicated in imprinting developmentally associated genes 90
14. LncRNAs regulating signaling pathways in ESCs 91
15. LncRNAs regulating organ development 92
16. LncRNAs affecting neural development 93
17. LncRNAs regulating organogenesis 94
18. Cellular localization and maturation of lncRNAs 95
19. LncRNAs regulating the stability and functions of other RNAs 95
20. LncRNAs functioning in protein modification pathways 96
21. Mechanisms of lncRNA:DNA/RNA interaction 96
22. Allosteric regulation of proteins by lncRNAs 98
23. Single cell analysis of lncRNA functions 99
24. LncRNAs in disease progression 100
25. LncRNA knockouts often show lack of phenotype: The importance of context and redundancy 101
26. Conclusions 102

References 102
Further reading 111

Current Topics in Developmental Biology, Volume 138
ISSN 0070-2153
https://doi.org/10.1016/bs.ctdb.2019.11.003

Abstract

Embryonic Stem cells are widely studied to elucidate the disease and developmental processes because of their capability to differentiate into cells of any lineage, Pervasive transcription is a distinct feature of all multicellular organisms and genomic elements such as enhancers and bidirectional or unidirectional promoters regulate these processes. Thousands of loci in each species produce a class of transcripts called noncoding RNAs (ncRNAs), that are well known for their influential regulatory roles in multiple biological processes including stem cell pluripotency and differentiation. The number of lncRNA species increases in more complex organisms highlighting the importance of RNA-based control in the evolution of multicellular organisms. Over the past decade, numerous studies have shed light on lncRNA biogenesis and functional significance in the cell and the organism. In this review, we focus primarily on lncRNAs affecting the stem cell state and developmental pathways.

1. Introduction

Discoveries elucidating critical features of embryonic and adult stem cells spanning several decades have revealed tremendous potential for experimental and regenerative medicine and advancements in biology. Stem cells can be classified either based on their origin or their ability to differentiate. For example, Embryonic Stem cells (ESCs) originate from the inner cell mass of the developing blastocyst while Induced Pluripotent Stem Cells (iPSCs) are obtained by a process of de-differentiation from adult cells. Similarly Adult Stem Cells (ASCs) are found in tissues and organs and has subtypes such as Mesenchymal stem cells (MSCs) or Hematopoietic stem cells (HSCs) depending on the organ or tissue of origin (Alison & Islam, 2009; Meirelles Lda & Nardi, 2009). Based on their ability to differentiate into other cell types, stem cells can be categorized into totipotent, pluripotent, multipotent or unipotent in decreasing order of their "stemness" or differentiation capability. While totipotent cells can give rise to the entire organism (such as the extraembryonic placenta), unipotent cells are restricted to only a single cell type (such as skin cells) (Ott et al., 2007)

The study of stem cells, particularly ESCs, is very important for modern Biology as their unique properties of self-renewal and pluripotency make them attractive models for therapeutic applications. These cells can be expanded as a pure population and maintained in their undifferentiated form for extended periods of time. This property of self-renewal ensures that ESCs can be frozen and cultured for multiple passages. By virtue of their pluripotent nature, ESCs can give rise to every cell type of the body contributing to the differentiation process of cells into multiples cell lineages of the adult tissue and germline. The ability of the ESCs to derive multiple lineages opens up opportunities for studying embryonic development and the related events during the earliest developmental stages and lineage specification. Deep understanding of the underlying molecular mechanisms of ESC differentiation and maintenance of pluripotency is required for designing cell based therapies that employ pluripotent stem cells (Yamanaka & Blau, 2010; Young, 2011).

ESCs can be maintained in their undifferentiated state in the presence of factors such as recombinant LIF or feeder cells. Through years of research, some of the factors which determine ESC fate have been identified. Dissecting these molecular signatures give important insights about developmental pathways active in an embryo as well as cues necessary for spatial and temporal switching of lineage-specific differentiation programs. It is now known that cellular factors regulating ESC fate are not exclusively of protein origin (Sutherland et al., 2018). One of the striking determinants that contribute immensely in various pathways and processes in stem cells is a group of nonprotein-coding RNAs that affect the landscape of transcriptome regulation, highlighting the interplay of epigenetic, transcriptional and post-translational mechanisms in determining cell fate and regulation of developmental processes (Batista & Chang, 2013; Rinn & Chang, 2012).

2. Origin of noncoding RNAs

The majority of the human genome is transcribed into RNAs without any coding potential. Based on their sizes, these RNAs can be classified as small or long noncoding RNAs (lncRNAs). Small ncRNAs are typically shorter than 200 nucleotides in length and can be further sub-categorized based on length, function and subcellular localization. Among these, micro RNAs (miRNAs), short interfering RNAs (siRNAs), Piwi-interacting RNAs (piRNAs), small nucleolar RNAs (snoRNAs) and short hairpin RNAs (shRNAs) have been widely studied and implicated in various aspects of embryonic stem cell biology (Fico, Fiorenzano, Pascale, Patriarca, & Minchiotti, 2019; Ozata, Gainetdinov, Zoch, O'Carroll, & Zamore, 2019). Long noncoding RNAs, however, form a distinct class that is poorly conserved among species and are generally expressed at low levels in cells. In fact, widespread interest in long noncoding RNA biology coincided with the advent of deep sequencing technologies that allowed scientists to curate, catalog and characterize them in depth. Today, long ncRNAs are been increasingly appreciated as regulatory molecules that play functional role in diverse cellular processes ranging from embryonic stem cells pluripotency, cell proliferation, neural processes, tumorigenesis and metastasis (Guttman et al., 2011; Pal & Rao, 2017). Expression of lncRNAs exhibit highly cell type specific or developmental stage-specific pattern, and are often dysregulated in disease conditions in different species. The spatiotemporal differences in lncRNA expression are intimately associated with how genes are expressed in response to different developmental cues. The molecular differences between cell types required to control cell identity and lineage commitment can be governed by the combined involvement of positive and negative feedback loops (Batista & Chang, 2013; Cabili et al., 2011).

Here, we discuss our latest understanding in exploring the functions of lncRNAs in embryonic stem cells (ESCs) and their molecular mechanisms of actions and roles of in cellular processes such as genomic imprinting, maintenance of pluripotency and organ development.

3. Core regulatory circuit in ESCs

The pluripotent state of ESCs is governed by a group of core transcription factors such as OCT4, SOX2 and NANOG and these are the main genes associated with establishing and maintaining the undifferentiated state in these cells. The auto-regulatory loop of OCT4, SOX2 and NANOG function in a concerted fashion and are interconnected to regulate their own promoters through a network of inhibitory and activating pathways. These factors can bind to the promoter regions of multiple genes and recruit various activators that act as positive regulators in maintaining the stemness of the ESCs. Similarly, they can negatively regulate genes driving lineage-specific differentiation and their repression allows ESCs to maintain a stable undifferentiated state. The loss of these core regulators leads to the rapid induction of a wide spectrum of lineage-specific regulators indicating that these genes are poised for activation. SETDB1 and Polycomb group (PcG) chromatin regulators have both been identified as key regulators involved in the repression of lineage-specific genes. OCT4 can bind to sumoylated SetDB1 that catalyzes the repressive histone modification H3K9me3 and associates with nucleosome complex thus further contributing to the repression. Thus, the core regulatory complex may recruit the SetDB1 and PcG complex through protein-protein interactions. The ability of the these factors to positively regulate the genes necessary to maintain ESC like state while repressing the genes that would help in exit from this state explains the ability of ESCs to self-renew in an undifferentiated state yet remain poised to differentiate into all cell types of the body in proper response to developmental cues. The PGRN is a complex association of multiple factors and cofactors that work in sync during the reprogramming and differentiation process. Other than OCT4, SOX2 and NANOG, the transcription factors TCF3, SMAD1, STAT3, ESRRB, KLF2, KLF4, SAL4, REX1, TBX3, ZFX, KLF5, RONIN, PRDM14, etc. have been shown to play important roles in the control of ESC state. These transcription factors can bind to promoter regions of multiple genes and regulate the pluripotent state by auto-regulatory feedback loops. ESCs can respond to small changes in their environments through a multitude of signaling pathways and these in turn activate or repress genes through control at the DNA level.

ESCs when removed from the appropriate culture conditions that maintain them as undifferentiated cells, have the tendency to differentiate into cells of all lineages. When appropriate factors are provided, ESCs can give rise to progeny consisting of derivatives of three germ layers: ectoderm, mesoderm and endoderm. In a developing embryo, the ectoderm gives rise to the central nervous system (brain and spinal cord), peripheral nervous system, sensory epithelia of the eye, ear and nose, epidermis and its appendages (nails and hair), mammary glands and the subcutaneous glands. The ectodermal development is called neurulation with respect to nervous tissue. The mesoderm give rise to the connective tissue, cartilage, and bones; blood, lymph vessels and cells; striated and smooth muscles, the kidneys, the gonads (ovaries and testes) and genital ducts and the serous membrane lining the body cavities. The endoderm gives rise to the epithelial lining of the gastrointestinal and respiratory tract, the parenchyma of the tonsils, the thymus, the thyroid, the parathyroid, the liver and the pancreas, epithelial lining of tympanic cavity, urinary bladder and urethra. Mouse ESCs in culture do not have the ability to give rise to trophoectoderm, thus reflecting the potential of the founder cells. Human ESCs, on the other hand can display cells with characteristics of trophoectoderm if induced with BMP4. Over the 20 years of research on ESC biology, different studies have come up with standardized protocols on differentiating ES cells into any particular cell type with definite factors and conditions. The lineage development protocols that are well established have explained different developmental stages and sequence of events leading to the generation of differentiated cells that very well recapitulates those in early embryos.

4. LncRNAs: New determinants of ES cell fate

ESCs can be configured to escape their regulatory circuitry and enter into differentiation process when cells are stimulated with suitable factors. Loss of important pluripotency markers through transcriptional and posttranscriptional mechanisms activate the differentiation process and activation of lineage-specific genes (Stein et al., 2012). The multiple signals that stimulate the ESCs differentiation cause changes in all classes of regulators like Transcription factors, cofactors, noncoding RNAs (ncRNAs), chromatin regulators shaping the genome architecture. The complex pluripotency gene regulatory network (PGRN) is influenced by multiple genetic and epigenetic factors that directly and/or indirectly lead to the regulation of the cellular processes in a systematic manner (Geng, Zhang, & Jiang, 2019; Wang & Li, 2017). Recent advancement in the field of lncRNAs indicates that lncRNAs mediate numerous biological processes in embryonic stem cell maintenance and differentiation. A substantial number of these lncRNAs are expressed at specific stages of embryonic development and regulates differentiation into diverse cell lineages (Luo et al., 2016). The underlying mechanism of ESCs differentiation and pluripotency maintenance and the events of somatic reprogramming is still inconclusive and remains elusive, but in recent years extensive studies on noncoding RNAs and small RNAs have unfolded a majority of processes and pathways that are more conducive to understand their role in biological phenomena.

5. Long noncoding RNAs (lncRNAs) and their biological function

The number of predicted protein-coding genes present in the mammalian genome is not on par with both the size of the genome as well as the abundance of the transcriptome. Approximately 5–10% of human genome undergo transcription, out of which only ~1% encode functional protein, the remaining ~4–9% are transcribed but their function is not yet known (Balas & Johnson, 2018; Ghosal, Das, & Chakrabarti, 2013). This ubiquitous transcription across genome has been demonstrated by various methods like whole transcriptome sequencing or tiling array. It had also been shown that such genome wide transcripts thus have prevalence and have diverse functions in eukaryotes ranging from plants, animals and fungi such as the Fission and the Budding yeast (Ponting, Oliver, & Reik, 2009). The complexity in whole genome transcription is exacerbated due to different combinations of exons and polyadenylation sites. Of the entire transcriptome data, >30% produce alternative transcripts and different exon combinations. Moreover, over half the mouse and human genes exhibit polyadenylation site among their transcripts (Carninci et al., 2005; Rinn et al., 2003). The fact that the numbers of noncoding transcripts are higher than coding ones suggests that noncoding RNAs might regulate complex developmental processes. Noncoding RNAs can be classified as Housekeeping ncRNAs and regulatory ncRNAs (Dinger, Pang, Mercer, & Mattick, 2008). Housekeeping ncRNAs are ribosomal, transfer, small nuclear and small nucleolar RNAs that are expressed constitutively. The regulatory ncRNAs are broadly classified into microRNAs, piwi-associated RNAs, small interfering RNAs. The transcripts may be localized in the nucleus or cytoplasm, and are often transcribed from either strand of a protein coding gene (Farazi, Juranek, & Tuschl, 2008; Kiyosawa, Mise, Iwase, Hayashizaki, & Abe, 2005).

LncRNAs are typically transcribed by RNA Polymerase II and generally do not possess an Open Reading Frame (ORF) sufficiently long enough for protein translation. LncRNAs are largely expressed in stem cells and progenitor cells that have open and active chromatin and often expressed in a spatially and temporally controlled manner. They are transcribed from large genomic regions flanking Transcription factor genes and other factors. Many lncRNAs possess features of protein coding RNAs such as 5′ Cap signal, Alternative Splicing and >60% of lncRNAs have Poly-A signal (Khawar, Mehmood, Abbasi, & Sheikh, 2018; Wang & Chang, 2011; Yan, Luo, Lu, & Shen, 2017). A lncRNA can be broadly placed into five categories: (a) Sense or (b) Anti-Sense when overlapping one or more exons of another transcript on the same or opposite strand, respectively, (c) Bidirectional when the expression of that transcript and a neighboring coding transcript is initiated within a close genomic proximity, (d) Intronic when it is derived wholly from within an intron or from a pre-mRNA sequence or (e) Intergenic when the transcript lies within the genomic interval between two genes. Most lncRNAs act in *trans* with their partner proteins forming large Ribonucleoprotein complexes and regulate the function of other genes (Kim & Shiekhattar, 2016; Kung, Colognori, & Lee, 2013; Vucicevic, Corradin, Ntini, Scacheri, & Orom, 2015). Intergenic lncRNAs are transcribed in close proximity to their protein coding genes and often act in *cis*. LncRNAs can also emerge due to insertion of transposable elements.

As transposable element sites often coincide with transcriptional start site, hence they might contribute to gene transcription repertoire (Chishima, Iwakiri, & Hamada, 2018; Ransohoff, Wei, & Khavari, 2018).

6. Discovery of lncRNAs: From sequences to function

With the completion of the Human Genome project it became evident that the complexity of different organisms cannot be defined by the number of genes present that give rise to protein coding transcripts. This simple understanding gave rise to advanced technologies like microarray and RNA sequencing at an unprecedented pace (Govindarajan, Duraiyan, Kaliyappan, & Palanisamy, 2012). Two independent studies had initially reported that the number of lncRNA genes is similar to protein coding genes. These studies had used DNA microarray with tiled or nested target sequences comprising of the entire chromosomal DNA sequence and allowed an unbiased survey of the transcribed genes. Some limitations of the tiling microarray studies are cross-hybridization and lack of information whether genes are transcribed or not. Nevertheless, these studies could confirm if transcription from noncoding loci was present using a range of techniques such as Northern Blot, Reverse transcription polymerase chain reactions, etc. However, the results were not completely conclusive since they could also represent transcriptional noise. The rapid rise of sequencing technologies meant that several groups could approach the problem of dissecting all transcripts emanating from the genome although assigning functions to all these transcripts has still remained technically challenging (Heather & Chain, 2016; Mockler et al., 2005).

7. Long noncoding RNAs and epigenetic regulation

Chromatin immunoprecipitation (ChIP) coupled with massively parallel sequencing has revealed complex DNA-protein signatures which regulate transcription by controlling the local chromatin environment around a gene. For example, Histone H3 Lysine 4 trimethylation (H3K4me3) or Histone H3 Lysine 36 trimethylation (H3K36me3) are well-known marks for actively transcribed genes and their presence across a large number of intergenic regions led to the discovery of novel lncRNAs also termed as long intergenic noncoding RNAs (lincRNAs). Among these lncRNAs, several show some level of evolutionary conservation and many of them have their own regulatory sequences such as promoter elements that also round up as binding sites of key pluripotency factors. Among their functions, cell cycle regulation, maintenance of pluripotency, regulation of DNA topology and coordinating immunogenic responses have been studied (Mattick & Rinn, 2015; Raha, Hong, & Snyder, 2010).

RNA sequencing, coupled with computational prediction of unannotated transcripts have led to the identification and subsequent validation of several lncRNAs at cell or tissue level resolution. By labeling mRNAs using metabolic markers, nascent transcription has been detected leading to identification of start and pause sites of actively transcribed genes and noncoding RNAs. Combining sequencing with biochemical assays such as Rapid Amplification of cDNA ends (RACE) has led to further characterization of full length lncRNAs (Lagarde et al., 2017).

8. Dissecting functional lncRNAs from transcriptional noise

The lncRNA field has fuelled a large number of computational prediction pipelines for the discovery and annotation of these transcripts. Among the well-known bioinformatic tools used for studying lncRNAs, TopHat and Cufflinks are used for mapping and assembling sequencing reads to the reference genome, BRAT is used for annotating sequences that need to consider gaps in reads. Cuffcompare is employed in detection and comparison of novel reads with reference annotations while Cuffmerge is used to merge several assemblies together. Once sequences are annotated, another set of tools can be used to predict their coding or noncoding potential. These include Coding-NonCoding Index (CNCI), Coding Potential Calculator (CPC) and iSeeRNA. Several databases are now available for curation and annotation of lncRNAs (Kim et al., 2013; Pertea, Kim, Pertea, Leek, & Salzberg, 2016; Sun, Liu, Zhang, & Meng, 2015, Table 1).

9. LncRNAs in ESC pluripotency and somatic cell reprogramming

Recent literature suggests that close to a thousand lncRNAs have been identified in mouse and human ESCs that might regulate stem cell renewal. Since ESCs represent a tightly controlled cellular state where a large number

Table 1 List of databases providing information about lncRNAs.

Name	**Function**	**Source**	**References**
LNCipedia	Comprehensive information about human lncRNA structure and sequence	https://lincipedia.org/	Volders et al. (2013)
LncRNAdb	LncRNAs that have regulatory roles in biological function	https://lncrnadb.org/	Amaral, Clark, Gascoigne, Dinger, and Mattick (2011)
LncACTdb	Interaction of ceRNAs with different types of RNAs	http://www.bio-bigdata.net/LncACTdb/	Wang et al. (2015)
LncReg	Comprehensive information of regulatory networks of lncRNAs	http://bioinformatics.ustc.edu.cn/lncreg/	Zhou, Shen, Khan, and Li (2015)
NONCODE	Collection of complete set of annotated lncRNAs	http://www.noncode.org/	Xiyuan et al. (2017)
LncRNome	Comprehensive knowledgebase of biologically significant and annotated lncRNAs	http://genome.igib.res.in/lncRNome	Bhartiya et al. (2013)
LncRNAtor	Functional investigation and conservation of lncRNAs between human and other organisms	http://lncrnator.ewha.ac.kr/	Park, Yu, Choi, Kim, and Lee (2014)
NPInter	Experimentally determined functional interaction of lncRNAs and protein biomolecules	http://www.bioinfo.org/NPInter/	Wu et al. (2006)
NRED	Collective database about gene expression information of lncRNAs	http://jsm-research.imb.uq.edu.au/NRED	Dinger et al. (2009)
TF2LncRNA	Information about any lncRNA-TF regulatory relationship in any specific cell line	https://omictools.com/tf2lncrna-tool	Jiang et al. (2014)

Table 1 List of databases providing information about lncRNAs.—cont'd

Name	Function	Source	References
LnCaNet, Lnc2Cancer	Comprehensive regulatory network of lncRNAs associated with various cancer types, experimentally supported	http://lncanet.bioinfo-minzhao.org/ http://www.bio-bigdata.com/lnc2cancer/	Liu and Zhao (2016) and Ning et al. (2016)
MiTranscriptome	List of lncRNA that are poly-adenylated and derived from analyzed RNA-Seq data	http://mitranscriptome.org/	Iyer et al. (2015)

of molecular pathways are operational, the importance of lncRNA mediated control of these processes has been underscored significantly. For example, a large number of studies have revealed a role of lncRNAs in controlling the epigenetic landscape of pluripotent cells either by directly regulating gene expression or altering the topology of DNA at sites of interaction. Several lncRNAs have been shown to interact with sites on DNA that are within 10 kb from the nearest gene and many of these directly bind to members of the pluripotency circuitry suggesting that these complexes may be assembled on their respective sites on DNA with the help of lncRNAs (Dinger et al., 2009; Jia, Chen, & Kang, 2013).

Studies on gene transcription from different stages of mouse development had identified lncRNAs co-expressed with different pluripotency and lineage-commitment genes suggesting their functional roles in these processes. To dissect the role of lncRNAs in pluripotency, Guttman et al. had identified the role of 26 lncRNAs potentially implicated in the maintenance of stem cell fate by performing a loss of function screen on 147 target lncRNAs using lentiviral based shRNAs in mouse ESCs. The depletion of these 26 lncRNAs led to a downregulation of NANOG and displayed characteristic features of differentiated cells suggesting that these lncRNAs are essential for the regulation of the pluripotent character of mouse ESCs. Similarly, a screen done using endoribonuclease prepared siRNAs (esiRNAs) identified three lncRNAs (Panct 1–3) as potential modulators of OCT4 expression in mouse ESCs (Chakraborty et al., 2012). In a different study, it was found that lincRNA-ROR acts as a sponge for miRNAs that target OCT4, SOX2 and NANOG thereby. Together these reports suggest that lncRNAs might control embryonic stem cell state by directly regulating the expression of transcription factors involved in maintenance of pluripotency (Hou et al., 2018).

LncRNAs have been shown to affect both pluripotency as well as the process of reprogramming (Table 2, Fig. 1). The global remodeling of the genome and the epigenome is a recurrent phenotype associated with

Table 2 Role of regulatory lncRNAs in stem cells.

LncRNA	Function	References
Panct 1–3	Potential mESCs pluripotency regulator by modulating OCT4 promoter activity	Chakraborty et al. (2012)
LincRNA-*ROR*	Acts as ceRNA by sponging miR-145 activity and prevents degradation of OCT4, SOX2, KLF4 mRNAs	Gupta et al. (2010)
ES1–3	Interacts with SOX2 and controls stem cell fate	Grammatikakis, Panda, Abdelmohsen, and Gorospe (2014)
Press1	Maintains stem cell state by recruiting Histone Acetylases on promoters of pluripotency genes	Jain et al. (2016)
N1–3	Promotes Neuronal differentiation by interacting with chromatin modulators and targeting miRNA	Barry, Guennewig, Fung, Kaczorowski, and Weickert (2015) and Sun et al. (2018)
Xist, Tsix	Crucial epigenetic modulators of X-Chromosome Inactivation (XCI) and dosage compensatory mechanism	Augui, Nora, and Heard (2011), Payer et al. (2013), and Pontier and Gribnau (2011)
Hottip	Impacts pluripotency in mESCs by regulating co-ordinate function of HoxA genes	Wang et al. (2011) and Yang et al. (2014)
Meg3, Hotair	Interacts with PRC2 complex and acts as transcriptional regulator of lineage markers	Terashima, Tange, Ishimura, and Suzuki (2017)

Table 2 Role of regulatory lncRNAs in stem cells.—cont'd

LncRNA	**Function**	**References**
Kcnq1ot1, *Airn*	Paternally expressed lncRNAs that work by repression of early embryonic development genes in mice	Latos et al. (2012) and Mancini-Dinardo, Steele, Levorse, Ingram, and Tilghman (2006)
Gas5	Highly expressed in hESCs and responsible for growth arrest	Xu et al. (2016)
Terra	Promotes self-renewal and maintains Pluripotent state	Fico et al. (2019)
Pnky	Localizes to NSCs and promotes neurogenesis and differentiation	Ramos et al. (2015)
Evf2	Promotes cortical brain development in association with Dlx5/6 loci	Bond et al. (2009)
Six3os	Interacts with epigenetic factors and regulates retinal development	Rapicavoli, Poth, Zhu, and Blackshaw (2011)
Tuna	Conserved lncRNA that regulates systematic neuronal development	Lin et al. (2014)
Braveheart, *Fendrr*	Impact the development of mature cardiomyocytes and maintains cardiac development	Ottaviani and da Costa Martins (2017)

the process of reprogramming and lncRNAs are shown to play a role in the process. Studies have revealed that the expression of lncRNAs in ES cells and iPSCs show a similar trend but are different from that seen in somatic cells such as hematopoietic stem cells or fibroblasts. Several lncRNAs have shown higher expression in iPSCs while not being expressed in ESCs suggesting that these lncRNAs might be involved in the reprogramming process or contributing to certain aspects of it. Certain lncRNAs such LincRNA *ROR*, lincRNA *SFMBT2*, lincRNA *VLDLR* are expressed at high levels in iPSCs and are also bound by the pluripotency factors OCT4, SOX2 and NANOG. Interestingly, downregulation of OCT4 leads to repression of these lncRNAs further strengthening the notion that lncRNAs expression is under the control of proteins implicated in the

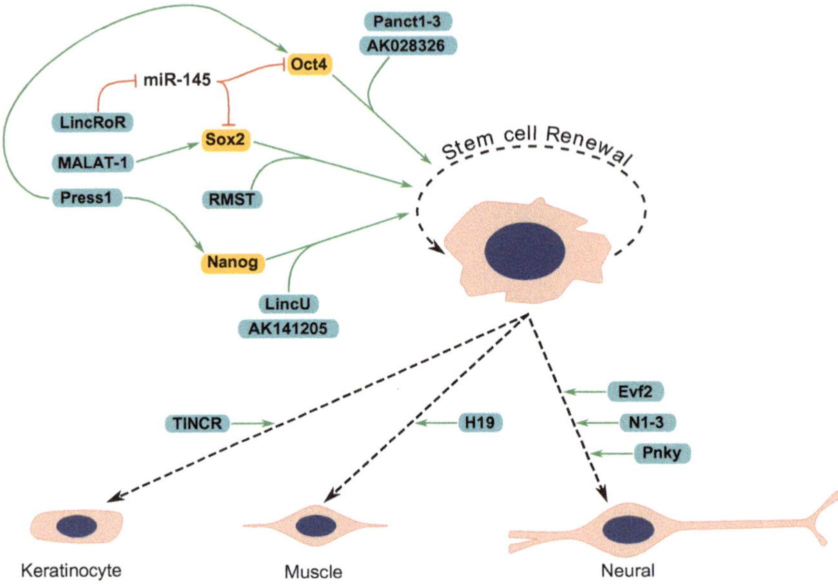

Fig. 1 LncRNAs regulate diverse pathways controlling ES cell pluripotency and differentiation. Selected lncRNAs identified from various studies are shown within context of their functional roles in these processes. *Adapted from Rinn, J. L., & Chang, H. Y. (2012). Genome regulation by long noncoding RNAs. Annual Review of Biochemistry, 81, 145–166. doi:10.1146/annurev-biochem-051410-092902.*

pluripotent pathway. Depletion of lincRNA *ROR* negatively impacts the formation of iPSC colonies while its overexpression increases colony formation. Notably, the depletion of these lncRNAs also impacts the levels of p53 expression which upon upregulation leads to oxidative stress and loss of cell viability. These results confirm that lncRNAs are important in inducing pluripotency by improving the survivability of resultant iPSC colonies (Gupta et al., 2010; Loewer et al., 2010).

LncRNA *AK028326* (OCT4 activated) and lncRNA *AK141205* (NANOG repressed) are examples of lncRNAs that are directly regulated by pluripotency factors and control ESC fate. Loss or gain of function studies with these lncRNAs show concomitant changes in the levels of OCT4 and NANOG suggesting a complex interplay between protein and lncRNA determinants of pluripotency (Fico et al., 2019). Similarly, SOX2 interacts with lncRNAs ES1, ES2 and ES3 and modulates stem cell state as well as neurogenesis (Grammatikakis et al., 2014).

Another lncRNA *Press1* was found to be overexpressed during the undifferentiated state in ESCs while being repressed in a p53 dependent manner during exit from pluripotency. Both lncRNA *Press1* and SIRT6, the histone H3 deacetylase that acts on Lysine 56 (H3K56ac) are expressed in hESCs. *Press1* regulates the expression of pluripotency genes by maintaining high levels of H3K56ac on chromatin on promoters, thereby contributing to the maintenance of stem cell state (Jain et al., 2016).

LncRNA *Panct1* is implicated in pluripotency pathway in mouse ESCs through its interaction with an uncharacterized protein TOBF1 which it recruits to promoters of pluripotency genes during the early G1 phase of the cell cycle. Loss of both *Panct1* and TOBF1 lead to downregulation of pluripotency factors suggesting that lncRNAs working in tandem with proteins regulate the ES cell state through a multitude of different molecular pathways (Chakraborty et al., 2017).

10. LncRNAs play a role in the differentiation of pluripotent stem cells

As mentioned earlier, reprogramming and differentiation are two processes encompassing numerous topological changes to the chromatin and consequent switching on or off of thousands of different noncoding RNAs (Pal & Rao, 2017). For example, lncRNA *Mistral* has been shown to recruit the MLL1 protein, an activator of transcription implicated in the differentiation of stem cells toward the hematopoietic lineage. Consequently, this leads to the expression of the homeotic genes *Hoxa6* and *Hoxa7* as cells differentiate toward the germ layers (Bertani, Sauer, Bolotin, & Sauer, 2011). A different lncRNA *N2* controls the differentiation of neuronal stem cells as well as controls neurogenesis by controlling the expression of miR125 (Sun et al., 2018). Similarly, lncRNA *N1* and *N3* interact with REST and SUZ12 complexes and regulates neuronal differentiation (Barry et al., 2015). In iPSCs, downregulation of lncRNA *Gtl2*, a member of the Dlk-Dio3 maternally imprinted cluster has been observed. In addition to this lncRNA, several other miRNAs belonging to the same cluster are all seen to be downregulated in undifferentiated cells. The cluster is also hypermethylated and hypoacetylated possibly leading to a general repression of transcription. Interestingly, when Valproic Acid, a Histone Deacetylase inhibitor is used, the entire cluster can be reactivated leading to normal differentiation and iPSC development (Dill & Naya, 2018).

Among the earliest noncoding RNAs studied is *Xist* which has a defined role in X-Chromosome Inactivation (XCI) during the differentiation of female ESCs. This phenomena ensures that in mammalian systems, female cells contain only one active X chromosome as a means of "dosage compensation" (Briggs & Reijo Pera, 2014; Minkovsky, Patel, & Plath, 2012). In undifferentiated cells, the transcription factors OCT4, SOX2 and NANOG bind to the Intron 1 of this lncRNA and prevent its expression. At the same time, the expression of *Tsix,* a lncRNA that functions antagonistic to *Xist* is activated by the combinatorial activities of REX1, KLF4, OCT4, SOX2 and NANOG. As differentiation proceeds, downregulation of the pluripotency factors leads to increased expression of *Xist* and this in turn recruits components of the Polycomb complex (PRC2) to initiate XCI. The process of XCI is indeed complex and other lncRNAS also play a role in maintaining the inactivation state. For example, *Xist, Tsix* and *Jpx* work in coordination during the process of XCI. *Jpx* shows a close LncRNA *Jpx* interacts closely with CTCF, a DNA binding protein implicated in chromosomal folding. CTCF mediated regulation of DNA topology affects *Xist* expression and the subsequent XIC process is initiated. At the same time, the canonical MAPK signaling pathway is downregulated leading the cells to exit from their pluripotent state. In the primed state when X Chromosome in inactivated, MAPK pathway genes are activated leading to differentiation of ESCs (Augui et al., 2011; Payer et al., 2013; Pontier & Gribnau, 2011).

11. LncRNAs regulating the epigenome

A hallmark of lncRNA activity is to function through repressive complexes with only a few cases of activating or enhancing lncRNAs characterized so far. An example of such a lncRNA is *Hottip*, which is expressed from the distal end of the HOXA gene cluster and which recruits WDR5 subunit of MLL1 thereby leading to transcriptional activation. This is an example of *cis*-regulation and its effect is on genes present in proximity in the nucleus. Interestingly, *Hottip* copy number is low and yet it manages to interact with multiple targets that are close together in the genomic cluster. Analysis of the WDR5-*Hottip* protein-RNA interaction reveals that WDR5 has a RNA binding pocket which when occupied by the lncRNA, makes WDR5 bind with higher stability to DNA. Since WDR5/MLL complex directly regulates chromatin conformation, performing pulldowns using this complex followed by sequencing the RNA portion revealed over 1400 WDR5 interacting RNAs, several of which were noncoding. Upon mutating the RNA binding pocket in HOTTIP, the temporal stability of the protein on DNA went down. Subsequently, the active chromatin mark H3K4Me3 was reduced leading to exit from ESC pluripotency (Song & Kingston, 2008;

Wang et al., 2011; Yang, Flynn, et al., 2014). This is a very unique example of *cis*-regulation by lncRNAs that directly impacts the pluripotent state in mouse ESCs.

Chromatin modifying complexes acting as repressors have been studied in greater detail and a large number of studies have revealed a direct link between the PRC2 complexes with a magnitude of lncRNAs that control the repressive state of chromatin, particularly at the promoters of genes. Several of these genes are important for the maintenance of pluripotency, XCI, development and lineage commitment. One of the components of this interaction is EZH2 which binds strongly with a large number of RNA molecules (Wang et al., 2018). Indeed, pull-down of PRC2 revealed thousands of such RNA targets in mouse ESCs. It is speculated that such interactions might be strengthened by additional protein components like JARID2. While EZH2 scans chromatin, actively transcribed genes advance EZH2 progression through elongating mRNAs while lncRNAs bound to silent regions maintain their transcriptionally inactive state in *trans*. Another example of such regulation is seen in the case of the DNA Methylase protein DNMT1 which through interaction with non-polyadenylated extra coding RNAs (ecRNAs) like *Cebpa* cause transcriptional repression across chromatin (Chase & Cross, 2011). JARID2 can stabilize PRC2 occupancy on chromatin thereby acting as a transcriptional regulator and several in-depth biochemical assays investigating protein-RNA binding has revealed its direct association with several lncRNAs. A large subset of lncRNAs are shared by EZH2 and JARID2 suggesting the complex interplay of function guided by lncRNAs (Yang, Flynn, et al., 2014). An example of such a lncRNA is *Meg3* which binds to both the PRC2 subunits but provides greater stability to the complex through its interaction with JARID2. This interaction in turn regulates PRC2 occupancy on chromatin and subsequent transcriptional regulation. LncRNA *Hotair* is another component of the JARID2-EZH2 interaction. Together these interactions between proteins, their lncRNA partners and DNA modulate how the chromatin activation or repression marks are established in the context of pluripotency and differentiation (Terashima et al., 2017).

12. The role of lncRNAs in dosage composition

The process of dosage compensation or equalization of the expression of genes between different biological sexes in an organism is facilitated through numerous pathways. It has been known for a long time that the

XIC facilitates this process through the expression of *Xist* which forms a repressive "cloud" over one of the active X chromosomes in female ES cells during differentiation thereby switching off the genes on this chromosome. This in turn determines which X chromosome will be targeted for formation of Barr Body, a transcriptionally inactive form of the chromosome. RNA *Fluorescent* in situ *Hybridization* (RNA FiSH) has been widely used to study the role of *Xist* in X chromosome inactivation in cells and offer insights about the timing of *Xist* expression and action. To understand the RNA: DNA interaction process in greater detail, techniques such as ChIRP (Chromatin Isolation by RNA Purification) and CHART (Capture Hybridization analysis of RNA targets) have been developed to identify chromatin binding sites of lncRNAs (Yang, Flynn, et al., 2014). These methods have revealed that beginning from the *Xist* locus, the transcriptional repression mark spreads to other regions of the X chromosome that are far away from the lncRNA expression locus. Active DNAaseI binding sites at these loci show that these might represent active chromatin prior to silencing and are rapidly silenced upon *Xist* activation. In a similar manner, when applied to the *Rox2* lncRNA in Drosophila, these techniques revealed a large number of genomic loci bound by *Rox2* on the X chromosome.

Characterization of the *Jpx* lncRNA pathway have shown strong association with the CTCF protein that regulates higher-order chromatin structure. By binding to CTCF, *Jpx* can sequester CTCF away from its binding sites on the *Xist* locus. This happens during the differentiation of ESCs when CTCF is lost from XIC leading to upregulation of *Xist*. Further to *Jpx* and CTCF interaction, a large number of other lncRNAs such as *Wrap53* have been shown to interact directly with CTCF. Together, these present a complex interplay of RNA binding proteins, lncRNAs and chromatin modulators at the center of transcriptional regulation of the X chromosome (Carmona, Lin, Chou, Arroyo, & Sun, 2018; Sun et al., 2013).

13. LncRNAs implicated in imprinting developmentally associated genes

During gametogenesis, certain imprinted genes are epigenetically marked in a sex-dependent manner. Subsequently these are silenced in one of the parent chromosomes in the developing embryo. It is now known that several lncRNAs are expressed from such imprinted regions which also regulate transcription. A number of these lncRNAs are very long, sometimes >100 kb in length. For example, lncRNA *Kcnq1ot1* and *Airn* are paternally expressed lncRNAs that work by repression of early embryonic development genes in mice (Kung et al., 2013). If the embryo inherits a loss of function version of these lncRNAs, its imprinting is lost and suffers from growth defects. Contrarily, maternal imprinting does not lead to similar features thereby establishing a tight regulatory control on lncRNA expression and transcription. The role of Kcnq1ot1 in repressing genes is different in the placenta and the embryo. In the placenta, the lncRNA recruits PRC2 and EHMT2, both repressor marks, on DNA far away from imprinted sites whereas in the embryo, the methylation marks are established on surrounding DNA. LncRNA *Airn* functions in a *cis* configuration and silences the paternal *Igf2r* allele. This silencing occurs because of continuous expression of *Airn* that inhibits RNA Polymerase II recruitment for transcription (Latos et al., 2012; Mancini-Dinardo et al., 2006; Nagano et al., 2008). Taken together these examples show how lncRNAs acting in *cis* and *trans* modulate transcription in different loci under different conditions although clearly in situ regulation at sites of lncRNA production produces a stronger effect on gene interaction networks through which lncRNAs work.

14. LncRNAs regulating signaling pathways in ESCs

A large number of external cues govern the intricate network of pluripotency and differentiation in stem cells and as mentioned before lncRNAs are one of the major players in these processes. LncRNA *Gas5* is responsible for growth arrest and is highly expressed in human ESCs. This lncRNA directly regulates the expression of OCT4 and SOX2 through the TGFβ signaling pathway where it maintains the expression of the signal receptor ligand NODAL by protecting it from miRNA mediating degradation (Xu et al., 2016). As a consequence, the pluripotent state of ESCs and iPSCs is maintained. In the case of the lncRNA *Digit* which is regulated by a enhancer element under the control of SMAD3, loss of function leads to Divergent to Goosecoid (GSC) activation that affects endodermal differentiation of both mouse and human ESCs (Daneshvar et al., 2016).

Telomeric RNA (*Terra*) is a well-known lncRNA that is expressed in high levels in ESCs and has been related to pluripotency maintenance. *Terra* is regulated by WNT/β-catenin pathway and upon overexpression can promote self-renewal. This lncRNA inhibits the function of TCF3 and consequently contributes to ESC pluripotency. In contrary, TCF3 overexpression prevents self-renewal of ESCs (Fico et al., 2019; Xu, Guo, Zhang, & Ye, 2018).

Among the metabolic pathway related lncRNAs, PTBP1 and HNRNPK are RNA binding proteins that physically interact with lncRNA *Lncenc1* and regulates the expression of genes implicated in the glycolytic pathway thereby leading to self-renewal in ESCs. Loss of function of *Lnccenc1* leads ESCs to exit from pluripotency (Sun et al., 2018).

15. LncRNAs regulating organ development

One of the most important gene clusters regulating embryonic body patterning and specification of cell lineages during the differentiation process is the HOX cluster. This contains 39 genes clustered into 4 groups (A–D) allowing gene expression to proceed in a coordinated manner. LncRNA *Hotair* is one of the widely studied lncRNAs that acts *in trans* to modulate the expression of *Hox* genes (Selleri et al., 2016). It gets transcribed from the HOXC cluster but represses the HOXD cluster located on a different chromosome by interacting with PRC2 and KDM1A-coREST-REST complexes and recruiting them to their site of action on specific gene targets in this cluster. Loss of function of *Hotair* in fibroblasts leads to activation of HOXD cluster genes due to release of repressive marks (McGann et al., 2014).

In an alternate manner, *cis* acting lncRNAs like *Hottip, Mistral* and *Hotairm1* regulate other HOX members by expression from proximity to targets, *Hottip* is expressed from 5′ end of the HOXA and activates specific genes responsible for distal limb development by interacting with members of the MLL1 activating complex. This complex functions by DNA looping bringing *Hottip* and distal HOX gene promoters in close proximity (Wang et al., 2011). LncRNA *Mistral* is expressed in mouse from the HOXA locus and is implicated in retinoic-acid (RA) mediated differentiations of ESCs. Loss of function of *Mistral* inhibits the expression of lineage commitment genes and maintains pluripotency of ESCs (Bertani et al., 2011).

The HOX cluster is implicated in several human diseases like cancer. *Hotairm1* is a lncRNA produced in the myeloid lineage and in leukemia cells it inhibits expression of HOXA genes (Zhang, Weissman, & Newburger, 2014). In addition to these, several other lncRNAs such as *Fendrr, Peril* and *Mdgt* with variable penetrance and lethality have been implicated in developmental processes through knockout studies.

16. LncRNAs affecting neural development

LncRNAs have also been implicated in neural development through loss of screens. One such screen identified several lncRNAs which prevent human ESCs from differentiating into mature neurons. Two lncRNAs, *N1* and *N3* were found to physically associate with SUZ12 and REST and are responsible for recruiting PRC2 complex proteins to genes specifying the glial lineage. LncRNA *N2* also functions as a precursor molecule for *miRNA-125b* and *Let-7* which are known to affect the differentiation pathway.

LncRNA *Pnky* localizes to neural stem cells (NSCs) and its knockdown in these cells promotes neurogenesis and neuronal differentiation. Under in vivo conditions, *Pnky* localizes to the nucleus and physically interacts with regulators of splicing like PTBP1. Together the lncRNA-protein complex coordinates the expression of several genes important for neuronal differentiation (Ramos et al., 2015).

Several lncRNAs have been associated with distinct brain cells in a tissue specific manner. These discoveries have been fuelled by genome wide expression studies and advanced imaging and biochemical technologies. Interestingly, although conservation of lncRNAs across species is considerably poor, significant conservation in seen in the case of lncRNAs expressed in the brain, particularly from birds to mammals. Some of these lncRNAs that are expressed from Ultra Conserved Regions (UCRs) often arise out of loci that are either overlapping or anti-sense to genes that function as molecular scaffolds for recruiting proteins to their targets (Polychronopoulos, King, Nash, Tan, & Lenhard, 2017). For example, the lncRNA *Dlx6os1* (also called *Evf2*) is located antisense to *Dlx6* and is responsible for cortical development, particularly in controlling the regions where projection neurons are found (Cajigas et al., 2018). It acts *in trans* to control the methylation at the *Dlx5/6* genomic loci and subsequent transcription from these sites which in turn regulate the activity of GABAergic neurons. Deletion of this lncRNA in mice reduced the number of GABAergic neurons in the hippocampus (Bond et al., 2009). Another lncRNA *Six3os* is implicated in retinal development by regulating expression of TUJ1 and OLIG2, markers for neurons and oligodendrocytes. Similar to a large number of other lncRNAs. *Six3os* interacts with EZH2 and mediates transcriptional repression of genes in the retina (Rapicavoli et al., 2011).

In Zebrafish, lncRNA *Tuna* has been implicated in development and loss of function of this lncRNA has led to neurodevelopment and locomotor response defects. This lncRNA is highly conserved across Zebrafish, humans and mouse and is implicated in the pluripotency pathway in addition to being expressed in the brain, spinal cord and eyes of adult tissues (Lin et al., 2014). Depletion of *Tuna* in mouse ESCs leads to a block of neuronal differentiation and this lncRNA is known to bind to promoters of NANOG, FGF4 and SOX2, all of which are important in the pluripotent pathway.

17. LncRNAs regulating organogenesis

Among the lncRNAs regulating organogenesis, *Braveheart* and *Fendrr* have been studied in the context of cardiac development (Ottaviani & da Costa Martins, 2017). These lncRNAs originate from the mesoderm which finally gives rise to the heart along with other organs. Knocking down *Braveheart* in mouse ESCs and cardiomyocytes impact the development of mature cardiomyocytes and expression of genes implicated in cardiac development. This establishes a role of *Braveheart* in the regeneration of cardiac cells following injury. The function of *Braveheart* is through epigenetic control of gene expression by interacting with the PRC2 complex (Klattenhoff et al., 2013).

Knockouts of lncRNA *Fendrr* is embryonic lethal and affects heart function although its loss of function did not result in any visible phenotype further fuelling the speculation that studies on lncRNAs might benefit from creating null mutants as opposed to loss of function versions (Grote et al., 2013). *Fendrr*, like *Braveheart* is also associated with PRC2 and functions through the recruitment of MLL1 in mouse embryos.

RNA sequencing of both progenitor and differentiating keratinocytes revealed lncRNA *Tincr* to be highly expressed during differentiation of keratinocytes. Upon knockout of *Tincr*, epidermis lacks distinctive features that accompany terminal differentiation and include intact lamellar bodies and keratohyalin granules. *Tincr* acts binding to the STAU1 protein that results in mRNA stabilization. One of the targets for this complex is the KRT80 which regulates keratinocyte differentiation (Kretz et al., 2013).

Similar to skin, lncRNAs also regulate adipogenesis and hematopoiesis and one of the lncRNAs identified during erythroid differentiation from mouse liver progenitor cells is lncRNA *Eps*. Knockdown of this lncRNA promotes apoptosis and inhibits differentiation (Hu, Yuan, Flygare, & Lodish, 2011).

18. Cellular localization and maturation of lncRNAs

Majority of lncRNAs are synthesized and processed in the nucleus and perform their activities in the cytoplasm. In the Allen Mouse brain atlas, it has been reported that over 800 lncRNAs showed diverse expression and localization sites in both the nucleus as well as cell bodies. Interestingly, certain lncRNAs showed novel subcellular compartments as well suggesting unknown functions (Mercer, Dinger, Sunkin, Mehler, & Mattick, 2008).

From transcriptomic studies, it has been observed that a majority of lncRNAs are transcribed in the 5′ to 3′ direction of a gene and most of these lncRNAs are expressed close to the initial introns and exons. Another important feature of lncRNAs is their secondary structural motifs which govern most of their functions. In the absence of major conservation across species, these secondary structures and their associated domains are postulated to mediate lncRNA:protein or lncRNA:DNA interactions (Delli Ponti, Armaos, Marti, & Tartaglia, 2018; Long, Wang, Youmans, & Cech, 2017).

19. LncRNAs regulating the stability and functions of other RNAs

LncRNAs can target miRNAs and other mRNAs and control their stability and half-life. Examples of such lncRNAs include STAU1-binding RNAs (1/2 sbs RNAs) and growth arrested DNA-damage inducible gene 7 (*Gadd7*) which decrease the stability of mRNAs, while *BACE1-AS* and *Tincr* increase mRNA stability (Rashid, Shah, & Shan, 2016).

At the translational level, LncRNA *P21* represses the translation of mRNAs while lncRNA *Uch1* promotes protein production (Hu et al., 2018).

LncRNA *H19* is a well-known lncRNA that plays important role in muscle differentiation by acting as a substrate for miRNA production and from its first exon miR-675-3p and miR-675-5p is production (Keniry et al., 2012).

The myoblast progenitor cells (MBs), differentiation regulates the production of active muscle cell which involves a highly regulated process that relies on Ying Yang 1 (YY1), a protein which binds to the promoter region of numerous lncRNAs. These lncRNAs are collectively named as YY1-associated muscle lncRNAs (*Yam*). Characterization of one of these lncRNAs, *Yam1*, which regulates myogenesis by inhibiting muscle differentiation factors Myogenin, Tnni2 and α-Actin. *Yam1* is thus a special lncRNA that promotes muscle differentiation by regulating the mechanism of miRNAs and mRNAs activity (Lu et al., 2013).

Multiple lncRNAs act as competing endogenous RNAs (ceRNAs) where these lncRNAs function by acting like miRNA sponges sequestering from their cognate targets. The ceRNA should be expressed in high amount to have sufficient numbers of miRNA binding site to substantially affect the pool of cellular miRNAs. LncRNA sequences that harbor different binding sites which are identical to miRNAs are called competing endogenous RNAs (ceRNAs) (Almeida, Reis, & Calin, 2012). ceRNAs are highly abundant inside cells to maintain the strong control over mRNA amounts in the cells. *HULC*, *linc-MD1*, *linc-ROR* act as a decoy for miR-372, miR-133/135 and miR-145, respectively (Yang et al., 2016).

20. LncRNAs functioning in protein modification pathways

A number of lncRNAs have been shown to function in pathways leading to modification of proteins through processes like ubiquitination, phosphorylation, etc. lnc-DC phosphorylates STAT3 leading to its activation while LincRNA-*p21* blocks the interaction of HIF-1 alpha with VHL protein under hypoxic conditions (Yang, Zhang, Mei, & Wu, 2014).

21. Mechanisms of lncRNA:DNA/RNA interaction

It is well appreciated in the lncRNA field that a large number of lncRNAs regulate chromatin level changes by direct or indirect contacts with DNA (Fig. 2).

In Drosophila, *roX1* and *roX2* ncRNA control dosage compensation by coordinating the interaction between members of the MSL complex and CLAMP for successful recruitment to cognate DNA site (Quinn et al., 2014; Yang, Wen, & Zhu, 2015).

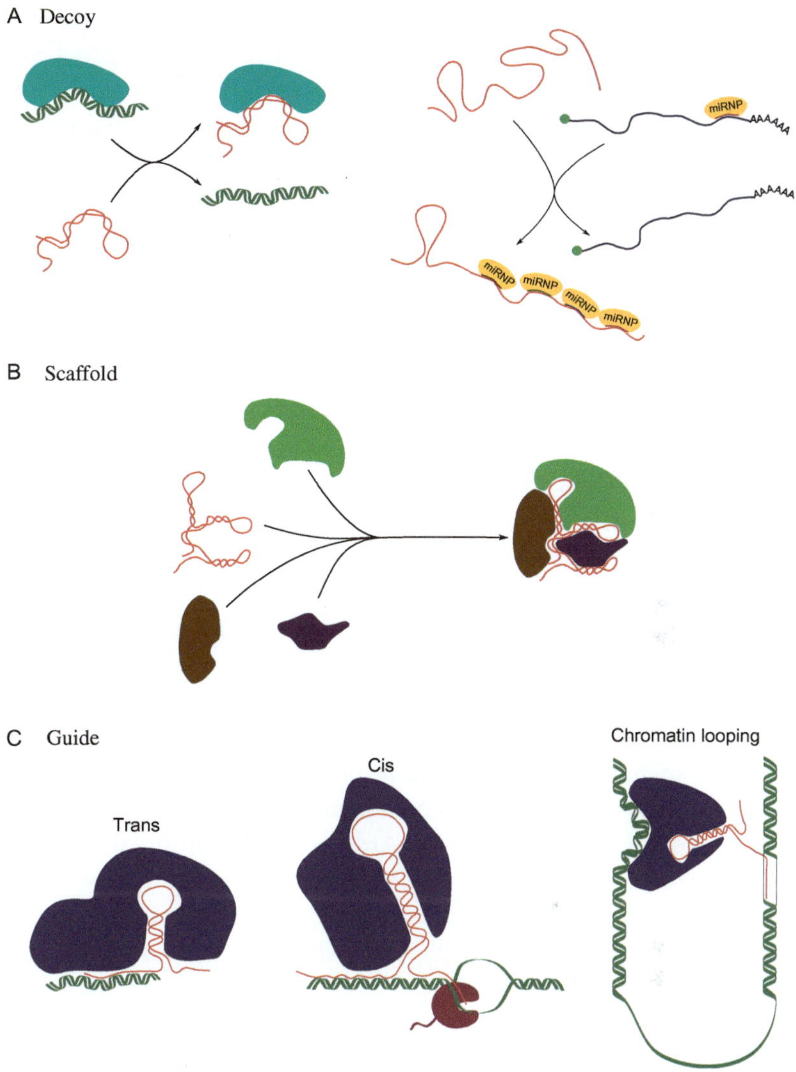

Fig. 2 Schematic showing the mechanism of action of lncRNAs through interactions with various binding partners. LncRNA is depicted in red, mRNA in black, DNA in green and protein binding partners are represented as colored objects. (A) LncRNAs acting as decoys, sequestering proteins from their cognate DNA or RNA binding sites. (B) LncRNAs acting as scaffold for the formation of protein complexes. (C) LncRNAs acting either in proximal (*cis*), distal (*trans*) or through chromosomal looping manner bringing DNA binding proteins to their sites of action.

During Neural differentiation, LncRNA *Dali* plays important role by interacting with DNMT1 which in turn recruits *Dali* to the target regions via interacting with other DNA binding proteins (Chalei et al., 2014).

Apart from forming complexes with protein, lncRNA also forms complex with DNA by Hoogsteen or Reverse-Hoogsteen base pairing between single stranded RNA and DNA molecules (Holland & Hoffman, 1996). LncRNA *Fendrr*, during mesodermal differentiation targets the promoter regions of *Foxf1* and *Pitx2* genes by forming a 3D structure through complementary base pairing (Marchese, Raimondi, & Huarte, 2017).

MEG3 targets the TGF-β pathway genes by forming triple helices with the chromatin and represses the genes by recruiting PRC2 complex (Mondal et al., 2015). Promoter associated RNAs (pRNA) induces de novo methylation by binding to complimentary DNA that is recognized by DNA Methyltransferase DNMT3b (Schmitz, Mayer, Postepska, & Grummt, 2010).

LncRNAs are also found to interact with other mRNAs such as in the case of lincRNA-p21 which initiates translation inhibitions of its target mRNAs by forming RNA:RNA interactions (Yoon et al., 2012).

LncRNA *Tincr* controls human epidermal differentiation by post-transcriptionally modifying large number of mRNA transcripts by directly interacting with influenced by Staufen1 (STAU1) protein (Kretz et al., 2013).

Many lncRNA and enhancer RNAs (eRNAs) have been known shown to play role in promoting enhancer-promoter looping, where Immediate Early Genes (IEGs) regulated by eRNAs in neurons appear to act downstream of such loops. In the induction of Arc gene in neurons, the enhancer-promoter interactions occur prior to eRNA synthesis in a stimulus-dependent manner and appears to be pre-requisite for the eRNA transcription to happen as this ERNA transcription only happens when Arc gene is present in WT type cells and not when Arc gene promoter is deleted despite of the wild-type levels of presence of RNAII binding at the enhancer regions. Consistently, Arc eRNA knockdown didn't affect the promoter-enhancer interactions, but instead promotes the Arc induction by facilitating the release of negative elongation factor NELF from paused RNAPII through its competition with other RNAs (Schaukowitch et al., 2014).

22. Allosteric regulation of proteins by lncRNAs

An increasing number of studies showed that lncRNAs can allosterically alter the function of the protein they interact with. TLS, an RNA binding protein regulates the transcription of certain genes by inhibiting the CBP HAT activity in an RNA dependent manner. The N-terminus has a strong inhibitory activity but binding of the TLS C-terminus prevents its inhibitory function. LncRNA *CCDN1* expressed from the 5′ regulatory region of a TLS target gene. *CCDN1* is known to allosterically modify TLS in *cis* to relieve the auto inhibitory configuration, thereby repressing *CCDN1* expression (Quan, Chen, & Zhang, 2015; Song, Wang, Arai, & Kurokawa, 2012).

LncRNAs can also modify the enzymatic activity of chromatin modulators. LncRNA *Evf2* is involved in neural development by regulating the expression of homeodomain transcription factor DLX5 and DLX6 in the developing mouse brain. It forms complex with the DLX homeodomain protein complex at the ultraconserved intergenic regions to repress gene expression (Berghoff et al., 2013). *Evf2*/DLX mass spectrometry results reveal the association of SWI/SNF chromatin remodelers Brahma related gene 1 (BRG1, SMARA4) and Brahma associated factor (BAF170, SMARCC2) in a developing mouse brain. Association of BRG1 with Evf2/DLX1is mediated by direct interaction with DLX1, but *Evf2* can increase the association of BRG1 binding to the DLX5/6 enhancers. *Evf2* can also inhibit BRG1 ATPase and chromatin remodeling activities, causing gene repression (Cajigas et al., 2015).

Recent reports have shown that the lncRNA transcribed from the regulatory elements such as eRNAs and promoter associated ncRNAs can modulate the gene expression by acting in *cis* to the regulatory region they occupy and the other transcription factors it associates with. The stable maintenance and spread across the genome of the transcription factor, YY, is regulated by the active presence of the regulatory element derived lncRNAs. These studies also suggest the positive feedback loop of the regulatory elements in maintaining genome stability and gene expression (Sigova et al., 2015).

23. Single cell analysis of lncRNA functions

Most of the lncRNA transcript profiling has been employed of bulk measurements, an average of thousands or millions of cells. Recent work at single cell level has revealed how much heterogeneity exists even within a "clonal" population of cells (Buganim et al., 2012). Recent characterization of lincHoxA1 located at the 3′ end of HoxA cluster brought to light the importance of studying the functions of lowly expressed lncRNAs at single cell levels (Maamar, Cabili, Rinn, & Raj, 2013). Earlier studies at the bulk cell level demonstrated a positive correlation between *lincHoxA1* and mRNA HoxA1. Surprisingly, at single cell level, it was observed that the mechanism is anticorrelated. The relationship was switch-like such that when the *lincHoxA1* number is more than 10 copies, then HOXA1 was repressed. Knockdown studies of lincHoxA1 via siRNAs and antisense oligonucleotides suggested that *lincHoxA1* partners with purine-rich element binding protein B (PURB) and exerts transcriptional silencing of HOXA1. The two depletion methods differ in their capacity to reduce *lincHoxA1* levels on the chromatin versus total levels. Thus, this study revealed two important points: first, the anticorrelation between lincHoxA1 and HOXA1 at a single cell level. Second, use of siRNAs which was effective in reducing the total *lincHoxA1* cellular level in bulk cells, might not reflect the function of certain lncRNAs where copy number is tightly linked to the roles they play inside the cell. This is because of the heterogeneity in silencing that RNA interference methods introduce in different cells of a population. As single cell analysis and antisense oligonucleotides technology become more robust and widely adopted, it is likely that many unknown features of lncRNAs will be revealed.

24. LncRNAs in disease progression

As seen from numerous lines of evidence, lncRNAs can serve as address codes for a large number of biological molecules to properly reach their intended targets. These influence the outcome of signaling pathways, developmental progression and in many cases reflect how a disease progresses. For example, among Mendelian disorders, the role of lncRNAs are being increasingly appreciated since only 7% of disease phenotypes are directly attributable to the underlying single nucleotide polymorphisms (SNPs) associated with the disease present in the protein coding region while the majority of these mutations lie in the non-protein coding regions (Batista & Chang, 2013; Hu et al., 2018). In the case of Facioscapulohumeral Muscular Dystrophy (FSHD), one of the most common myopathies, there is a reduction in the copy numbers of D4Z4 repeats which in turn results in activation of lncRNA *Dbe-T*, a *cis* acting lncRNA which localizes to FSHD locus. This lncRNA recruits a component of the MLL/TRX complex called ASH1L to chromatin and derepresses target genes thereby performing functions of an activator. Promoter mutations affecting the lncRNA expression results in the disease phenotype (Cabianca et al., 2012; Lemmers et al., 2018).

In the case of HELLP syndrome (hemolysis, elevated liver enzymes, low platelets), a 205 kb capped and polyadenylated lncRNA is associated with the important genes responsible for the disease progression. This correlation was established through linkage analysis and the function of the lncRNA was attributed to its role in allowing cells to transit from G2 to M phase during the cell cycle. Rescuing the disease associated mutations in the lncRNA using morpholinos was able to revert some of the features of gene expression and cell invasion defects associated with the disease (McAninch, Roberts, & Bianco-Miotto, 2017).

25. LncRNA knockouts often show lack of phenotype: The importance of context and redundancy

It is now known that certain lncRNAs despite exhibiting higher expression levels and conservation in different tissue types and species, might not show any persistent phenotype when knocked out. It is interesting to observe that a lot of functional redundancy exists between lncRNA classes and individual lncRNA function. On many cases knockout phenotypes are subtle or non-existent during developmental process, whereas in vitro studies suggest their essential role by stark changes pointing to defined functional roles. An example is lncRNA *MALAT1*, which is one of the most abundant and widely expressed and conserved lncRNAs in vertebrates. *MALAT1* localizes to the nuclear speckles and functions by modulating a subset of genes that play essential roles in formation of synapse and nucleus to regulate metastasis potential in cancer cells and regulate several splicing events in mRNAs (Gutschner et al., 2013). In nuclear speckles, *MALAT1* is found to interact with E3 SUMO-protein ligase CBX4, which is a component of PRC2 (Ma, Zhang, Sun, & Cheng, 2014). Three independent *MALAT1* knockout mice studies, however, led to different functional observations. Unexpectedly, these mice were viable and fertile, showing no apparent gross phenotypes. One study using *MALAT1* knockout strains showed that the mice exhibit enhanced tumor differentiation and reduced tendency to form mammary tumor in a background of transgenic expression of strong oncogene. Another *MALAT1* knockout study showed retinal vascularization phenotype. A third study showed that *MALAT1* knockout affects several other genes, including a *MALAT1* neighboring gene indicating the *cis*-regulatory role of *MALAT1* in gene transcription (Zhang et al., 2012). Despite the high conservation and expression between different organisms underscoring *MALAT1* as an important candidate for important cellular functions, it does not appear to regulate physiological processes linked to organismal development.

26. Conclusions

LncRNAs are no more considered as by-products of pervasive transcription or "transcriptional noise." A decade of studies on lncRNAs have revealed features of transcriptional control and functional significance that put them at the forefront of cellular decisions. However, their low expression and difficulties in annotation make it difficult to assign defined cellular roles. Thus, although several thousand lncRNAs have been identified through sequencing efforts, only a few of them have been functionally characterized. In several cases, the overlap of lncRNAs with closely associated protein coding genes makes it complicated to assign functions with confidence. This is made more complicated due to the problems of off-targeting that most perturbation studies encompass. It is thus becoming more common to investigate lncRNA roles using a combination of multiple knockdown/knockout and overexpression strategies. The revolutionary CRISPR technology has addressed some of these features due to the ease of performing many of these experiments as well as allowing innovative ways of visualizing lncRNAs in real time, a major step forward for attributing functional roles to lncRNAs. The processes of stem cell pluripotency and differentiation represent crucial phases in the development of an organism and as more and more lncRNAs are implicated in regulating these pathways, harnessing next generation strategies to study them in a defined and systematic manner will become increasingly important.

References

Alison, M. R., & Islam, S. (2009). Attributes of adult stem cells. *The Journal of Pathology*, *217*(2), 144–160. https://doi.org/10.1002/path.2498.

Almeida, M. I., Reis, R. M., & Calin, G. A. (2012). Decoy activity through microRNAs: The therapeutic implications. *Expert Opinion on Biological Therapy*, *12*(9), 1153–1159. https://doi.org/10.1517/14712598.2012.693470.

Amaral, P. P., Clark, M. B., Gascoigne, D. K., Dinger, M. E., & Mattick, J. S. (2011). lncRNAdb: A reference database for long noncoding RNAs. *Nucleic Acids Research*, *39*(Database issue), D146–D151. https://doi.org/10.1093/nar/gkq1138.

Augui, S., Nora, E. P., & Heard, E. (2011). Regulation of X-chromosome inactivation by the X-inactivation centre. *Nature Reviews. Genetics, 12*(6), 429–442. https://doi.org/10.1038/nrg2987.

Balas, M. M., & Johnson, A. M. (2018). Exploring the mechanisms behind long noncoding RNAs and cancer. *Non-Coding RNA Research, 3*(3), 108–117. https://doi.org/10.1016/j.ncrna.2018.03.001.

Barry, G., Guennewig, B., Fung, S., Kaczorowski, D., & Weickert, C. S. (2015). Long noncoding RNA expression during aging in the human subependymal zone. *Frontiers in Neurology, 6*, 45. https://doi.org/10.3389/fneur.2015.00045.

Batista, P. J., & Chang, H. Y. (2013). Long noncoding RNAs: Cellular address codes in development and disease. *Cell, 152*(6), 1298–1307. https://doi.org/10.1016/j.cell.2013.02.012.

Berghoff, E. G., Clark, M. F., Chen, S., Cajigas, I., Leib, D. E., & Kohtz, J. D. (2013). Evf2 (Dlx6as) lncRNA regulates ultraconserved enhancer methylation and the differential transcriptional control of adjacent genes. *Development, 140*(21), 4407–4416. https://doi.org/10.1242/dev.099390.

Bertani, S., Sauer, S., Bolotin, E., & Sauer, F. (2011). The noncoding RNA mistral activates Hoxa6 and Hoxa7 expression and stem cell differentiation by recruiting MLL1 to chromatin. *Molecular Cell, 43*(6), 1040–1046. https://doi.org/10.1016/j.molcel.2011.08.019.

Bhartiya, D., Pal, K., Ghosh, S., Kapoor, S., Jalali, S., Panwar, B., et al. (2013). lncRNome: A comprehensive knowledgebase of human long noncoding RNAs. *Database (Oxford), 2013*. https://doi.org/10.1093/database/bat034bat034.

Bond, A. M., Vangompel, M. J., Sametsky, E. A., Clark, M. F., Savage, J. C., Disterhoft, J. F., et al. (2009). Balanced gene regulation by an embryonic brain ncRNA is critical for adult hippocampal GABA circuitry. *Nature Neuroscience, 12*(8), 1020–1027. https://doi.org/10.1038/nn.2371.

Briggs, S. F., & Reijo Pera, R. A. (2014). X chromosome inactivation: Recent advances and a look forward. *Current Opinion in Genetics & Development, 28*, 78–82. https://doi.org/10.1016/j.gde.2014.09.010.

Buganim, Y., Faddah, D. A., Cheng, A. W., Itskovich, E., Markoulaki, S., Ganz, K., et al. (2012). Single-cell expression analyses during cellular reprogramming reveal an early stochastic and a late hierarchic phase. *Cell, 150*(6), 1209–1222. https://doi.org/10.1016/j.cell.2012.08.023.

Cabianca, D. S., Casa, V., Bodega, B., Xynos, A., Ginelli, E., Tanaka, Y., et al. (2012). A long ncRNA links copy number variation to a polycomb/trithorax epigenetic switch in FSHD muscular dystrophy. *Cell, 149*(4), 819–831. https://doi.org/10.1016/j.cell.2012.03.035.

Cabili, M. N., Trapnell, C., Goff, L., Koziol, M., Tazon-Vega, B., Regev, A., et al. (2011). Integrative annotation of human large intergenic noncoding RNAs reveals global properties and specific subclasses. *Genes & Development, 25*(18), 1915–1927. https://doi.org/10.1101/gad.17446611.

Cajigas, I., Chakraborty, A., Swyter, K. R., Luo, H., Bastidas, M., Nigro, M., et al. (2018). The Evf2 ultraconserved enhancer lncRNA functionally and spatially organizes Megabase distant genes in the developing forebrain. *Molecular Cell, 71*(6). https://doi.org/10.1016/j.molcel.2018.07.024956–972.e959.

Cajigas, I., Leib, D. E., Cochrane, J., Luo, H., Swyter, K. R., Chen, S., et al. (2015). Evf2 lncRNA/BRG1/DLX1 interactions reveal RNA-dependent inhibition of chromatin remodeling. *Development, 142*(15), 2641–2652. https://doi.org/10.1242/dev.126318.

Carmona, S., Lin, B., Chou, T., Arroyo, K., & Sun, S. (2018). LncRNA Jpx induces Xist expression in mice using both trans and cis mechanisms. *PLoS Genetics, 14*(5), e1007378. https://doi.org/10.1371/journal.pgen.1007378.

Carninci, P., Kasukawa, T., Katayama, S., Gough, J., Frith, M. C., Maeda, N., et al. (2005). The transcriptional landscape of the mammalian genome. *Science*, *309*(5740), 1559–1563. https://doi.org/10.1126/science.1112014.

Chakraborty, D., Kappei, D., Theis, M., Nitzsche, A., Ding, L., Paszkowski-Rogacz, M., et al. (2012). Combined RNAi and localization for functionally dissecting long noncoding RNAs. *Nature Methods*, *9*(4), 360–362. https://doi.org/10.1038/nmeth.1894.

Chakraborty, D., Paszkowski-Rogacz, M., Berger, N., Ding, L., Mircetic, J., Fu, J., et al. (2017). lncRNA Panct1 maintains mouse embryonic stem cell identity by regulating TOBF1 recruitment to Oct-sox sequences in early G1. *Cell Reports*, *21*(11), 3012–3021. https://doi.org/10.1016/j.celrep.2017.11.045.

Chalei, V., Sansom, S. N., Kong, L., Lee, S., Montiel, J. F., Vance, K. W., et al. (2014). The long non-coding RNA Dali is an epigenetic regulator of neural differentiation. *eLife*, *3*, e04530. https://doi.org/10.7554/eLife.04530.

Chase, A., & Cross, N. C. (2011). Aberrations of EZH2 in cancer. *Clinical Cancer Research*, *17*(9), 2613–2618. https://doi.org/10.1158/1078-0432.CCR-10-2156.

Chishima, T., Iwakiri, J., & Hamada, M. (2018). Identification of transposable elements contributing to tissue-specific expression of long non-coding RNAs. *Genes (Basel)*, *9*(1), 23. https://doi.org/10.3390/genes9010023.

Daneshvar, K., Pondick, J. V., Kim, B. M., Zhou, C., York, S. R., Macklin, J. A., et al. (2016). DIGIT is a conserved long noncoding RNA that regulates GSC expression to control definitive endoderm differentiation of embryonic stem cells. *Cell Reports*, *17*(2), 353–365. https://doi.org/10.1016/j.celrep.2016.09.017.

Delli Ponti, R., Armaos, A., Marti, S., & Tartaglia, G. G. (2018). A method for RNA structure prediction shows evidence for structure in lncRNAs. *Frontiers in Molecular Biosciences*, *5*, 111. https://doi.org/10.3389/fmolb.2018.00111.

Dill, T. L., & Naya, F. J. (2018). A hearty dose of noncoding RNAs: The imprinted DLK1-DIO3 locus in cardiac development and disease. *Journal of Cardiovascular Development and Disease*, *5*(3), 37. https://doi.org/10.3390/jcdd5030037.

Dinger, M. E., Pang, K. C., Mercer, T. R., Crowe, M. L., Grimmond, S. M., & Mattick, J. S. (2009). NRED: A database of long noncoding RNA expression. *Nucleic Acids Research*, *37*(Database issue), D122–D126. https://doi.org/10.1093/nar/gkn617.

Dinger, M. E., Pang, K. C., Mercer, T. R., & Mattick, J. S. (2008). Differentiating protein-coding and noncoding RNA: Challenges and ambiguities. *PLoS Computational Biology*, *4*(11). e1000176https://doi.org/10.1371/journal.pcbi.1000176.

Farazi, T. A., Juranek, S. A., & Tuschl, T. (2008). The growing catalog of small RNAs and their association with distinct Argonaute/Piwi family members. *Development*, *135*(7), 1201–1214. https://doi.org/10.1242/dev.005629.

Fico, A., Fiorenzano, A., Pascale, E., Patriarca, E. J., & Minchiotti, G. (2019). Long non-coding RNA in stem cell pluripotency and lineage commitment: Functions and evolutionary conservation. *Cellular and Molecular Life Sciences*, *76*(8), 1459–1471. https://doi.org/10.1007/s00018-018-3000-z.

Geng, T., Zhang, D., & Jiang, W. (2019). Concise review: Epigenetic regulation of transition among different pluripotent states. *Stem Cells*, *37*, 1372–1380. https://doi.org/10.1002/stem.3064.

Ghosal, S., Das, S., & Chakrabarti, J. (2013). Long noncoding RNAs: New players in the molecular mechanism for maintenance and differentiation of pluripotent stem cells. *Stem Cells and Development*, *22*(16), 2240–2253. https://doi.org/10.1089/scd.2013.0014.

Govindarajan, R., Duraiyan, J., Kaliyappan, K., & Palanisamy, M. (2012). Microarray and its applications. *Journal of Pharmacy & Bioallied Sciences*, *4*(Suppl. 2), S310–S312. https://doi.org/10.4103/0975-7406.100283.

Grammatikakis, I., Panda, A. C., Abdelmohsen, K., & Gorospe, M. (2014). Long noncoding RNAs(lncRNAs) and the molecular hallmarks of aging. *Aging (Albany NY)*, *6*(12), 992–1009. https://doi.org/10.18632/aging.100710.

Grote, P., Wittler, L., Hendrix, D., Koch, F., Wahrisch, S., Beisaw, A., et al. (2013). The tissue-specific lncRNA Fendrr is an essential regulator of heart and body wall development in the mouse. *Developmental Cell*, *24*(2), 206–214. https://doi.org/10.1016/j.devcel.2012.12.012.

Gupta, R. A., Shah, N., Wang, K. C., Kim, J., Horlings, H. M., Wong, D. J., et al. (2010). Long non-coding RNA HOTAIR reprograms chromatin state to promote cancer metastasis. *Nature*, *464*(7291), 1071–1076. https://doi.org/10.1038/nature08975.

Gutschner, T., Hammerle, M., Eissmann, M., Hsu, J., Kim, Y., Hung, G., et al. (2013). The noncoding RNA MALAT1 is a critical regulator of the metastasis phenotype of lung cancer cells. *Cancer Research*, *73*(3), 1180–1189. https://doi.org/10.1158/0008-5472.CAN-12-2850.

Guttman, M., Donaghey, J., Carey, B. W., Garber, M., Grenier, J. K., Munson, G., et al. (2011). lincRNAs act in the circuitry controlling pluripotency and differentiation. *Nature*, *477*(7364), 295–300. https://doi.org/10.1038/nature10398.

Heather, J. M., & Chain, B. (2016). The sequence of sequencers: The history of sequencing DNA. *Genomics*, *107*(1), 1–8. https://doi.org/10.1016/j.ygeno.2015.11.003.

Holland, J. A., & Hoffman, D. W. (1996). Structural features and stability of an RNA triple helix in solution. *Nucleic Acids Research*, *24*(14), 2841–2848. https://doi.org/10.1093/nar/24.14.2841.

Hou, L., Tu, J., Cheng, F., Yang, H., Yu, F., Wang, M., et al. (2018). Long noncoding RNA ROR promotes breast cancer by regulating the TGF-beta pathway. *Cancer Cell International*, *18*, 142. https://doi.org/10.1186/s12935-018-0638-4.

Hu, G., Niu, F., Humburg, B. A., Liao, K., Bendi, S., Callen, S., et al. (2018). Molecular mechanisms of long noncoding RNAs and their role in disease pathogenesis. *Oncotarget*, *9*(26), 18648–18663. https://doi.org/10.18632/oncotarget.24307.

Hu, W., Yuan, B., Flygare, J., & Lodish, H. F. (2011). Long noncoding RNA-mediated anti-apoptotic activity in murine erythroid terminal differentiation. *Genes & Development*, *25*(24), 2573–2578. https://doi.org/10.1101/gad.178780.111.

Iyer, M. K., Niknafs, Y. S., Malik, R., Singhal, U., Sahu, A., Hosono, Y., et al. (2015). The landscape of long noncoding RNAs in the human transcriptome. *Nature Genetics*, *47*(3), 199–208. https://doi.org/10.1038/ng.3192.

Jain, A. K., Xi, Y., McCarthy, R., Allton, K., Akdemir, K. C., Patel, L. R., et al. (2016). LncPRESS1 is a p53-regulated LncRNA that safeguards pluripotency by disrupting SIRT6-mediated De-acetylation of histone H3K56. *Molecular Cell*, *64*(5), 967–981. https://doi.org/10.1016/j.molcel.2016.10.039.

Jia, W., Chen, W., & Kang, J. (2013). The functions of microRNAs and long non-coding RNAs in embryonic and induced pluripotent stem cells. *Genomics, Proteomics & Bioinformatics*, *11*(5), 275–283. https://doi.org/10.1016/j.gpb.2013.09.004.

Jiang, Q., Wang, J., Wang, Y., Ma, R., Wu, X., & Li, Y. (2014). TF2LncRNA: Identifying common transcription factors for a list of lncRNA genes from ChIP-Seq data. *BioMed Research International*, *2014*, 317642. https://doi.org/10.1155/2014/317642.

Keniry, A., Oxley, D., Monnier, P., Kyba, M., Dandolo, L., Smits, G., et al. (2012). The H19 lincRNA is a developmental reservoir of miR-675 that suppresses growth and Igf1r. *Nature Cell Biology*, *14*(7), 659–665. https://doi.org/10.1038/ncb2521.

Khawar, M. B., Mehmood, R., Abbasi, M. H., & Sheikh, N. (2018). Multifactorial role of long non-coding RNAs (LncRNAs) in hematopoiesis. *Frontiers in Bioscience (Scholar Edition)*, *10*, 119–126.

Kim, D., Pertea, G., Trapnell, C., Pimentel, H., Kelley, R., & Salzberg, S. L. (2013). TopHat2: Accurate alignment of transcriptomes in the presence of insertions, deletions and gene fusions. *Genome Biology, 14*(4), R36. https://doi.org/10.1186/gb-2013-14-4-r36.

Kim, T. K., & Shiekhattar, R. (2016). Diverse regulatory interactions of long noncoding RNAs. *Current Opinion in Genetics & Development, 36*, 73–82. https://doi.org/10.1016/j.gde.2016.03.014.

Kiyosawa, H., Mise, N., Iwase, S., Hayashizaki, Y., & Abe, K. (2005). Disclosing hidden transcripts: Mouse natural sense-antisense transcripts tend to be poly(A) negative and nuclear localized. *Genome Research, 15*(4), 463–474. https://doi.org/10.1101/gr.3155905.

Klattenhoff, C. A., Scheuermann, J. C., Surface, L. E., Bradley, R. K., Fields, P. A., Steinhauser, M. L., et al. (2013). Braveheart, a long noncoding RNA required for cardiovascular lineage commitment. *Cell, 152*(3), 570–583. https://doi.org/10.1016/j.cell.2013.01.003.

Kretz, M., Siprashvili, Z., Chu, C., Webster, D. E., Zehnder, A., Qu, K., et al. (2013). Control of somatic tissue differentiation by the long non-coding RNA TINCR. *Nature, 493*(7431), 231–235. https://doi.org/10.1038/nature11661.

Kung, J. T., Colognori, D., & Lee, J. T. (2013). Long noncoding RNAs: Past, present, and future. *Genetics, 193*(3), 651–669. https://doi.org/10.1534/genetics.112.146704.

Lagarde, J., Uszczynska-Ratajczak, B., Carbonell, S., Perez-Lluch, S., Abad, A., Davis, C., et al. (2017). High-throughput annotation of full-length long noncoding RNAs with capture long-read sequencing. *Nature Genetics, 49*(12), 1731–1740. https://doi.org/10.1038/ng.3988.

Latos, P. A., Pauler, F. M., Koerner, M. V., Senergin, H. B., Hudson, Q. J., Stocsits, R. R., et al. (2012). Airn transcriptional overlap, but not its lncRNA products, induces imprinted Igf2r silencing. *Science, 338*(6113), 1469–1472. https://doi.org/10.1126/science.1228110.

Lemmers, R., van der Vliet, P. J., Vreijling, J. P., Henderson, D., van der Stoep, N., Voermans, N., et al. (2018). Cis D4Z4 repeat duplications associated with facioscapulohumeral muscular dystrophy type 2. *Human Molecular Genetics, 27*(20), 3488–3497. https://doi.org/10.1093/hmg/ddy236.

Lin, N., Chang, K. Y., Li, Z., Gates, K., Rana, Z. A., Dang, J., et al. (2014). An evolutionarily conserved long noncoding RNA TUNA controls pluripotency and neural lineage commitment. *Molecular Cell, 53*(6), 1005–1019. https://doi.org/10.1016/j.molcel.2014.01.021.

Liu, Y., & Zhao, M. (2016). lnCaNet: Pan-cancer co-expression network for human lncRNA and cancer genes. *Bioinformatics, 32*(10), 1595–1597. https://doi.org/10.1093/bioinformatics/btw017.

Loewer, S., Cabili, M. N., Guttman, M., Loh, Y. H., Thomas, K., Park, I. H., et al. (2010). Large intergenic non-coding RNA-RoR modulates reprogramming of human induced pluripotent stem cells. *Nature Genetics, 42*(12), 1113–1117. https://doi.org/10.1038/ng.710.

Long, Y., Wang, X., Youmans, D. T., & Cech, T. R. (2017). How do lncRNAs regulate transcription? *Science Advances, 3*(9), eaao2110. https://doi.org/10.1126/sciadv.aao2110.

Lu, L., Sun, K., Chen, X., Zhao, Y., Wang, L., Zhou, L., et al. (2013). Genome-wide survey by ChIP-seq reveals YY1 regulation of lincRNAs in skeletal myogenesis. *The EMBO Journal, 32*(19), 2575–2588. https://doi.org/10.1038/emboj.2013.182.

Luo, S., Lu, J. Y., Liu, L., Yin, Y., Chen, C., Han, X., et al. (2016). Divergent lncRNAs regulate gene expression and lineage differentiation in pluripotent cells. *Cell Stem Cell, 18*(5), 637–652. https://doi.org/10.1016/j.stem.2016.01.024.

Ma, R. G., Zhang, Y., Sun, T. T., & Cheng, B. (2014). Epigenetic regulation by polycomb group complexes: Focus on roles of CBX proteins. *Journal of Zhejiang University. Science. B, 15*(5), 412–428. https://doi.org/10.1631/jzus.B1400077.

Maamar, H., Cabili, M. N., Rinn, J., & Raj, A. (2013). Linc-HOXA1 is a noncoding RNA that represses Hoxa1 transcription in cis. *Genes & Development, 27*(11), 1260–1271. https://doi.org/10.1101/gad.217018.113.

Mancini-Dinardo, D., Steele, S. J., Levorse, J. M., Ingram, R. S., & Tilghman, S. M. (2006). Elongation of the Kcnq1ot1 transcript is required for genomic imprinting of neighboring genes. *Genes & Development, 20*(10), 1268–1282. https://doi.org/10.1101/gad.1416906.

Marchese, F. P., Raimondi, I., & Huarte, M. (2017). The multidimensional mechanisms of long noncoding RNA function. *Genome Biology, 18*(1), 206. https://doi.org/10.1186/s13059-017-1348-2.

Mattick, J. S., & Rinn, J. L. (2015). Discovery and annotation of long noncoding RNAs. *Nature Structural & Molecular Biology, 22*(1), 5–7. https://doi.org/10.1038/nsmb.2942.

McAninch, D., Roberts, C. T., & Bianco-Miotto, T. (2017). Mechanistic insight into long noncoding RNAs and the placenta. *International Journal of Molecular Sciences, 18*(7), 1371. https://doi.org/10.3390/ijms18071371.

McGann, J. C., Oyer, J. A., Garg, S., Yao, H., Liu, J., Feng, X., et al. (2014). Polycomb- and REST-associated histone deacetylases are independent pathways toward a mature neuronal phenotype. *eLife, 3*, e04235. https://doi.org/10.7554/eLife.04235.

Meirelles Lda, S., & Nardi, N. B. (2009). Methodology, biology and clinical applications of mesenchymal stem cells. *Frontiers in Bioscience (Landmark edition), 14*, 4281–4298.

Mercer, T. R., Dinger, M. E., Sunkin, S. M., Mehler, M. F., & Mattick, J. S. (2008). Specific expression of long noncoding RNAs in the mouse brain. *Proceedings of the National Academy of Sciences of the United States of America, 105*(2), 716–721. https://doi.org/10.1073/pnas.0706729105.

Minkovsky, A., Patel, S., & Plath, K. (2012). Concise review: Pluripotency and the transcriptional inactivation of the female mammalian X chromosome. *Stem Cells, 30*(1), 48–54. https://doi.org/10.1002/stem.755.

Mockler, T. C., Chan, S., Sundaresan, A., Chen, H., Jacobsen, S. E., & Ecker, J. R. (2005). Applications of DNA tiling arrays for whole-genome analysis. *Genomics, 85*(1), 1–15. https://doi.org/10.1016/j.ygeno.2004.10.005.

Mondal, T., Subhash, S., Vaid, R., Enroth, S., Uday, S., Reinius, B., et al. (2015). MEG3 long noncoding RNA regulates the TGF-beta pathway genes through formation of RNA-DNA triplex structures. *Nature Communications, 6*, 7743. https://doi.org/10.1038/ncomms8743.

Nagano, T., Mitchell, J. A., Sanz, L. A., Pauler, F. M., Ferguson-Smith, A. C., Feil, R., et al. (2008). The air noncoding RNA epigenetically silences transcription by targeting G9a to chromatin. *Science, 322*(5908), 1717–1720. https://doi.org/10.1126/science.1163802.

Ning, S., Zhang, J., Wang, P., Zhi, H., Wang, J., Liu, Y., et al. (2016). Lnc2Cancer: A manually curated database of experimentally supported lncRNAs associated with various human cancers. *Nucleic Acids Research, 44*(D1), D980–D985. https://doi.org/10.1093/nar/gkv1094.

Ott, H. C., Matthiesen, T. S., Brechtken, J., Grindle, S., Goh, S. K., Nelson, W., et al. (2007). The adult human heart as a source for stem cells: Repair strategies with embryonic-like progenitor cells. *Nature Clinical Practice Cardiovascular Medicine, 4*(Suppl. 1), S27–S39. https://doi.org/10.1038/ncpcardio0771.

Ottaviani, L., & da Costa Martins, P. A. (2017). Non-coding RNAs in cardiac hypertrophy. *The Journal of Physiology, 595*(12), 4037–4050. https://doi.org/10.1113/JP273129.

Ozata, D. M., Gainetdinov, I., Zoch, A., O'Carroll, D., & Zamore, P. D. (2019). PIWI-interacting RNAs: Small RNAs with big functions. *Nature Reviews Genetics*, *20*(2), 89–108. https://doi.org/10.1038/s41576-018-0073-3.

Pal, D., & Rao, M. R. S. (2017). Long noncoding RNAs in pluripotency of stem cells and cell fate specification. *Advances in Experimental Medicine and Biology*, *1008*, 223–252. https://doi.org/10.1007/978-981-10-5203-3_8.

Park, C., Yu, N., Choi, I., Kim, W., & Lee, S. (2014). lncRNAtor: A comprehensive resource for functional investigation of long non-coding RNAs. *Bioinformatics*, *30*(17), 2480–2485. https://doi.org/10.1093/bioinformatics/btu325.

Payer, B., Rosenberg, M., Yamaji, M., Yabuta, Y., Koyanagi-Aoi, M., Hayashi, K., et al. (2013). Tsix RNA and the germline factor, PRDM14, link X reactivation and stem cell reprogramming. *Molecular Cell*, *52*(6), 805–818. https://doi.org/10.1016/j.molcel.2013.10.023.

Pertea, M., Kim, D., Pertea, G. M., Leek, J. T., & Salzberg, S. L. (2016). Transcript-level expression analysis of RNA-seq experiments with HISAT, StringTie and Ballgown. *Nature Protocols*, *11*(9), 1650–1667. https://doi.org/10.1038/nprot.2016.095.

Polychronopoulos, D., King, J. W. D., Nash, A. J., Tan, G., & Lenhard, B. (2017). Conserved non-coding elements: developmental gene regulation meets genome organization. *Nucleic Acids Research*, *45*(22), 12611–12624. https://doi.org/10.1093/nar/gkx1074.

Pontier, D. B., & Gribnau, J. (2011). Xist regulation and function explored. *Human Genetics*, *130*(2), 223–236. https://doi.org/10.1007/s00439-011-1008-7.

Ponting, C. P., Oliver, P. L., & Reik, W. (2009). Evolution and functions of long noncoding RNAs. *Cell*, *136*(4), 629–641. https://doi.org/10.1016/j.cell.2009.02.006.

Quan, M., Chen, J., & Zhang, D. (2015). Exploring the secrets of long noncoding RNAs. *International Journal of Molecular Sciences*, *16*(3), 5467–5496. https://doi.org/10.3390/ijms16035467.

Quinn, J. J., Ilik, I. A., Qu, K., Georgiev, P., Chu, C., Akhtar, A., et al. (2014). Revealing long noncoding RNA architecture and functions using domain-specific chromatin isolation by RNA purification. *Nature Biotechnology*, *32*(9), 933–940. https://doi.org/10.1038/nbt.2943.

Raha, D., Hong, M., & Snyder, M. (2010). ChIP-Seq: A method for global identification of regulatory elements in the genome. *Current Protocols in Molecular Biology*. Chapter 21, Unit 21 19 21–14. https://doi.org/10.1002/0471142727.mb2119s91.

Ramos, A. D., Andersen, R. E., Liu, S. J., Nowakowski, T. J., Hong, S. J., Gertz, C., et al. (2015). The long noncoding RNA Pnky regulates neuronal differentiation of embryonic and postnatal neural stem cells. *Cell Stem Cell*, *16*(4), 439–447. https://doi.org/10.1016/j.stem.2015.02.007.

Ransohoff, J. D., Wei, Y., & Khavari, P. A. (2018). The functions and unique features of long intergenic non-coding RNA. *Nature Reviews Molecular Cell Biology*, *19*(3), 143–157. https://doi.org/10.1038/nrm.2017.104.

Rapicavoli, N. A., Poth, E. M., Zhu, H., & Blackshaw, S. (2011). The long noncoding RNA Six3OS acts in trans to regulate retinal development by modulating Six3 activity. *Neural Development*, *6*, 32. https://doi.org/10.1186/1749-8104-6-32.

Rashid, F., Shah, A., & Shan, G. (2016). Long non-coding RNAs in the cytoplasm. *Genomics, Proteomics & Bioinformatics*, *14*(2), 73–80. https://doi.org/10.1016/j.gpb.2016.03.005.

Rinn, J. L., & Chang, H. Y. (2012). Genome regulation by long noncoding RNAs. *Annual Review of Biochemistry*, *81*, 145–166. https://doi.org/10.1146/annurev-biochem-051410-092902.

Rinn, J. L., Euskirchen, G., Bertone, P., Martone, R., Luscombe, N. M., Hartman, S., et al. (2003). The transcriptional activity of human chromosome 22. *Genes & Development*, *17*(4), 529–540. https://doi.org/10.1101/gad.1055203.

Schaukowitch, K., Joo, J. Y., Liu, X., Watts, J. K., Martinez, C., & Kim, T. K. (2014). Enhancer RNA facilitates NELF release from immediate early genes. *Molecular Cell*, *56*(1), 29–42. https://doi.org/10.1016/j.molcel.2014.08.023.

Schmitz, K. M., Mayer, C., Postepska, A., & Grummt, I. (2010). Interaction of noncoding RNA with the rDNA promoter mediates recruitment of DNMT3b and silencing of rRNA genes. *Genes & Development*, *24*(20), 2264–2269. https://doi.org/10.1101/gad.590910.

Selleri, L., Bartolomei, M. S., Bickmore, W. A., He, L., Stubbs, L., Reik, W., et al. (2016). A Hox-embedded Long noncoding RNA: Is it all hot air? *PLoS Genetics*, *12*(12). e1006485https://doi.org/10.1371/journal.pgen.1006485.

Sigova, A. A., Abraham, B. J., Ji, X., Molinie, B., Hannett, N. M., Guo, Y. E., et al. (2015). Transcription factor trapping by RNA in gene regulatory elements. *Science*, *350*(6263), 978–981. https://doi.org/10.1126/science.aad3346.

Song, J. J., & Kingston, R. E. (2008). WDR5 interacts with mixed lineage leukemia (MLL) protein via the histone H3-binding pocket. *The Journal of Biological Chemistry*, *283*(50), 35258–35264. https://doi.org/10.1074/jbc.M806900200.

Song, X., Wang, X., Arai, S., & Kurokawa, R. (2012). Promoter-associated noncoding RNA from the CCND1 promoter. *Methods in Molecular Biology*, *809*, 609–622. https://doi.org/10.1007/978-1-61779-376-9_39.

Stein, G. S., Stein, J. L., van, J. W. A., Lian, J. B., Montecino, M., Medina, R., et al. (2012). The architectural organization of human stem cell cycle regulatory machinery. *Current Pharmaceutical Design*, *18*(13), 1679–1685. https://doi.org/10.2174/138161212799859639.

Sun, S., Del Rosario, B. C., Szanto, A., Ogawa, Y., Jeon, Y., & Lee, J. T. (2013). Jpx RNA activates Xist by evicting CTCF. *Cell*, *153*(7), 1537–1551. https://doi.org/10.1016/j.cell.2013.05.028.

Sun, L., Liu, H., Zhang, L., & Meng, J. (2015). lncRScan-SVM: A tool for predicting long non-coding RNAs using support vector machine. *PLoS One*, *10*(10), e0139654. https://doi.org/10.1371/journal.pone.0139654.

Sun, Z., Zhu, M., Lv, P., Cheng, L., Wang, Q., Tian, P., et al. (2018). The Long noncoding RNA Lncenc1 maintains naive states of mouse ESCs by promoting the glycolysis pathway. *Stem Cell Reports*, *11*(3), 741–755. https://doi.org/10.1016/j.stemcr.2018.08.001.

Sutherland, L., Ruhe, M., Gattegno-Ho, D., Mann, K., Greaves, J., Koscielniak, M., et al. (2018). LIF-dependent survival of embryonic stem cells is regulated by a novel palmitoylated Gab1 signalling protein. *Journal of Cell Science*, *131*(18). https://doi.org/10.1242/jcs.222257.

Terashima, M., Tange, S., Ishimura, A., & Suzuki, T. (2017). MEG3 long noncoding RNA contributes to the epigenetic regulation of epithelial-mesenchymal transition in lung Cancer cell lines. *The Journal of Biological Chemistry*, *292*(1), 82–99. https://doi.org/10.1074/jbc.M116.750950.

Volders, P. J., Helsens, K., Wang, X., Menten, B., Martens, L., Gevaert, K., et al. (2013). LNCipedia: A database for annotated human lncRNA transcript sequences and structures. *Nucleic Acids Research*, *41*(Database issue), D246–D251. https://doi.org/10.1093/nar/gks915.

Vucicevic, D., Corradin, O., Ntini, E., Scacheri, P. C., & Orom, U. A. (2015). Long ncRNA expression associates with tissue-specific enhancers. *Cell Cycle*, *14*(2), 253–260. https://doi.org/10.4161/15384101.2014.977641.

Wang, K. C., & Chang, H. Y. (2011). Molecular mechanisms of long noncoding RNAs. *Molecular Cell*, *43*(6), 904–914. https://doi.org/10.1016/j.molcel.2011.08.018.

Wang, R., & Li, T. (2017). DNA methylation is correlated with pluripotency of stem cells. *Current Stem Cell Research & Therapy*, *12*(6), 442–446. https://doi.org/10.2174/1574888X11666161226145432.

Wang, P., Ning, S., Zhang, Y., Li, R., Ye, J., Zhao, Z., et al. (2015). Identification of lncRNA-associated competing triplets reveals global patterns and prognostic markers for cancer. *Nucleic Acids Research*, *43*(7), 3478–3489. https://doi.org/10.1093/nar/gkv233.

Wang, Y., Xie, Y., Li, L., He, Y., Zheng, D., Yu, P., et al. (2018). EZH2 RIP-seq identifies tissue-specific long non-coding RNAs. *Current Gene Therapy*, *18*(5), 275–285. https://doi.org/10.2174/1566523218666181008125010.

Wang, K. C., Yang, Y. W., Liu, B., Sanyal, A., Corces-Zimmerman, R., Chen, Y., et al. (2011). A long noncoding RNA maintains active chromatin to coordinate homeotic gene expression. *Nature*, *472*(7341), 120–124. https://doi.org/10.1038/nature09819.

Wu, T., Wang, J., Liu, C., Zhang, Y., Shi, B., Zhu, X., et al. (2006). NPInter: The noncoding RNAs and protein related biomacromolecules interaction database. *Nucleic Acids Research*, *34*(Database issue), D150–D152. https://doi.org/10.1093/nar/gkj025.

Xiyuan, L., Dechao, B., Liang, S., Yang, W., Shuangsang, F., Hui, L., et al. (2017). Using the NONCODE database resource. *Current Protocols in Bioinformatics*, 58, 12 16 11–12 16 19. https://doi.org/10.1002/cpbi.25.

Xu, X., Guo, M., Zhang, N., & Ye, S. (2018). Telomeric noncoding RNA promotes mouse embryonic stem cell self-renewal through inhibition of TCF3 activity. *American Journal of Physiology Cell Physiology*, *314*(6), C712–C720. https://doi.org/10.1152/ajpcell.00292.2017.

Xu, C., Zhang, Y., Wang, Q., Xu, Z., Jiang, J., Gao, Y., et al. (2016). Long non-coding RNA GAS5 controls human embryonic stem cell self-renewal by maintaining NODAL signalling. *Nature Communications*, 7, 13287. https://doi.org/10.1038/ncomms13287.

Yamanaka, S., & Blau, H. M. (2010). Nuclear reprogramming to a pluripotent state by three approaches. *Nature*, *465*(7299), 704–712. https://doi.org/10.1038/nature09229.

Yan, P., Luo, S., Lu, J. Y., & Shen, X. (2017). Cis- and trans-acting lncRNAs in pluripotency and reprogramming. *Current Opinion in Genetics & Development*, *46*, 170–178. https://doi.org/10.1016/j.gde.2017.07.009.

Yang, Y. W., Flynn, R. A., Chen, Y., Qu, K., Wan, B., Wang, K. C., et al. (2014). Essential role of lncRNA binding for WDR5 maintenance of active chromatin and embryonic stem cell pluripotency. *eLife*, *3*, e02046. https://doi.org/10.7554/eLife.02046.

Yang, Y., Wen, L., & Zhu, H. (2015). Unveiling the hidden function of long non-coding RNA by identifying its major partner-protein. *Cell & Bioscience*, *5*, 59. https://doi.org/10.1186/s13578-015-0050-x.

Yang, C., Wu, D., Gao, L., Liu, X., Jin, Y., Wang, D., et al. (2016). Competing endogenous RNA networks in human cancer: Hypothesis, validation, and perspectives. *Oncotarget*, 7(12), 13479–13490. https://doi.org/10.18632/oncotarget.7266.

Yang, F., Zhang, H., Mei, Y., & Wu, M. (2014). Reciprocal regulation of HIF-1alpha and lincRNA-p21 modulates the Warburg effect. *Molecular Cell*, *53*(1), 88–100. https://doi.org/10.1016/j.molcel.2013.11.004.

Yoon, J. H., Abdelmohsen, K., Srikantan, S., Yang, X., Martindale, J. L., De, S., et al. (2012). LincRNA-p21 suppresses target mRNA translation. *Molecular Cell*, *47*(4), 648–655. https://doi.org/10.1016/j.molcel.2012.06.027.

Young, R. A. (2011). Control of the embryonic stem cell state. *Cell*, *144*(6), 940–954. https://doi.org/10.1016/j.cell.2011.01.032.

Zhang, B., Arun, G., Mao, Y. S., Lazar, Z., Hung, G., Bhattacharjee, G., et al. (2012). The lncRNA MALAT1 is dispensable for mouse development but its transcription plays a cis-regulatory role in the adult. *Cell Reports*, *2*(1), 111–123. https://doi.org/10.1016/j.celrep.2012.06.003.

Zhang, X., Weissman, S. M., & Newburger, P. E. (2014). Long intergenic non-coding RNA HOTAIRM1 regulates cell cycle progression during myeloid maturation in NB4 human promyelocytic leukemia cells. *RNA Biology*, *11*(6), 777–787. https://doi.org/10.4161/rna.28828.

Zhou, Z., Shen, Y., Khan, M. R., & Li, A. (2015). LncReg: A reference resource for lncRNA-associated regulatory networks. *Database (Oxford)*, *2015*. https://doi.org/10.1093/database/bav083.

Further reading

Boyer, L. A., Lee, T. I., Cole, M. F., Johnstone, S. E., Levine, S. S., Zucker, J. P., et al. (2005). Core transcriptional regulatory circuitry in human embryonic stem cells. *Cell*, *122*(6), 947–956. https://doi.org/10.1016/j.cell.2005.08.020.

Brunner, A. L., Beck, A. H., Edris, B., Sweeney, R. T., Zhu, S. X., Li, R., et al. (2012). Transcriptional profiling of long non-coding RNAs and novel transcribed regions across a diverse panel of archived human cancers. *Genome Biology*, *13*(8), R75. https://doi.org/10.1186/gb-2012-13-8-r75.

de Lara, J. C., Arzate-Mejia, R. G., & Recillas-Targa, F. (2019). Enhancer RNAs: Insights into their biological role. *Epigenet Insights*, *12*. https://doi.org/10.1177/2516865719846093 2516865719846093.

Ding, M., Liu, Y., Liao, X., Zhan, H., Liu, Y., & Huang, W. (2018). Enhancer RNAs (eRNAs): New insights into gene transcription and disease treatment. *Journal of Cancer*, *9*(13), 2334–2340. https://doi.org/10.7150/jca.25829.

Gutschner, T., & Diederichs, S. (2012). The hallmarks of cancer: A long non-coding RNA point of view. *RNA Biology*, *9*(6), 703–719. https://doi.org/10.4161/rna.20481.

Heninger, A. K., & Buchholz, F. (2007). Production of endoribonuclease-prepared short interfering RNAs (esiRNAs) for specific and effective gene silencing in mammalian cells. *CSH Protocols*, *2007*, pdb prot4824. https://doi.org/10.1101/pdb.prot4824pdb.

Keller, G. (2005). Embryonic stem cell differentiation: Emergence of a new era in biology and medicine. *Genes & Development*, *19*(10), 1129–1155. https://doi.org/10.1101/gad.1303605.

Maass, P. G., Rump, A., Schulz, H., Stricker, S., Schulze, L., Platzer, K., et al. (2012). A misplaced lncRNA causes brachydactyly in humans. *The Journal of Clinical Investigation*, *122*(11), 3990–4002. https://doi.org/10.1172/JCI65508.

Navarro, P., Page, D. R., Avner, P., & Rougeulle, C. (2006). Tsix-mediated epigenetic switch of a CTCF-flanked region of the Xist promoter determines the Xist transcription program. *Genes & Development*, *20*(20), 2787–2792. https://doi.org/10.1101/gad.389006.

Odorico, J. S., Kaufman, D. S., & Thomson, J. A. (2001). Multilineage differentiation from human embryonic stem cell lines. *Stem Cells*, *19*(3), 193–204. https://doi.org/10.1634/stemcells.19-3-193.

Salehi, S., Taheri, M. N., Azarpira, N., Zare, A., & Behzad-Behbahani, A. (2017). State of the art technologies to explore long non-coding RNAs in cancer. *Journal of Cellular and Molecular Medicine*, *21*(12), 3120–3140. https://doi.org/10.1111/jcmm.13238.

Schor, I. E., Bussotti, G., Males, M., Forneris, M., Viales, R. R., Enright, A. J., et al. (2018). Non-coding RNA expression, function, and variation during Drosophila embryogenesis. *Current Biology*, *28*(22). 3547–3561.e3549. https://doi.org/10.1016/j.cub.2018.09.026.

Tatsumi, D., Hayashi, Y., Endo, M., Kobayashi, H., Yoshioka, T., Kiso, K., et al. (2018). DNMTs and SETDB1 function as co-repressors in MAX-mediated repression of germ cell-related genes in mouse embryonic stem cells. *PLoS One*, *13*(11), e0205969. https://doi.org/10.1371/journal.pone.0205969.

Tsankov, A. M., Gu, H., Akopian, V., Ziller, M. J., Donaghey, J., Amit, I., et al. (2015). Transcription factor binding dynamics during human ES cell differentiation. *Nature*, *518*(7539), 344–349. https://doi.org/10.1038/nature14233.

Wang, F., Tang, Z., Shao, H., Guo, J., Tan, T., Dong, Y., et al. (2018). Long noncoding RNA HOTTIP cooperates with CCCTC-binding factor to coordinate HOXA gene expression. *Biochemical and Biophysical Research Communications*, *500*(4), 852–859. https://doi.org/10.1016/j.bbrc.2018.04.173.

Xu, Y., Wu, W., Han, Q., Wang, Y., Li, C., Zhang, P., et al. (2019). Post-translational modification control of RNA-binding protein hnRNPK function. *Open Biology*, *9*(3), 180239. https://doi.org/10.1098/rsob.180239.

CHAPTER THREE

Regulation of pluripotency and reprogramming by RNA binding proteins

Dan Li[a,b], Mohamed S. Kishta[c,d,e], Jianlong Wang[a,b,e,*]

[a]Department of Cell, Developmental and Regenerative Biology; The Black Family Stem Cell Institute; Icahn School of Medicine at Mount Sinai, New York, NY, United States

[b]The Graduate School of Biomedical Sciences, Icahn School of Medicine at Mount Sinai, New York, NY, United States

[c]Hormones Department, Medical Research Division, National Research Centre, Cairo, Egypt

[d]Stem Cell Lab., Center of Excellence for Advanced Sciences, National Research Centre, Cairo, Egypt

[e]Department of Medicine, Columbia Center for Human Development, Columbia University Irving Medical Center, New York, NY, United States

[*]Corresponding author: e-mail address: jw3925@cumc.columbia.edu

Contents

1. Pluripotency and reprogramming 114
2. RNA binding proteins 115
   2.1 Epigenetic regulation 119
   2.2 RNA modification 119
   2.3 Alternative splicing 121
   2.4 Alternative polyadenylation 122
   2.5 Nuclear retention and export of RNAs 123
   2.6 Translation 123
   2.7 mRNA stability and degradation 124
3. RNA helicases and DEAD-box helicase family 124
   3.1 DDX3 125
   3.2 DDX5/DDX17 127
   3.3 DDX6 128
   3.4 DDX18 129
   3.5 DDX21 129
   3.6 DDX47 and DDX52 129
4. Conclusions 130
Acknowledgments 130
References 130

Abstract

Embryonic stem cells have the capacities of self-renewal and pluripotency. Pluripotency establishment (somatic cell reprogramming), maintenance, and execution (differentiation) require orchestrated regulatory mechanisms of a cell's molecular machinery,

Current Topics in Developmental Biology, Volume 138
ISSN 0070-2153
https://doi.org/10.1016/bs.ctdb.2020.01.003

including signaling pathways, epigenetics, transcription, translation, and protein degradation. RNA binding proteins (RBPs) take part in every process of RNA regulation and recent studies began to address their important functions in the regulation of pluripotency and reprogramming. Here, we discuss the roles of RBPs in key regulatory steps in the control of pluripotency and reprogramming. Among RNA binding proteins are a group of RNA helicases that are responsible for RNA structure remodeling with important functional implications. We highlight the largest family of RNA helicases, DDX (DEAD-box) helicase family and our current understanding of their functions specifically in the regulation of pluripotency and reprogramming.

1. Pluripotency and reprogramming

Embryonic stem cells (ESCs) are derived from the inner cell mass (ICM) of the blastocyst and they can be maintained indefinitely in culture. ESCs have two main characteristics: self-renewal, the ability of a cell to propagate indefinitely in the same state; and pluripotency, the potential of a single ESC to develop into any cell types of an embryo or an adult animal (Young, 2011).

During the mouse embryo development, at around embryonic day 3.5 (E3.5), the blastomeres compact into a blastocyst and the blastocyst has two different cell populations: the outer layer cells or the trophectoderm which will develop into the extra embryonic tissues; and the ICM which will develop into the primitive endoderm (hypoblast) and primitive ectoderm (epiblast). The primitive endoderm will give rise to the secondary extra embryonic tissues while the primitive ectoderm will produce the three germ layers of the embryo: ectoderm, mesoderm and endoderm (Morris et al., 2010).

At around mouse embryonic day E4.5, the blastocyst implants into the uterus to undergo the further development. ESCs are derived from the ICM of the pre-implantation blastocyst. It is also found that some of the post-implantation epiblast cells are capable of giving rise to all three embryonic germ layers, like ESCs. Based on the definition of pluripotency, these cells would be also considered pluripotent (Young, 2011). However, there are many differences between the cells derived from the ICM of the pre-implantation blastocyst and the cells from the post-implantation epiblast, such as the capacity to contribute to the chimeras and germ line transmission, the signaling to support cell's pluripotency. Besides, human ESCs, which are also derived from the ICM of human pre-implantation embryos (Thomson et al., 1998), display characteristics much closer to the mouse post-implantation epiblast stem cells (EpiSCs), than to the mouse ICM-derived mESCs. This observation suggests that hESCs correspond to a more differentiated developmental stage, or a primed pluripotency state (Brons et al., 2007; Tesar et al., 2007).

Research into molecules that control pluripotency has led to the landmark discovery in 2006 by Yamanaka's group who found a way to convert the mouse fibroblasts into a pluripotent ESC-like state through over expression of four transcription factors *Oct4*, *Sox2*, *Klf4*, and *c-Myc*. These reprogrammed cells are called induced pluripotent stem cells (iPSCs) and are highly similar to ESCs (Takahashi & Yamanaka, 2006). In 2007, Yamanaka's group also reported the reprogramming of human somatic cells to a pluripotent state with the same set of factors (Takahashi et al., 2007).

Human ESCs and iPSCs have tremendous therapeutic and regenerative potentials by providing a precious resource for drug testing, disease modeling, and cell replacement. A better understanding of the molecular regulatory mechanisms underlying pluripotency and reprogramming is a prerequisite for ESCs and iPSCs to be applied in disease therapeutics and regenerative medicine. An interplay of transcription factors, epigenetic factors, and signal transduction pathways are crucially important in the regulation of establishment, maintenance, and execution of pluripotency. While epigenetic and transcriptional regulation of pluripotency and reprogramming has been extensively studied and reviewed (reviewed by Chambers & Tomlinson, 2009; Gökbuget & Blelloch, 2019; Theunissen & Jaenisch, 2017; Yeo & Ng, 2013), post-transcriptional encompassing translational and posttranslational controls are relatively under-explored and are becoming the subjects of an ever increasing number of recent publications in the field of stem cell biology understanding pluripotency and reprogramming (Di Stefano et al., 2019; Freimer, Hu, & Blelloch, 2018; Li et al., 2017a; Yoffe et al., 2016; Zhang et al., 2020).

2. RNA binding proteins

RNA binding proteins (RBPs) are key factors in gene expression regulation by participating in every RNA-involved process, from transcription, RNA maturation, transport, stability, to translation and RNA degradation (reviewed by Guallar & Wang, 2014; Ye & Blelloch, 2014). RBPs are defined as proteins that contain one or multiple well-known RNA-binding domains (RBDs); or less commonly, proteins that reside within the ribonucleoproteins even if they don't directly interact with RNA (Gerstberger, Hafner, & Tuschl, 2014).

RBPs can be classified based on their target RNAs: mRNA-binding, tRNA-binding, pre-rRNA-binding, small nucleolar RNA (snoRNA)-binding, small nuclear RNA (snRNA)-binding, and other non-coding RNA (ncRNA)-binding. Notably, some RBPs can interact with different RNA types, such as the RNA exosome that regulates general RNA turnover. In these cases, researchers usually group the RBPs into their predominant target groups. Also, some RBPs with well-known RBDs are without available RNA target information (Gerstberger et al., 2014). Many published studies are focused on the mRNA-binding proteins (mRBPs), and relatively less is known about other RBPs subclasses such as the ncRNA-binding proteins. In the Online Mendelian Inheritance in Man (OMIM) database that links the known diseases to the relevant genes in the human genome, there are around 150 RBPs listed. In this list, only one-third of the RBPs are mRNA-binding, with the rest mostly targeting diverse ncRNAs (Hamosh, 2004), supporting the significance of studying the latter group.

Owing to the important roles that RBPs play in the gene expression regulation, it is not surprising that RBP families are well conserved across eukaryotes. Previous studies show that there are at least 200 distinct RBPs which are also present in the lowest common animal ancestor (Anantharaman, Koonin, & Aravind, 2002). Gerstberger et al. reported that in human, 50% of the RBP families are conserved in *S. cerevisiae* and even more are conserved in higher eukaryotes. The relative percentage of each RBP subclass based on its RNA targets is also maintained across phylogenies, 38% for mRBPs and 12% for tRNA-binding proteins (Gerstberger et al., 2014). Gerstberger et al. also showed that in human, 98% of paralogous RBP families are ubiquitously expressed across tissues while only 2% of paralogous families have tissue-specific expression patterns (Gerstberger et al., 2014). As evolutionary origin and tissue-specificity of gene expression often correlate with the protein function, highly evolutionary conservation and ubiquitous expression of RBPs support their critical roles in basic cellular functions (Freilich et al., 2005; Ramsköld, Wang, Burge, & Sandberg, 2009; Winter, Goodstadt, & Ponting, 2004).

RBD determines the specificity of binding of a certain RBP to its targets (MacKay, Font, & Segal, 2011). The following RBDs are some of the best characterized domains described in the literature:

RNA-Recognition Motif (RRM): 90–100 amino acids in length, present in up to six copies per protein, the most abundant and the most extensively studied RBD in higher vertebrates (Maris, Dominguez, & Allain, 2005).

The RRM-RNA interaction is specific to single-stranded RNA, with low sequence-specificity. RRM has been shown to be also capable to interact with DNA and proteins.

K-Homology Domain (KH): around 70 amino acids in length. KH can recognize four single-stranded nucleotides with rather weak affinity, the stronger affinity or longer than four nucleotides target can be achieved by synergy in multiple copies (Beuth, Pennell, Arnvig, Martin, & Taylor, 2005).

Double-Stranded RNA-Binding Domain (dsRBD): around 70–75 amino acids, present in up to five copies per protein. The dsRBD recognizes double-stranded RNA in a sequence-independent way. The recognition covers 15 nucleotides with two minor grooves separated by a major groove. The additional functional domains modulate the binding specificity for various RNA shapes (Stefl, Skrisovska, & Allain, 2005).

DEAD-Box Domain: the name of DEAD-box is coming from their characteristic Asp-Glu-Ala-Asp (DEAD) motifs. DEAD-box proteins form the largest helicase family and they utilize ATP to bind or remodel RNA and ribonucleoproteins (Linder & Jankowsky, 2011). A major focus of this review (see more in Section 3).

PUF RNA-Binding Repeats: the PUF (formed by Pumilio and FBF) domain is around 36 amino acids, present six to eight tandem repeats per protein, packed in a curved structure, bind to single-stranded RNAs (Guallar & Wang, 2014).

PAZ Domain: the PAZ (Piwi/Argonaute/Zwille) domain is around 110 amino acids, recognizes the two-base 3′ overhang of dsRNA and ssRNA. The PAZ-domain RBPs function in the post-transcriptional gene silencing (Tian, Simanshu, Ma, & Patel, 2011).

Zinc-Finger Domains (ZnF): Znf domain is a classical DNA-binding domain, but it is also able to interact with RNA (Teplova & Patel, 2008). ZnFs present alone or in multiple copies per protein, they can also work in combination with other RBDs.

Even though above RBDs are well characterized in their association with RNA, many proteins have now been shown to interact with RNA in the absence of known RBDs. It is still necessary to determine whether those candidates directly interact with their RNA targets, and to characterize potential new RBDs. The high number of proteins that have been shown to bind RNA underscores the importance of both RBPs and RNA regulation in cellular function (Guallar & Wang, 2014).

Because RBPs are involved in every process of RNA regulation, from transcriptional to post- transcriptional as well as translational regulation, it is not surprising that they also play important roles in pluripotency and reprogramming (reviewed by Guallar & Wang, 2014; Ye & Blelloch, 2014). Therefore, categorization of RBPs in pluripotent stem cells provides an inroad to understanding their biology in pluripotency and reprogramming. Kwon et al. identified 555 proteins, including 283 novel RBP candidates, to constitute the mESC mRNA interactome. In this interactome, 68 proteins are preferentially expressed in ESCs by comparison to differentiated cells (Kwon et al., 2013). Bao et al. developed an approach to capture the newly transcribed RNA interactome using click chemistry (RICK) and applied it in mESCs. They identified 518 high-confidence proteins, 160 of which are overlapped with Kwon et al.'s interactome and the rest 358 are defined as RICK-exclusive mESC RBPs with RNA binding and polyA-RNA binding capacities. Among these 358 proteins, expression levels of 95 proteins are higher in mESCs than in differentiated cells, suggesting their specific roles in ESC self-renewal and pluripotency (Bao et al., 2018). He et al. performed proteomic identification of RNA-binding regions in mESCs, and identified 803 nuclear RBPs, many of which are well-known transcriptional regulators and chromatin modifiers, such as NANOG and TET2 (He et al., 2016).

Mechanistically, RBPs participate in the regulation of pluripotency and reprogramming in many different regulatory layers (Fig. 1), which are discussed in detail below.

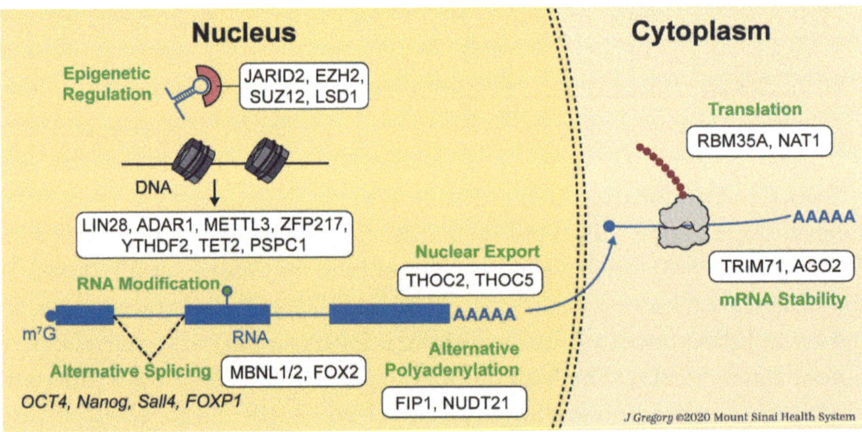

Fig. 1 RBPs participate in multiple regulatory layers to control pluripotency and reprogramming. See Sections 2.1–2.7 for details. *Illustration by Jill K Gregory. Used with permission of ©Mount Sinai Health System.*

2.1 Epigenetic regulation

RBPs can also interact with ncRNAs to control chromatin activation or repression, the epigenetic control that serves as another important regulatory layer in the embryonic development. Examples of such RBPs are JARID2, an Xist-interacting RBP that promotes PRC2 recruitment for X chromosome inactivation in early female development and also during female ESC differentiation *in vitro* (da Rocha et al., 2014; Kaneko et al., 2014); and EZH2 and SUZ12, two catalytic subunits of PRC2 (polycomb repressive complex 2) that interact with lncRNAs (such as *HOTAIR* lncRNA (Brockdorff, 2013)) to function during embryonic development. *HOTAIR* lncRNA is also bound by an epigenetic modifier LSD1, a histone demethylase. LSD1 plays important roles in ESC differentiation through its H3 demethylase activity (Adamo et al., 2011; Whyte et al., 2012). Studies by Tsai et al. showed that *HOTAIR* promoted the bridging between PRC2 and LSD1 to facilitate their cooperation in regulating gene repression (Kaya & Higuchi, 2010). In addition, *HOTAIR* is induced during differentiation and its expression is also required in epithelial-to-mesenchymal transition (EMT) and metastasis in cancer cell lines (Gupta et al., 2010; Pádua Alves et al., 2013).

2.2 RNA modification

Posttranscriptional RNA modification provides a new layer of gene regulation at the RNA level. RNA modifications can be separated into two types: the addition of untemplated nucleotides and the chemical modification of the template nucleotides. One example for the former is the uridylation of *pre-let-7* miRNA: *let-7* is an important miRNA in facilitating ESC differentiation and repressing reprogramming of somatic cells (Melton, Judson, & Blelloch, 2010). Its formation can be regulated through uridylation in two opposite ways: LIN28A can direct 3′ terminal uridylyl transferases (TUTases) ZCCHC11 and ZCCHC6 to add a string of around 11 uridines to the *pre-let-7* miRNA (Hagan, Piskounova, & Gregory, 2009; Heo et al., 2008), then the oligouridylated *pre-let-7* would be targeted and degraded by the DIS3L2 exoribonuclease (Ustianenko et al., 2013); on the contrary, the addition of only one uridine to *pre-let-7* would facilitate the maturation of *let-7* (Heo et al., 2012).

The latter type of RNA modification can take several forms. One major class of RNA modifications is editing by deamination (Bass, 2002). Classical examples are Adenosine-to-inosine, A-to-I, catalyzed by the adenosine deaminase acting on RNA (ADAR) family (Eggington, Greene, & Bass, 2011), and cytidine-to-uridine, C-to-U, catalyzed by the AID-APOBEC enzyme (Powell et al., 1987). A-to-I RNA editing catalyzed by ADAR1 is important in human embryogenesis and ADAR1 is required for hESC differentiation and neural induction (Chen et al., 2015; Shtrichman et al., 2012).

A second class of RNA modifications is methylation of adenosine to form *N*6-methyladenosine (m^6A), the most abundant modification of eukaryotic mRNA which is critical for pluripotency and reprogramming. METTL3, a m^6A methyltransferase, is required for m^6A in mRNAs of ESCs. While ESCs without *Mettl3* can preserve their naïve pluripotent identity, *Mettl3* knockout (KO) naïve ESCs cannot be transferred to primed state, and they lose differentiation competence, staying in a hyper-naïve pluripotency state. Such resistance to differentiation is because during the transition from naïve to primed pluripotent states, m^6A is required to timely destabilize the transcripts of pluripotency factors, which is necessary for proper lineage differentiation (Batista et al., 2014; Geula et al., 2015). In iPSC reprogramming and naïve ESCs, ZFP217 interacts with and sequesters METTL3, inhibiting m^6A deposition on the transcripts of the core stem cell network, such as *Nanog*, *Sox2*, and *c-Myc* (Aguilo et al., 2015). Apart from transcripts of these core pluripotency factors, a recent paper shows that in human pluripotent stem cells, m^6A is also important in the regulation of R-loops, the tripartite nucleic acid structures that are formed during transcription with an RNA:DNA hybrid and a non-hybridized single-stranded DNA. During cell cycles, m^6A-containing R-loops accumulate during G_2/M phases and are drastically depleted during G_0/G_1 phases. An m^6A reader, YTHDF2, interacts with RNA:DNA hybrids. The depletion of YTHDF2 or METTL3 leads to accumulation of RNA:DNA hybrids and increases γH2AX, a marker of DNA double-strand breaks, indicating genome instability (Abakir et al., 2019).

A third class of RNA modifications is oxidation of 5-methylcytidine (5mC) to 5-hydroxymethylcytidine (5hmC) on RNA that can be catalyzed by *Tet* enzymes (Delatte et al., 2016; Fu et al., 2014; Masiello & Biggiogera, 2017; Miao et al., 2016; Zhang, Xiong, Qi, Feng, & Yuan, 2016). In mESCs, TET2 can be recruited to actively transcribed *MERVL* RNAs through its physical association with another RBP PSPC1 and deposit 5hmC modification on *MERVL* RNAs, contributing to *MERVL* destabilization in mESCs (Guallar et al., 2018). Besides TET2, PSPC1 can also recruit HDAC1/2 (histone deacetylase complex) to silence *MERVL* transcriptionally (Guallar et al., 2018). Readers are encouraged to read this review (Frye & Blanco, 2016) to gain additional information on RNA modifications in development and stem cells.

2.3 Alternative splicing

In mouse and human, more than half of the genes can generate different transcripts through alternative splicing (Modrek, 2001). Many pluripotency-associated transcripts, including *OCT4* and *Nanog*, two of the most important pluripotency factors, have different isoforms generated through alternative splicing: in human, *OCT4A* is the key pluripotency transcription factor in ESCs while *OCT4B* expresses in nonpluripotent cells without known functions (Wang & Dai, 2010); in mouse, the three isoforms of *Nanog* contributes with various efficacies to maintaining ESC pluripotency (Atlasi, Mowla, Ziaee, Gokhale, & Andrews, 2008; Das, Jena, & Levasseur, 2011). Another example is *Sall4*, a transcription factor essential for pluripotency. It has two isoforms, *Sall4a* and *Sall4b*, which can form either homodimers or a heterodimer with each other. The genomic binding loci of *Sall4a* and *Sall4b* are overlapped but not identical. *Sall4b* is relatively more important than *Sall4a* in the regulation of pluripotency as *Sall4b*, but not *Sall4a*, can partially rescue the loss-of-function phenotype of both isoforms (Rao et al., 2010). A fourth example is *FOXP1*. It has an ESC-specific isoform that promotes iPSC reprogramming and ESC maintenance by stimulating the expression of pluripotency factors (Gabut et al., 2011).

Apart from the RBPs involved in the core machinery of the spliceosome, specific RBPs are also needed to generate the ESC-specific splicing signature. For example, MBNL1 and MBNL2 are conserved negative regulators of cassette exon alternative splicing events that are differentially controlled among cell types. The alternative splicing event of aforementioned ESC-specific *FOXP1* isoform is inhibited by MBNL1/2 and consistent with such inhibitory control of ESC-specific *FOXP1* isoform, the depletion of *MBNL1/2* enhances iPSC reprogramming (Han et al., 2013). Another splicing regulator example is *FOX2*, which is critical for pluripotency in hESCs as its depletion drives hESCs into differentiation and death. The CLIP-seq (crosslinking immunoprecipitation with high-throughput sequencing) from Yeo et al. showed that FOX2 binding to one intron induced the inclusion of the upstream flanking exon and the exclusion of the downstream flanking exon. In addition, FOX2 also acts as an upstream splicing master regulator because it targets and regulates the alternative splicing of several other splicing regulators, such as *LIN28*, *FOX2* itself and serine/threonine kinases (Yeo et al., 2009). All these highlight the important functions of FOX2 in the splicing program to maintain the hESC pluripotency. For more examples, we direct readers to two related reviews of the subject (Chen, 2015; Cheong & Lufkin, 2011).

2.4 Alternative polyadenylation

Around 70% of the mammalian RNAs are subjected to alternative polyadenylation (APA), leading to different 3′UTR lengths of transcripts (Derti et al., 2012). The various 3′UTR lengthening can affect the stability, localization and translation of transcripts, leading to differential protein expression. Previous studies show that APA is closely related with cell states: somatic cell reprogramming is associated with 3′UTR shortening (Ji & Tian, 2009; Sandberg, Neilson, Sarma, Sharp, & Burge, 2008), whereas embryonic development and exit from pluripotency are accompanied by 3′UTR lengthening (Ji, Lee, Pan, Jiang, & Tian, 2009; Shepard et al., 2011). In transcript cleavage and polyadenylation, cleavage and polyadenylation specificity factor (CPSF) complex recognizes the polyadenylation signal flanking upstream of the cleavage site (Lackford et al., 2014). Lackford et al. demonstrated that *Fip1*, one subunit of CPSF, functioned as an mRNA 3′ processing factor in establishing ESC-specific APA profile. *Fip1* knockdown resulted in partial differentiation in mESCs and inhibited MEF reprogramming. Deep sequencing showed that *Fip1* depletion changed the APA profile of 374 genes with 3′UTR lengthening (Lackford et al., 2014). Further studies are needed to investigate how *Fip1* regulates the 3′UTR length in contributing to pluripotency. Another protein complex, known as cleavage factor Im (CFIm) complex, acts as an activator of transcript cleavage and polyadenylation. *Nudt21* (also called *Cpsf5*), a component of CFIm, regulates cell fates by manipulating alternative polyadenylation. *Nudt21* is a barrier to reprogramming as its depletion dramatically increases iPSC reprogramming efficiency. Its depletion also enhances the transdifferentiation of MEFs to induced trophoblast stem cells but impairs ESC differentiation. Mechanistically, *Nudt21* knockdown facilitates alternative polyadenylation of chromatin regulators, such as *Rybp*, *Chd1*, and *Wdr5*, that play important roles in reprogramming (Ang et al., 2011; Brumbaugh et al., 2018; Gaspar-Maia et al., 2009; Li et al., 2017b). For additional information on APA in stem cell biology, readers are referred to this review (Mueller, Cheung, & Rando, 2013).

2.5 Nuclear retention and export of RNAs

Most RNAs need to be exported from nucleus to cytoplasm to function, so gene expression can also be regulated through controlling the access of RNA to the cytoplasmic machineries (*e.g.*, translation machinery). A study from Wang et al. showed this regulatory level played important roles in ESCs. The THO complex is a conserved complex regulating mRNA export from the nucleus to the cytoplasm. The depletion of two subunits of the THO complex, namely *Thoc2* or *Thoc5*, didn't change the overall transcripts level, however, resulted in the nuclear accumulation of a subset of pluripotency-related transcripts, including *Nanog*, *Esrrb*, *Klf4*, and *Sox2*. The interaction of THOC2 with these pluripotency-related mRNAs is THOC5-dependent. THOC5 is an adaptor protein, which is down-regulated in normal development. The knockdown of *Thoc5* promotes ESC differentiation and inhibits somatic cell reprogramming, while over-expression of *Thoc5* delays the differentiation in ESCs (Wang et al., 2013). This example emphasizes the important role of RNA nuclear export control in pluripotency regulation (Saunders & Wang, 2014).

2.6 Translation

RBPs can also adjust the RNA/ribonucleoprotein structures to control the accessibility of the RNA to ribosomes or the movement of ribosomes along the mRNA to control protein synthesis. At this regulatory level, RBPs often bind to the 5′UTR of RNA and such 5′UTR-RBP interactions have been reported to regulate ESC proliferation and differentiation (Ye & Blelloch, 2014). For example, RBM35A was found to target the 5′UTR of *Sox2* and *Oct4* transcripts to prevent their loading into the polysomes, as demonstrated through RBM35A immunoprecipitation and polysome profiling. And *Rbm35a* depletion blocks ESC differentiation and facilitates somatic cell reprograming through promoting the expression of key pluripotency transcription factors, including *Oct4* and *Sox2* (Fagoonee et al., 2013). Another example is *NAT1* (also known as *EIF4G2*, *DAP5*, and *p97*), which is homologous to the C-terminal of eukaryotic translation initiation factor 4G (*EIF4G1*). In both mESCs and hESCs, its depletion results in the resistance to differentiation induction due partly to the translational block of NAT1-mediated translation of a specific subgroup of proteins that are critical for ESC differentiation (Sugiyama et al., 2017; Yoffe et al., 2016). Translational control in pluripotency and reprogramming is being increasingly recognized (Tahmasebi, Amiri, & Sonenberg, 2019), although much more work needs to be done to further unravel this important regulatory layer.

2.7 mRNA stability and degradation

During quality surveillance of RNA, RBPs can bind to aberrant RNAs and export them in the cytoplasm for degradation (Reed & Hurt, 2002), as well as modify RNA in nucleus for degradation (Houseley, LaCava, & Tollervey, 2006; LaCava et al., 2005). Some RBPs that function in RNA quality control have been shown to be important for pluripotency and reprogramming. For example, TRIM71 can interact with miRNA-containing AGO2 and cooperate with ESC-specific *miR-290* and *miR-302* to target the 3′UTR of *Cdkn1a*, a repressor of the G_1-S transition, inhibiting its activity to promote the cell cycle process for optimal ESC self-renewal (Chang et al., 2012). Loedige et al. also demonstrated the binding of TRIM71 to the 3′UTRs of a subset of prodifferentiation genes, leading to the downregulation of mRNA levels in an AGO2-independent way (Loedige, Gaidatzis, Sack, Meister, & Filipowicz, 2013). In addition, Worringer et al. showed that overexpression of *TRIM71* promoted human somatic cell reprogramming, which was partly due to the post-transcriptional inhibition of the fibroblast-enriched *EGR1* transcripts to which TRIM71 binds and negatively regulates (Worringer et al., 2014).

In sum, RBPs can function in multiple regulatory layers to control pluripotency and reprogramming. Some RBPs can even work multifunctionally by controlling various molecular layers of RNA regulation. One example is coming from the study by Dardenne et al., demonstrating the multiple functions of two RNA helicases DDX5 and DDX17 in various regulatory layers controlling myogenesis and EMT (epithelial-to-mesenchymal transition) (Dardenne et al., 2014). We will further discuss this particular family of RBPs below.

3. RNA helicases and DEAD-box helicase family

RNA is one of the most important biological macromolecules that function in many biological processes. To be functional, RNA must fold into specific secondary or tertiary structures in three dimensions, and many proteins are involved in the RNA folding/remodeling to regulate the physical characteristics of RNA or form ribonucleoprotein complexes for further function (Jarmoskaite & Russell, 2014). For example, in the spliceosome assembly, Sub2 and Prp5, two DEAD-box helicases, are required to promote the rearrangements allowing the recognition base-pairing between the branchpoint and U2 snRNA (Ruby, Chang, & Abelson, 1993).

Helicases are the enzymes responsible for nucleic acids remodeling by using the energy from nucleoside triphosphate binding and hydrolysis (Hardwick & Luisi, 2013). Helicases function in almost every cellular process in which nucleic acids are involved. Until now, at least two mechanisms have been reported: canonical translocation-based duplex unwinding and duplex unwinding by local strand separation, which are employed by some viral RNA helicases and DEAD-box helicases, respectively (Jankowsky, 2011). Helicases are classified into six superfamilies (SFs) and all eukaryotic RNA helicases are found in six families belonging to SFs 1 and 2; the remaining families in eukaryotes are composed of DNA helicases. Some families consist of both RNA and DNA helicases and some helicases work on both DNA and RNA (Fairman-Williams, Guenther, & Jankowsky, 2010; Putnam & Jankowsky, 2013). The DEAD-box (DDX) is the largest family of RNA helicases, belonging to SF2 (Hardwick & Luisi, 2013). The name of DEAD-box reflects their characteristic Asp-Glu-Ala-Asp (DEAD) motifs. This family is present in all eukaryotes and also in many Archaea and bacteria. These highly conserved helicases are involved in virtually every RNA metabolism step, from ribosome biogenesis, to transcription, RNA maturation, microRNA processing, translation, and RNA degradation. DDX proteins generally function as part of large multicomponent complex, like the spliceosome (Linder & Jankowsky, 2011). Because DDXs are widely involved in the RNA metabolism, it is not surprising that some DDX members also play important roles in pluripotency and reprogramming. To date, direct implication of DDX family members in stem cell pluripotency and somatic cell reprogramming came from the studies of following DDX factors.

3.1 DDX3

The expression level of DDX3 is highly enriched in human undifferentiated stem cells compared to differentiated cells. Inhibition of DDX3 reduces cellular proliferation in hESCs but doesn't decrease proliferation of human embryonic fibroblast cells. In hESCs, inhibition of DDX3 also downregulates critical pluripotency markers (OCT4, SOX2, and NANOG) and facilitates differentiation (Kerr, Bol, Vesuna, & Raman, 2019) (Fig. 2A). Interestingly, Cruciat et al. reported that DDX3 could bind

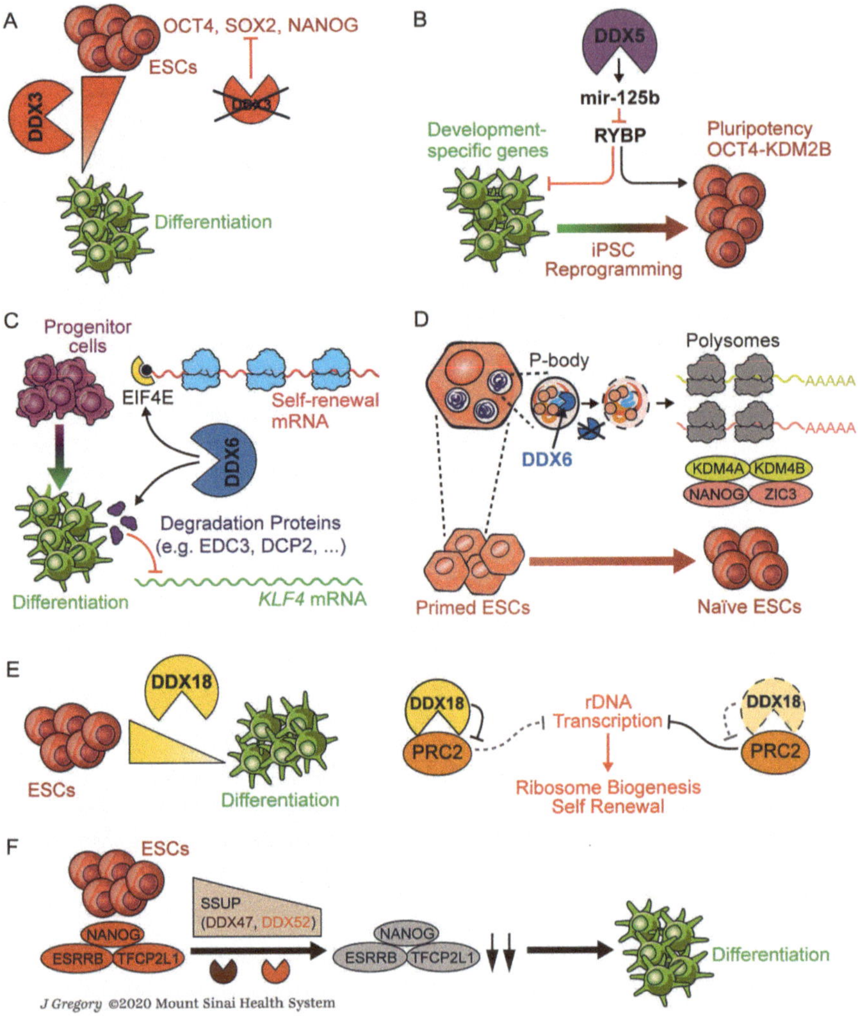

Fig. 2 DDX proteins that are reported to be important in pluripotency and reprogramming. (A) DDX3 expression is highly enriched in human undifferentiated stem cells and decreases during differentiation. The inhibition of DDX3 in hESCs downregulates core pluripotency factors. (B) DDX5 inhibits iPSC reprogramming. Depletion of *Ddx5* downregulates miRNA-125b, leading to the increase of RYBP. RYBP upregulation inhibits development-specific gene expression, and facilitates pluripotency through the activation of OCT4-KDM2B network. (C) DDX6 is necessary for the maintenance of adult progenitor cell functions, through facilitating the translation of proliferation and self-renewal transcripts, and the degradation of differentiation-inducing *KLF4* mRNA. (D) Depletion of *DDX6* promotes the reprogramming of primed hESCs to a naïve state. DDX6 interacts with P-body proteins and suppresses the target transcripts *(Continued)*

and activate the casein kinase 1 isoform epsilon (CK1ε), which leads to the Wnt-dependent phosphorylation of Disheveled, enabling β-catenin's function in activating its target genes (Cruciat et al., 2013). As Wnt signaling plays important roles in both pluripotency and reprogramming, it remains to be determined whether DDX3's regulatory role as the subunit of CK1ε defined by Cruciat et al. is part of the mechanism.

3.2 DDX5/DDX17

A notable example for DDX's multi-functionality in the stem cell field comes from the study of DDX5 and DDX17 paralogs (Dardenne et al., 2014). Dardenne et al. show that DDX5 and DDX17 can cooperate with hnRNP (heterogeneous nuclear ribonucleoprotein) H/F splicing factors to express an epithelial- and myoblast-specific splicing subprogram. Also, DDX5 and DDX17 serve as transcriptional coregulators of key differentiation transcription factors to drive the transcription programs specific to the myogenesis and EMT (epithelial-to-mesenchymal transition), which in turn can produce differentiation-specific miRNAs resulting in the down-regulation of DDX5 and DDX17. Another example is the association of DDX3 and DDX5 with the Wnt/β-catenin signaling pathway (Cruciat et al., 2013; Yang, Lin, & Liu, 2006). This pathway is very important in the embryonic development and is directly linked to the pluripotency core transcription factors, playing essential roles in pluripotency and self-renewal regulation (Kim & Kimmel, 2006). In Yang et al. (2006) showed that the stimulated DDX5 could displace the inhibitor Axin from β-catenin, facilitating the transfer of β-catenin to the nucleus to activate the target gene expression instead of being phosphorylated and degraded.

Fig. 2—cont'd in P-bodies. Depletion of *DDX6* leads to dissolvement of P-bodies, releasing target mRNAs that encode key cell fate transcription regulators and chromatin factors. The released mRNAs are translated to promote pluripotency and reprogramming. (E) *Ddx18* expression is highly enriched in mESCs and decreases along differentiation. DDX18 interacts with PRC2, preventing it from accessing and marking rDNA with H3K27me3. DDX18 downregulation leads to inhibition of rDNA transcription and reduced ribosomal protein level as well as global translation level. (F) DDX47 and DDX52, as the components of SSUP, are highly expressed in ESCs and important for maintaining the protein levels of pluripotency factors. Downregulation of DDX47 or DDX52 leads degradation of pluripotency factors and consequent differentiation of mESCs. The references for each group are listed in the main text. *Illustration by Jill K Gregory. Used with permission of ©Mount Sinai Health System.*

DDX5 also inhibits iPSC reprogramming. The depletion of *Ddx5* results in the dysregulation of dozens of miRNAs, including downregulation of *microRNA-125b*, which inhibits the expression of non-canonical polycomb complex 1 (PRC1) subunit *Rybp*. RYBP upregulation upon *Ddx5* depletion not only facilitates the deposition of inhibitory H2AK119ub1 at lineage-specific genes through PRC1 but also activates the OCT4-KDM2B network to enhance pluripotency-associated gene expression independently of PRC1 (Fig. 2B) (Li et al., 2017b).

3.3 DDX6

DDX6 has been shown to be necessary for the maintenance of adult progenitor cell functions (Wang, Arribas-Layton, Chen, Lykke-Andersen, & Sen, 2015). On one hand, to maintain self-renewal, DDX6 facilitates the translation of proliferation and self-renewal transcripts by recruiting them to translation initiation factor EIF4E; on the other hand, to prevent differentiation of progenitor cells, through association with mRNA degradation proteins, DDX6 targets and destabilizes the mRNA of *KLF4*, a differentiation-inducing transcription factor that is required for the activation of the epidermal differentiation and the conversion of fibroblasts to keratinocyte-like cells (Fig. 2C) (Chen, Mistry, & Sen, 2014; Mistry, Chen, Wang, Zhang, & Sen, 2014; Segre, Bauer, & Fuchs, 1999; Wang et al., 2015). In mESCs, *Ddx6* is required to maintain normal mESC cell morphology and proliferation. The loss of *Ddx6* produces a similar downstream consequence as the depletion of *Dgcr8*, which is essential for miRNA biogenesis. Instead of miRNA-induced RNA degradation, *Ddx6* is important in miRNA-induced translational repression in mESCs (Freimer et al., 2018). Recently, Di Stefano et al. showed that DDX6 is an important regulator of pluripotency in both human and mouse ESCs as its depletion leads ESCs to a differentiation-resistant state. Suppression of *DDX6* also promotes the reprogramming of primed hESCs to a naïve state. DDX6 was also found to regulate the differentiation potential of adult somatic progenitors in a context-dependent manner. Mechanistically, DDX6 is associated with critical P-body proteins and mediates the translational suppression of the target transcripts in P-bodies. DDX6 loss results in dissolvement of P-bodies, which releases mRNAs encoding key cell fate transcription regulators and chromatin factors to the translational machinery in promoting pluripotency and reprogramming (Fig. 2D) (Di Stefano et al., 2019).

3.4 DDX18

Zuo et al. (2009) constructed a protein interaction network encompassing hESC-enriched proteins in hESCs and found that DDX18 is among the top 5% highly connected nodes, suggesting that DDX18 may have important functions in hESCs. Very recently, *Ddx18* was reported to be required for mESC maintenance and embryonic development. DDX18 directly interacts with PRC2 and modulates the formation of PRC2 complex. Such interaction prevents PRC2 from accessing and marking ribosomal DNA (rDNA) with repressive H3K27me3. rRNA (ribosomal RNA, the product of rDNA transcription) is highly expressed in ESCs and becomes downregulated upon differentiation. *Ddx18* depletion increases PRC2 occupancy at rDNA loci to inhibit rDNA transcription, leading to reduced ribosomal protein level and global translation level (Fig. 2E) (Zhang et al., 2020). Owing to the alternative pluripotent states between mouse and human ESCs, it remains to be addressed whether the human ortholog DDX18 may play a similar or distinct role in regulating human pluripotency and reprogramming.

3.5 DDX21

RNA helicase DDX21 functions in multiple steps of ribosome biogenesis by coordinating transcription and rRNA processing. DDX21 was found associated with actively transcribed ribosomal genes as well as rRNAs and snoRNAs, facilitating rRNA modification (Calo et al., 2015). The multifaceted function of DDX21 in ribosome biogenesis suggests its potential role in ESCs as both ribosomal genes and rRNAs are highly expressed in ESCs and properly downregulated/repressed during early differentiation (Ingolia, Lareau, & Weissman, 2011; Savić et al., 2014; Woolnough, Atwood, Liu, Zhao, & Giles, 2016), although a definite functional contribution to pluripotency and reprogramming has yet to be tested.

3.6 DDX47 and DDX52

DDX47 and DDX52 are subunits of small subunit processome (SSUP), which mediates 18S rRNA biogenesis (Phipps, Charette, & Baserga, 2011; Tafforeau et al., 2013). The components of SSUP are highly expressed in stem cells and important to maintain the protein levels of pluripotency factors. As SSUP subunits, both DDX47 and DDX52 are validated to be necessary for ESC maintenance and efficient iPSC reprogramming: (1) depletion of either of them in mESCs induced differentiation; (2) they help to sustain the protein levels of labile pluripotency factors NANOG and OCT4 in mESCs; and (3) Both are required for efficient reprogramming of iPSCs (Fig. 2F) (You, Park, & Kim, 2015).

4. Conclusions

Understanding molecular mechanisms underlying pluripotency and reprograming is highly significant both scientifically and clinically. The post-transcriptional regulation by RBPs constitutes an important regulatory layer for controlling pluripotency and reprogramming. Although RBPs have been studied widely because of their involvement in a broad range of cellular processes, their regulatory functions in stem cell field are only just beginning to be appreciated. As post-transcriptional regulation enables cells to quickly respond by adjusting protein abundance, future studies are warranted to dissect mechanistic actions of RBPs during cell fate transitions and further our understanding of their roles in pluripotency and reprogramming. The potential multifaceted functions of RBPs on both RNA and DNA targets at transcriptome/epitranscriptome and genome/epigenome levels should be more carefully examined in light of their dual DNA/RNA binding capacities. Finally, as many RBPs contain intrinsically disordered regions, the roles of RBPs in the regulation of phase separations and gene expression, which are only recently recognized (A & Weber, 2019; Shorter, 2019; Xiao et al., 2019; Youn et al., 2019), await more future investigations at both physiological and pathological conditions.

Together, RBP studies would provide a platform and new framework for better understanding of molecular mechanisms underlying pluripotency and reprogramming, which would bring us closer to the practical applications of pluripotent ESCs/iPSCs for regenerative medicine, tissue engineering, and disease therapeutics.

Acknowledgments

We are grateful to Jill Gregory for the illustrations of Figs. 1 and 2 and the members of Wang lab for helpful discussions. We thank the following funding sources for support: National Institutes of Health (GM129157, HD095938, HD097268, and HL146664) and New York State Stem Cell Science (NYSTEM#C32583GG and C32569GG).

References

A, & Weber. (2019). Evidence for and against liquid-liquid phase separation in the nucleus. *Non-Coding RNA*, *5*(4), 50. https://doi.org/10.3390/ncrna5040050.

Abakir, A., Giles, T. C., Cristini, A., Foster, J. M., Dai, N., Starczak, M., et al. (2019). N 6-methyladenosine regulates the stability of RNA:DNA hybrids in human cells. *Nature Genetics*, *52*, 48–55. https://doi.org/10.1038/s41588-019-0549-x.

Adamo, A., Sesé, B., Boue, S., Castaño, J., Paramonov, I., Barrero, M. J., et al. (2011). LSD1 regulates the balance between self-renewal and differentiation in human embryonic stem cells. *Nature Cell Biology*, *13*(6), 652–661. https://doi.org/10.1038/ncb2246.

Aguilo, F., Zhang, F., Sancho, A., Fidalgo, M., Di Cecilia, S., Vashisht, A., et al. (2015). Coordination of m 6 a mRNA methylation and gene transcription by ZFP217 regulates pluripotency and reprogramming. *Cell Stem Cell*, *17*(6), 689–704. https://doi.org/10.1016/j.stem.2015.09.005.

Anantharaman, V., Koonin, E. V., & Aravind, L. (2002). Comparative genomics and evolution of proteins involved in RNA metabolism. in survey and summary. *Nucleic Acids Research*, *30*(7), 1427–1464. https://doi.org/10.1093/nar/30.7.1427.

Ang, Y. S., Tsai, S. Y., Lee, D. F., Monk, J., Su, J., Ratnakumar, K., et al. (2011). Wdr5 mediates self-renewal and reprogramming via the embryonic stem cell core transcriptional network. *Cell*, *145*(2), 183–197. https://doi.org/10.1016/j.cell.2011.03.003.

Atlasi, Y., Mowla, S. J., Ziaee, S. A. M., Gokhale, P. J., & Andrews, P. W. (2008). OCT4 spliced variants are differentially expressed in human pluripotent and nonpluripotent cells. *Stem Cells*, *26*(12), 3068–3074. https://doi.org/10.1634/stemcells.2008-0530.

Bao, X., Guo, X., Yin, M., Tariq, M., Lai, Y., Kanwal, S., et al. (2018). Capturing the interactome of newly transcribed RNA. *Nature Methods*, *15*(3), 213–220. https://doi.org/10.1038/nmeth.4595.

Bass, B. L. (2002). RNA editing by adenosine deaminases that act on RNA. *Annual Review of Biochemistry*, *71*(1), 817–846. https://doi.org/10.1146/annurev.biochem.71.110601.135501.

Batista, P. J., Molinie, B., Wang, J., Qu, K., Zhang, J., Li, L., et al. (2014). m6A RNA modification controls cell fate transition in mammalian embryonic stem cells. *Cell Stem Cell*, *15*(6), 707–719. https://doi.org/10.1016/J.STEM.2014.09.019.

Beuth, B., Pennell, S., Arnvig, K. B., Martin, S. R., & Taylor, I. A. (2005). Structure of a mycobacterium tuberculosis NusA–RNA complex. *The EMBO Journal*, *24*(20), 3576–3587. https://doi.org/10.1038/sj.emboj.7600829.

Brockdorff, N. (2013). Noncoding RNA and Polycomb recruitment. *RNA*, *19*, 429–442. https://doi.org/10.1261/rna.037598.112.

Brons, I. G. M., Smithers, L. E., Trotter, M. W. B., Rugg-Gunn, P., Sun, B., Chuva De Sousa Lopes, S. M., et al. (2007). Derivation of pluripotent epiblast stem cells from mammalian embryos. *Nature*, *448*(7150), 191–195. https://doi.org/10.1038/nature05950.

Brumbaugh, J., Di Stefano, B., Wang, X., Borkent, M., Forouzmand, E., Clowers, K. J., et al. (2018). Nudt21 controls cell fate by connecting alternative polyadenylation to chromatin signaling. *Cell*, *172*(1-2), 106–120.e21. https://doi.org/10.1016/j.cell.2017.11.023.

Calo, E., Flynn, R. A., Martin, L., Spitale, R. C., Chang, H. Y., & Wysocka, J. (2015). RNA helicase DDX21 coordinates transcription and ribosomal RNA processing. *Nature*, *518*(7538), 249–253. https://doi.org/10.1038/nature13923.

Chambers, I., & Tomlinson, S. R. (2009, July 15). The transcriptional foundation of pluripotency. *Development*, *136*, 2311–2322. https://doi.org/10.1242/dev.024398.

Chang, H. M., Martinez, N. J., Thornton, J. E., Hagan, J. P., Nguyen, K. D., & Gregory, R. I. (2012). Trim71 cooperates with microRNAs to repress Cdkn1a expression and promote embryonic stem cell proliferation. *Nature Communications*, *3*, 923. https://doi.org/10.1038/ncomms1909.

Chen, K. (2015). Alternative splicing: An important mechanism in stem cell biology. *World Journal of Stem Cells*, 7(1), 1. https://doi.org/10.4252/wjsc.v7.i1.1.

Chen, Y., Mistry, D. S., & Sen, G. L. (2014). Highly rapid and efficient conversion of human fibroblasts to keratinocyte-like cells. *Journal of Investigative Dermatology*, *134*(2), 335–344. https://doi.org/10.1038/jid.2013.327.

Chen, T., Xiang, J.-F., Zhu, S., Chen, S., Yin, Q.-F., Zhang, X.-O., et al. (2015). ADAR1 is required for differentiation and neural induction by regulating microRNA processing in a catalytically independent manner. *Cell Research*, *25*, 459–476. https://doi.org/10.1038/cr.2015.24.

Cheong, C. Y., & Lufkin, T. (2011). Alternative splicing in self-renewal of embryonic stem cells. *Stem Cells International, 2011*, 560261, https://doi.org/10.4061/2011/560261.

Cruciat, C. M., Dolde, C., De Groot, R. E. A., Ohkawara, B., Reinhard, C., Korswagen, H. C., et al. (2013). RNA helicase DDX3 is a regulatory subunit of casein kinase 1 in Wnt-β-catenin signaling. *Science, 339*(6126), 1436–1441. https://doi.org/10.1126/science.1231499.

da Rocha, S. T., Boeva, V., Escamilla-Del-Arenal, M., Ancelin, K., Granier, C., Matias, N. R., et al. (2014). Jarid2 Is implicated in the initial Xist-induced targeting of PRC2 to the inactive X chromosome. *Molecular Cell, 53*(2), 301–316. https://doi.org/10.1016/j.molcel.2014.01.002.

Dardenne, E., PolayEspinoza, M., Fattet, L., Germann, S., Lambert, M. P., Neil, H., et al. (2014). RNA helicases DDX5 and DDX17 dynamically orchestrate transcription, miRNA, and splicing programs in cell differentiation. *Cell Reports*, 7(6), 1900–1913. https://doi.org/10.1016/j.celrep.2014.05.010.

Das, S., Jena, S., & Levasseur, D. N. (2011). Alternative splicing produces nanog protein variants with different capacities for self-renewal and pluripotency in embryonic stem cells. *Journal of Biological Chemistry, 286*(49), 42690–42703. https://doi.org/10.1074/jbc.M111.290189.

Delatte, B., Wang, F., Ngoc, L. V., Collignon, E., Bonvin, E., Deplus, R., et al. (2016). Transcriptome-wide distribution and function of RNA hydroxymethylcytosine. *Science, 351*(6270), 282–285. https://doi.org/10.1126/science.aac5253.

Derti, A., Garrett-Engele, P., MacIsaac, K. D., Stevens, R. C., Sriram, S., Chen, R., et al. (2012). A quantitative atlas of polyadenylation in five mammals. *Genome Research, 22*(6), 1173–1183. https://doi.org/10.1101/gr.132563.111.

Di Stefano, B., Luo, E. C., Haggerty, C., Aigner, S., Charlton, J., Brumbaugh, J., et al. (2019). The RNA helicase DDX6 controls cellular plasticity by modulating P-body homeostasis. *Cell Stem Cell, 25*(5), 622–638.e13. https://doi.org/10.1016/j.stem.2019.08.018.

Eggington, J. M., Greene, T., & Bass, B. L. (2011). Predicting sites of ADAR editing in double-stranded RNA. *Nature Communications, 2*, 319. https://doi.org/10.1038/ncomms1324.

Fagoonee, S., Bearzi, C., Di Cunto, F., Clohessy, J. G., Rizzi, R., Reschke, M., et al. (2013). The RNA binding protein ESRP1 fine-tunes the expression of pluripotency-related factors in mouse embryonic stem cells. *PLoS One, 8*(8), e72300. https://doi.org/10.1371/journal.pone.0072300.

Fairman-Williams, M. E., Guenther, U. P., & Jankowsky, E. (2010). SF1 and SF2 helicases: Family matters. *Current Opinion in Structural Biology, 20*(3), 313–324. https://doi.org/10.1016/j.sbi.2010.03.011.

Freilich, S., Massingham, T., Bhattacharyya, S., Ponsting, H., Lyons, P. A., Freeman, T. C., et al. (2005). Open access relationship between the tissue-specificity of mouse gene expression and the evolutionary origin and function of the proteins. *Genome Biology, 6*(7), 56. https://doi.org/10.1186/gb-2005-6-7-r56.

Freimer, J. W., Hu, T. J., & Blelloch, R. (2018). Decoupling the impact of microRNAs on translational repression versus RNA degradation in embryonic stem cells. *eLife* 7, e38014. https://doi.org/10.7554/eLife.38014.

Frye, M., & Blanco, S. (2016). Post-transcriptional modifications in development and stem cells. *Development (Cambridge, England), 143*(21), 3871–3881. https://doi.org/10.1242/dev.136556.

Fu, L., Guerrero, C. R., Zhong, N., Amato, N. J., Liu, Y., Liu, S., et al. (2014). Tet-mediated formation of 5-hydroxymethylcytosine in RNA. *Journal of the American Chemical Society, 136*(33), 11582–11585. https://doi.org/10.1021/ja505305z.

Gabut, M., Samavarchi-Tehrani, P., Wang, X., Slobodeniuc, V., O'Hanlon, D., Sung, H. K., et al. (2011). An alternative splicing switch regulates embryonic stem cell pluripotency and reprogramming. *Cell, 147*(1), 132–146. https://doi.org/10.1016/j.cell.2011.08.023.

Gaspar-Maia, A., Alajem, A., Polesso, F., Sridharan, R., Mason, M. J., Heidersbach, A., et al. (2009). Chd1 regulates open chromatin and pluripotency of embryonic stem cells. *Nature, 460*(7257), 863–868. https://doi.org/10.1038/nature08212.

Gerstberger, S., Hafner, M., & Tuschl, T. (2014). A census of human RNA-binding proteins. *Nature Reviews Genetics, 15*(12), 829–845. https://doi.org/10.1038/nrg3813.

Geula, S., Moshitch-Moshkovitz, S., Dominissini, D., Mansour, A. A., Kol, N., Salmon-Divon, M., et al. (2015). m ^{6} A mRNA methylation facilitates resolution of naïve pluripotency toward differentiation. *Science, 347*(6225), 1002–1006. https://doi.org/10.1126/science.1261417.

Gökbuget, D., & Blelloch, R. (2019). Epigenetic control of transcriptional regulation in pluripotency and early differentiation. *Development (Cambridge), 146*, 1–16. https://doi.org/10.1242/dev.164772.

Guallar, D., Bi, X., Pardavila, J. A., Huang, X., Saenz, C., Shi, X., et al. (2018). RNA-dependent chromatin targeting of TET2 for endogenous retrovirus control in pluripotent stem cells. *Nature Genetics, 50*(3), 443–451. https://doi.org/10.1038/s41588-018-0060-9.

Guallar, D., & Wang, J. (2014). RNA-binding proteins in pluripotency, differentiation, and reprogramming. *Frontiers in Biology, 9*(5), 389–409. https://doi.org/10.1007/s11515-014-1326-y.

Gupta, R. A., Shah, N., Wang, K. C., Kim, J., Horlings, H. M., Wong, D. J., et al. (2010). Long non-coding RNA HOTAIR reprograms chromatin state to promote cancer metastasis. *Nature, 464*(7291), 1071–1076. https://doi.org/10.1038/nature08975.

Hagan, J. P., Piskounova, E., & Gregory, R. I. (2009). Lin28 recruits the TUTase Zcchc11 to inhibit let-7 maturation in mouse embryonic stem cells. *Nature Structural and Molecular Biology, 16*(10), 1021–1025. https://doi.org/10.1038/nsmb.1676.

Hamosh, A. (2004). Online Mendelian inheritance in man (OMIM), a knowledgebase of human genes and genetic disorders. *Nucleic Acids Research, 33*(Database issue), D514–D517. https://doi.org/10.1093/nar/gki033.

Han, H., Irimia, M., Ross, P. J., Sung, H. K., Alipanahi, B., David, L., et al. (2013). MBNL proteins repress ES-cell-specific alternative splicing and reprogramming. *Nature, 498*(7453), 241–245. https://doi.org/10.1038/nature12270.

Hardwick, S. W., & Luisi, B. F. (2013). Rarely at rest: RNA helicases and their busy contributions to RNA degradation, regulation and quality control. *RNA Biology, 10*, 56–70. https://doi.org/10.4161/rna.22270.

He, C., Sidoli, S., Warneford-Thomson, R., Tatomer, D. C., Wilusz, J. E., Garcia, B. A., et al. (2016). High-resolution mapping of RNA-binding regions in the nuclear proteome of embryonic stem cells. *Molecular Cell, 64*(2), 416–430. https://doi.org/10.1016/j.molcel.2016.09.034.

Heo, I., Ha, M., Lim, J., Yoon, M. J., Park, J. E., Kwon, S. C., et al. (2012). Mono-uridylation of pre-microRNA as a key step in the biogenesis of group II let-7 microRNAs. *Cell, 151*(3), 521–532. https://doi.org/10.1016/j.cell.2012.09.022.

Heo, I., Joo, C., Cho, J., Ha, M., Han, J., & Kim, V. N. (2008). Lin28 mediates the terminal Uridylation of let-7 precursor MicroRNA. *Molecular Cell, 32*(2), 276–284. https://doi.org/10.1016/j.molcel.2008.09.014.

Houseley, J., LaCava, J., & Tollervey, D. (2006, July 17). RNA-quality control by the exosome. *Nature Reviews Molecular Cell Biology*, 7, 529–539. https://doi.org/10.1038/nrm1964.

Ingolia, N. T., Lareau, L. F., & Weissman, J. S. (2011). Ribosome profiling of mouse embryonic stem cells reveals the complexity and dynamics of mammalian proteomes. *Cell, 147*(4), 789–802. https://doi.org/10.1016/J.CELL.2011.10.002.

Jankowsky, E. (2011). RNA helicases at work: Binding and rearranging. *Trends in Biochemical Sciences, 36*(1), 19–29. https://doi.org/10.1016/j.tibs.2010.07.008.

Jarmoskaite, I., & Russell, R. (2014). RNA helicase proteins as chaperones and remodelers. *Annual Review of Biochemistry, 83*(1), 697–725. https://doi.org/10.1146/annurev-biochem-060713-035546.

Ji, Z., Lee, J. Y., Pan, Z., Jiang, B., & Tian, B. (2009). Progressive lengthening of 3′ untranslated regions of mRNAs by alternative polyadenylation during mouse embryonic development. *Proceedings of the National Academy of Sciences of the United States of America, 106*(17), 7028–7033. https://doi.org/10.1073/pnas.0900028106.

Ji, Z., & Tian, B. (2009). Reprogramming of 3′ untranslated regions of mRNAs by alternative polyadenylation in generation of pluripotent stem cells from different cell types. *PLoS One, 4*(12), e8419. https://doi.org/10.1371/journal.pone.0008419.

Kaneko, S., Bonasio, R., Saldaña-Meyer, R., Yoshida, T., Son, J., Nishino, K., et al. (2014). Interactions between JARID2 and noncoding RNAs regulate PRC2 recruitment to chromatin. *Molecular Cell, 53*(2), 290–300. https://doi.org/10.1016/j.molcel.2013.11.012.

Kaya, M., & Higuchi, H. (2010). Nonlinear elasticity and an 8-nm working stroke of single myosin molecules in myofilaments. *Science, 329*(5992), 686–689. https://doi.org/10.1126/science.1191484.

Kerr, C. L., Bol, G. M., Vesuna, F., & Raman, V. (2019). Targeting RNA helicase DDX3 in stem cell maintenance and teratoma formation. www.Genes&Cancer.com, *Genes & Cancer, 10*(1–2), 11–20. https://doi.org/10.18632/genesandcancer.187.

Kim, L., & Kimmel, A. (2006). GSK3 at the edge: Regulation of developmental specification and cell polarization. *Current Drug Targets,* 7(11), 1411–1419. https://doi.org/10.2174/1389450110607011411.

Kwon, S. C., Yi, H., Eichelbaum, K., Föhr, S., Fischer, B., You, K. T., et al. (2013). The RNA-binding protein repertoire of embryonic stem cells. *Nature Structural and Molecular Biology, 20*(9), 1122–1130. https://doi.org/10.1038/nsmb.2638.

LaCava, J., Houseley, J., Saveanu, C., Petfalski, E., Thompson, E., Jacquier, A., et al. (2005). RNA degradation by the exosome is promoted by a nuclear polyadenylation complex. *Cell, 121*(5), 713–724. https://doi.org/10.1016/j.cell.2005.04.029.

Lackford, B., Yao, C., Charles, G. M., Weng, L., Zheng, X., Choi, E.-A., et al. (2014). Fip1 regulates mRNA alternative polyadenylation to promote stem cell self-renewal. *The EMBO Journal, 33*(8), 878–889. https://doi.org/10.1002/embj.201386537.

Li, H., Lai, P., Jia, J., Song, Y., Xia, Q., Huang, K., et al. (2017a). Erratum: RNA helicase DDX5 inhibits reprogramming to pluripotency by miRNA-based repression of RYBP and its PRC1-dependent and -independent functions (Cell Stem Cell (2017) 20(4) (462–477) (S1934590916304581) (10.1016/j.stem.2016.12.002)). *Cell Stem Cell, 20*(4), 571. https://doi.org/10.1016/j.stem.2017.03.014.

Li, H., Lai, P., Jia, J., Song, Y., Xia, Q., Huang, K., et al. (2017b). RNA helicase DDX5 inhibits reprogramming to pluripotency by miRNA-based repression of RYBP and its PRC1-dependent and -independent functions. *Cell Stem Cell, 20*(4). https://doi.org/10.1016/j.stem.2016.12.002, 462–477.e6.

Linder, P., & Jankowsky, E. (2011). From unwinding to clamping—The DEAD box RNA helicase family. *Nature Reviews Molecular Cell Biology, 12*, 505–516. https://doi.org/10.1038/nrm3154.

Loedige, I., Gaidatzis, D., Sack, R., Meister, G., & Filipowicz, W. (2013). The mammalian TRIM-NHL protein TRIM71/LIN-41 is a repressor of mRNA function. *Nucleic Acids Research, 41*(1), 518–532. https://doi.org/10.1093/nar/gks1032.

MacKay, J. P., Font, J., & Segal, D. J. (2011, March). The prospects for designer single-stranded RNA-binding proteins. *Nature Structural and Molecular Biology, 18*, 256–261. https://doi.org/10.1038/nsmb.2005.

Maris, C., Dominguez, C., & Allain, F. H.-T. (2005). The RNA recognition motif, a plastic RNA-binding platform to regulate post-transcriptional gene expression. *FEBS Journal, 272*(9), 2118–2131. https://doi.org/10.1111/j.1742-4658.2005.04653.x.

Masiello, I., & Biggiogera, M. (2017). Ultrastructural localization of 5-methylcytosine on DNA and RNA. *Cellular and Molecular Life Sciences*, *74*(16), 3057–3064. https://doi.org/10.1007/s00018-017-2521-1.

Melton, C., Judson, R. L., & Blelloch, R. (2010). Opposing microRNA families regulate self-renewal in mouse embryonic stem cells. *Nature*, *463*(7281), 621–626. https://doi.org/10.1038/nature08725.

Miao, Z., Xin, N., Wei, B., Hua, X., Zhang, G., Leng, C., et al. (2016). 5-hydroxymethylcytosine is detected in RNA from mouse brain tissues. *Brain Research*, *1642*, 546–552. https://doi.org/10.1016/j.brainres.2016.04.055.

Mistry, D. S., Chen, Y., Wang, Y., Zhang, K., & Sen, G. L. (2014). SNAI2 controls the undifferentiated state of human epidermal progenitor cells. *Stem Cells*, *32*(12), 3209–3218. https://doi.org/10.1002/stem.1809.

Modrek, B. (2001). Genome-wide detection of alternative splicing in expressed sequences of human genes. *Nucleic Acids Research*, *29*(13), 2850–2859. https://doi.org/10.1093/nar/29.13.2850.

Morris, S. A., Teo, R. T. Y., Li, H., Robson, P., Glover, D. M., & Zernicka-Goetz, M. (2010). Origin and formation of the first two distinct cell types of the inner cell mass in the mouse embryo. *Proceedings of the National Academy of Sciences of the United States of America*, *107*(14), 6364–6369. https://doi.org/10.1073/pnas.0915063107.

Mueller, A. A., Cheung, T. H., & Rando, T. A. (2013). All's well that ends well: Alternative polyadenylation and its implications for stem cell biology. *Current Opinion in Cell Biology*, *25*, 222–232. https://doi.org/10.1016/j.ceb.2012.12.008.

Pádua Alves, C., Fonseca, A. S., Muys, B. R., de Barros e Lima Bueno, R., Bürger, M. C., de Souza, J. E. S., et al. (2013). Brief report: The lincRNA Hotair is required for epithelial-to-mesenchymal transition and stemness maintenance of cancer cell lines. *Stem Cells*, *31*(12), 2827–2832. https://doi.org/10.1002/stem.1547.

Phipps, K. R., Charette, J. M., & Baserga, S. J. (2011). The small subunit processome in ribosome biogenesis-progress and prospects. *Wiley Interdisciplinary Reviews: RNA*, *2*(1), 1–21. https://doi.org/10.1002/wrna.57.

Powell, L. M., Wallis, S. C., Pease, R. J., Edwards, Y. H., Knott, T. J., & Scott, J. (1987). A novel form of tissue-specific RNA processing produces apolipoprotein-B48 in intestine. *Cell*, *50*(6), 831–840. https://doi.org/10.1016/0092-8674(87)90510-1.

Putnam, A. A., & Jankowsky, E. (2013). DEAD-box helicases as integrators of RNA, nucleotide and protein binding. *Biochimica et Biophysica Acta—Gene Regulatory Mechanisms*, *1829*(8), 884–893. https://doi.org/10.1016/j.bbagrm.2013.02.002.

Ramsköld, D., Wang, E. T., Burge, C. B., & Sandberg, R. (2009). An abundance of ubiquitously expressed genes revealed by tissue transcriptome sequence data. *PLoS Computational Biology*, *5*(12), e1000598. https://doi.org/10.1371/journal.pcbi.1000598.

Rao, S., Zhen, S., Roumiantsev, S., McDonald, L. T., Yuan, G.-C., & Orkin, S. H. (2010). Differential roles of Sall4 isoforms in embryonic stem cell pluripotency. *Molecular and Cellular Biology*, *30*(22), 5364–5380. https://doi.org/10.1128/mcb.00419-10.

Reed, R., & Hurt, E. (2002, February 22). A conserved mRNA export machinery coupled to pre-mRNA splicing. *Cell*, *108*, 523–531. https://doi.org/10.1016/S0092-8674(02)00627-X.

Ruby, S. W., Chang, T. H., & Abelson, J. (1993). Four yeast spliceosomal proteins (PRP5, PRP9, PRP11, and PRP21) interact to promote U2 snRNP binding to pre-mRNA. *Genes and Development*, 7(10), 1909–1925. https://doi.org/10.1101/gad.7.10.1909.

Sandberg, R., Neilson, J. R., Sarma, A., Sharp, P. A., & Burge, C. B. (2008). Proliferating cells express mRNAs with shortened 3′ untranslated regions and fewer microRNA target sites. *Science*, *320*(5883), 1643–1647. https://doi.org/10.1126/science.1155390.

Saunders, A., & Wang, J. (2014, May 1). Export and expression: MRNAs deliver new messages for controlling pluripotency. *Cell Stem Cell, 14*, 549–550. https://doi.org/10.1016/j.stem.2014.04.009.

Savić, N., Bär, D., Leone, S., Frommel, S. C., Weber, F. A., Vollenweider, E., et al. (2014). LncRNA maturation to initiate heterochromatin formation in the nucleolus is required for exit from pluripotency in ESCs. *Cell Stem Cell, 15*(6), 720–734. https://doi.org/10.1016/j.stem.2014.10.005.

Segre, J. A., Bauer, C., & Fuchs, E. (1999). Klf4 is a transcription factor required for establishing the barrier function of the skin. *Nature Genetics, 22*(4), 356–360. https://doi.org/10.1038/11926.

Shepard, P. J., Choi, E. A., Lu, J., Flanagan, L. A., Hertel, K. J., & Shi, Y. (2011). Complex and dynamic landscape of RNA polyadenylation revealed by PAS-Seq. *RNA, 17*(4), 761–772. https://doi.org/10.1261/rna.2581711.

Shorter, J. (2019). Phase separation of RNA-binding proteins in physiology and disease: An introduction to the JBC reviews thematic series. *Journal of Biological Chemistry, 294*, 7113–7114. https://doi.org/10.1074/jbc.REV119.007944.

Shtrichman, R., Germanguz, I., Mandel, R., Ziskind, A., Nahor, I., Safran, M., et al. (2012). Altered A-to-I RNA editing in human embryogenesis. *PLoS One*, 7(7), e41576. https://doi.org/10.1371/journal.pone.0041576.

Stefl, R., Skrisovska, L., & Allain, F. H. (2005). RNA sequence- and shape-dependent recognition by proteins in the ribonucleoprotein particle. *EMBO Reports, 6*(1), 33–38. https://doi.org/10.1038/sj.embor.7400325.

Sugiyama, H., Takahashi, K., Yamamoto, T., Iwasaki, M., Narita, M., Nakamura, M., et al. (2017). Nat1 promotes translation of specific proteins that induce differentiation of mouse embryonic stem cells. *Proceedings of the National Academy of Sciences of the United States of America, 114*(2), 340–345. https://doi.org/10.1073/pnas.1617234114.

Tafforeau, L., Zorbas, C., Langhendries, J. L., Mullineux, S. T., Stamatopoulou, V., Mullier, R., et al. (2013). The complexity of human ribosome biogenesis revealed by systematic nucleolar screening of pre-rRNA processing factors. *Molecular Cell, 51*(4), 539–551. https://doi.org/10.1016/j.molcel.2013.08.011.

Tahmasebi, S., Amiri, M., & Sonenberg, N. (2019). Translational control in stem cells. *Frontiers in Genetics, 9*, 709. https://doi.org/10.3389/fgene.2018.00709.

Takahashi, K., Tanabe, K., Ohnuki, M., Narita, M., Ichisaka, T., Tomoda, K., et al. (2007). Induction of pluripotent stem cells from adult human fibroblasts by defined factors. *Cell, 131*(5), 861–872. https://doi.org/10.1016/j.cell.2007.11.019.

Takahashi, K., & Yamanaka, S. (2006). Induction of pluripotent stem cells from mouse embryonic and adult fibroblast cultures by defined factors. *Cell, 126*(4), 663–676. https://doi.org/10.1016/j.cell.2006.07.024.

Teplova, M., & Patel, D. J. (2008). Structural insights into RNA recognition by the alternative-splicing regulator muscleblind-like MBNL1. *Nature Structural and Molecular Biology, 15*(12), 1343–1351. https://doi.org/10.1038/nsmb.1519.

Tesar, P. J., Chenoweth, J. G., Brook, F. A., Davies, T. J., Evans, E. P., Mack, D. L., et al. (2007). New cell lines from mouse epiblast share defining features with human embryonic stem cells. *Nature, 448*(7150), 196–199. https://doi.org/10.1038/nature05972.

Theunissen, T. W., & Jaenisch, R. (2017). Mechanisms of gene regulation in human embryos and pluripotent stem cells. *Development (Cambridge, England), 144*(24), 4496–4509. https://doi.org/10.1242/dev.157404.

Thomson, J. A., Itskovitz-Eldor, J., Shapiro, S. S., Waknitz, M. A., Swiergiel, J. J., Marshall, V. S., et al. (1998). Embryonic stem cell lines derived from human blastocysts. *Science (New York, N.Y.), 282*(5391), 1145–1147. https://doi.org/10.1126/SCIENCE.282.5391.1145.

Tian, Y., Simanshu, D. K., Ma, J. B., & Patel, D. J. (2011). Structural basis for piRNA 2′-O-methylated 3′-end recognition by Piwi PAZ (Piwi/Argonaute/Zwille) domains. *Proceedings of the National Academy of Sciences of the United States of America, 108*(3), 903–910. https://doi.org/10.1073/pnas.1017762108.

Ustianenko, D., Hrossova, D., Potesil, D., Chalupnikova, K., Hrazdilova, K., Pachernik, J., et al. (2013). Mammalian DIS3L2 exoribonuclease targets the uridylated precursors of let-7 miRNAs. *RNA, 19*(12), 1632–1638. https://doi.org/10.1261/rna.040055.113.

Wang, Y., Arribas-Layton, M., Chen, Y., Lykke-Andersen, J., & Sen, G. L. (2015). DDX6 orchestrates mammalian progenitor function through the mRNA degradation and translation pathways. *Molecular Cell, 60*(1), 118–130. https://doi.org/10.1016/J.MOLCEL.2015.08.014.

Wang, X., & Dai, J. (2010). Isoforms of *OCT4* contribute to the confusing diversity in stem cell biology. *Stem Cells, 28*(5), 885–893. https://doi.org/10.1002/stem.419.

Wang, L., Miao, Y. L., Zheng, X., Lackford, B., Zhou, B., Han, L., et al. (2013). The THO complex regulates pluripotency gene mRNA export and controls embryonic stem cell self-renewal and somatic cell reprogramming. *Cell Stem Cell, 13*(6), 676–690. https://doi.org/10.1016/j.stem.2013.10.008.

Whyte, W. A., Bilodeau, S., Orlando, D. A., Hoke, H. A., Frampton, G. M., Foster, C. T., et al. (2012). Enhancer decommissioning by LSD1 during embryonic stem cell differentiation. *Nature, 482*(7384), 221–225. https://doi.org/10.1038/nature10805.

Winter, E. E., Goodstadt, L., & Ponting, C. P. (2004). Elevated rates of protein secretion, evolution, and disease among tissue-specific genes. *Genome Research, 14*(1), 54–61. https://doi.org/10.1101/gr.1924004.

Woolnough, J. L., Atwood, B. L., Liu, Z., Zhao, R., & Giles, K. E. (2016). The regulation of rRNA gene transcription during directed differentiation of human embryonic stem cells. *PLoS One, 11*(6), e0157276. https://doi.org/10.1371/journal.pone.0157276.

Worringer, K. A., Rand, T. A., Hayashi, Y., Sami, S., Takahashi, K., Tanabe, K., et al. (2014). The let-7/LIN-41 pathway regulates reprogramming to human induced pluripotent stem cells by controlling expression of prodifferentiation genes. *Cell Stem Cell, 14*(1), 40–52. https://doi.org/10.1016/j.stem.2013.11.001.

Xiao, Y., Chen, J., Wan, Y., Gao, Q., Jing, N., Zheng, Y., et al. (2019). Regulation of zebrafish dorsoventral patterning by phase separation of RNA-binding protein Rbm14. *Cell Discovery, 5*, 37. https://doi.org/10.1038/s41421-019-0106-x.

Yang, L., Lin, C., & Liu, Z. R. (2006). P68 RNA helicase mediates PDGF-induced epithelial mesenchymal transition by displacing axin from β-catenin. *Cell, 127*(1), 139–155. https://doi.org/10.1016/j.cell.2006.08.036.

Ye, J., & Blelloch, R. (2014). Regulation of pluripotency by RNA binding proteins. *Cell Stem Cell, 15*, 271–280. https://doi.org/10.1016/j.stem.2014.08.010.

Yeo, G. W., Coufal, N. G., Liang, T. Y., Peng, G. E., Fu, X. D., & Gage, F. H. (2009). An RNA code for the FOX2 splicing regulator revealed by mapping RNA-protein interactions in stem cells. *Nature Structural and Molecular Biology, 16*(2), 130–137. https://doi.org/10.1038/nsmb.1545.

Yeo, J. C., & Ng, H. H. (2013, January). The transcriptional regulation of pluripotency. *Cell Research, 23*, 20–32. https://doi.org/10.1038/cr.2012.172.

Yoffe, Y., David, M., Kalaora, R., Povodovski, L., Friedlander, G., Feldmesser, E., et al. (2016). Cap-independent translation by DAP5 controls cell fate decisions in human embryonic stem cells. *Genes & Development, 30*(17), 1991–2004. https://doi.org/10.1101/gad.285239.116.

You, K. T., Park, J., & Kim, V. N. (2015). Role of the small subunit processome in the maintenance of pluripotent stem cells. *Genes & Development, 29*(19), 2004–2009. https://doi.org/10.1101/gad.267112.115.

Youn, J. Y., Dyakov, B. J. A., Zhang, J., Knight, J. D. R., Vernon, R. M., Forman-Kay, J. D., et al. (2019, October 17). Properties of stress granule and P-body proteomes. *Molecular Cell*, *76*, 286–294. https://doi.org/10.1016/j.molcel.2019.09.014.

Young, R. A. (2011, March 18). Control of the embryonic stem cell state. *Cell*, *144*, 940–954. https://doi.org/10.1016/j.cell.2011.01.032.

Zhang, H., Wu, Z., Lu, J. Y., Huang, B., Zhou, H., Xie, W., et al. (2020). DEAD-box helicase 18 counteracts PRC2 to safeguard ribosomal DNA in pluripotency regulation. *Cell Reports*, *30*(1), 81–97.e7. https://doi.org/10.1016/j.celrep.2019.12.021.

Zhang, H. Y., Xiong, J., Qi, B. L., Feng, Y. Q., & Yuan, B. F. (2016). The existence of 5-hydroxymethylcytosine and 5-formylcytosine in both DNA and RNA in mammals. *Chemical Communications*, *52*(4), 737–740. https://doi.org/10.1039/c5cc07354e.

Zuo, C., Liang, S., Wang, Z., Li, H., Zheng, W., & Ma, W. (2009). Enriching protein-protein and functional interaction networks in human embryonic stem cells. *International Journal of Molecular Medicine*, *23*(6), 811–819. https://doi.org/10.3892/ijmm_00000197.

CHAPTER FOUR

Generating primed pluripotent epiblast stem cells: A methodology chapter

Milan Samanta, Sundeep Kalantry*
Department of Human Genetics, University of Michigan Medical School, Ann Arbor, MI, United States
*Corresponding author: e-mail address: kalantry@umich.edu

Contents

1. Introduction 140
2. Materials 141
   2.1 Mouse embryonic fibroblasts (MEFs) 141
   2.2 Mouse embryonic stem cells (mESCs) 142
   2.3 Mouse epiblast stem cells (EpiSCs) 145
   2.4 Epiblast like stem cells (EpiLCs) 147
   2.5 Equipment 149
3. Methods 149
   3.1 Mouse embryonic fibroblasts (MEFs) 150
   3.2 Mouse embryonic stem cells (mESCs) 153
   3.3 Mouse epiblast stem cells (mEpiSCs) 158
   3.4 EpiLCs 162
4. Discussion 166
5. Recipes 166
6. Notes 171
Acknowledgments 173
References 173

Abstract

At least two distinct pluripotent states, referred to as naïve and primed, define the early mammalian embryo. In the mouse, the pluripotent epiblast cells in the pre/peri-implantation embryo are the source of naïve embryonic stem cells (ESCs). After the embryo implants, the epiblast lineage generates a restricted or primed population of stem cells, referred to as epiblast stem cells (EpiSCs). ESCs can be cultured in EpiSC media to generate epiblast-like cells (EpiLCs). The differentiation of naive ESCs into primed EpiLCs permits insights into the development and differentiation of the pluripotent epiblast lineage. This chapter describes the generation and characterization of EpiSCs as well as EpiLCs.

Current Topics in Developmental Biology, Volume 138
ISSN 0070-2153
https://doi.org/10.1016/bs.ctdb.2020.01.005

1. Introduction

Mammalian embryogenesis has been studied most intensively in the mouse. All cells of the very early mouse embryo are totipotent and can give rise to both extraembryonic (placental and yolk sac) and embryonic (epiblast and its derivatives) cells (Rossant, 1976; Tarkowski, 1959; Tarkowski & Wroblewska, 1967). The first morphological difference during embryogenesis is the formation of the trophectoderm in the blastocyst stage embryo. The remaining cells in the blastocyst form the inner cell mass, which subsequently separates into the primitive endoderm and the pluripotent epiblast cells. The trophectoderm and the primitive endoderm lineages generate the extra-embryonic cells of the placenta and the yolk sac. The epiblast, on the other hand, gives rise to somatic tissues and germ cells. The *in vitro* culture of epiblast-derived pluripotent stem cells has therefore garnered considerable interest both to glean insights into the early embryonic developmental program and as substrates for regenerative medicine.

Two types of stem cell lines have been derived from the mouse epiblast. The first is embryonic stem cells (ESCs) (Evans & Kaufman, 1981; Martin, 1981) and the second is epiblast stem cells (EpiSCs) (Brons et al., 2007; Guo et al., 2009; Tesar et al., 2007). Both ESCs and EpiSCs can be differentiated *in vitro* into the mesoderm, endoderm, ectoderm, as well as germ cells (Brons et al., 2007; Evans & Kaufman, 1981; Martin, 1981; Tesar et al., 2007). When injected into the blastocyst, ESCs can generate chimeric animals, but EpiSCs rarely can (Bradley, Evans, Kaufman, & Robertson, 1984; Brons et al., 2007; Tesar et al., 2007). EpiSCs express many (e.g., *Oct4*, *Nanog*, and *Sox2*), but not all (e.g., *Rex1*, *Klf2*, and *Klf4*) pluripotency-associated genes (Hayashi, Ohta, Kurimoto, Aramaki, & Saitou, 2011; Tesar et al., 2007). In the embryo, ESCs are *in vitro* analogs of the pre- or peri-implantation embryonic epiblast (Brook & Gardner, 1997; Evans & Kaufman, 1981; Martin, 1981). EpiSCs, on the other hand, are representatives of the early post-implantation epiblast (Gayen, Maclary, Buttigieg, Hinten, & Kalantry, 2015). In agreement, female ESCs contain two active X-chromosomes (Rastan, 1983; Rastan & Robertson, 1985), like in the epiblast cells of peri-implantation embryos (Gardner & Lyon, 1971; Mak et al., 2004), and female EpiSCs harbor an inactivated X-chromosome (Bao et al., 2009; Gayen et al., 2015; Guo et al., 2009), as in the epiblast cells of post-implantation embryos (McMahon, Fosten, & Monk, 1983; Monk & Harper, 1979). Due to their similarities to ESCs, EpiSCs are also considered pluripotent. But, due to their differences, ESCs are referred to as naïve pluripotent stem cells and EpiSCs as primed pluripotent stem cells, which are slightly differentiated cells relative to their ESC counterparts.

ESCs can be differentiated into cells that resemble epiblast-like stem cells (EpiLCs), which morphologically and transcriptionally mimic EpiSCs (Hackett & Surani, 2014; Hayashi et al., 2011; Nakamura et al., 2016). The derivation of EpiSCs and the stereotyped conversion of ESCs into EpiLCs provide a molecular dissection of naïve to primed pluripotency.

2. Materials

2.1 Mouse embryonic fibroblasts (MEFs)

2.1.1 MEF isolation

1. Pregnant mouse at E(15–17) (see Note 1).
2. Sterile dissection tools: scissors, blunt forceps, small blunt forceps, scalpels or oil-free razor blades.
3. Phosphate-buffered saline (PBS) without calcium and magnesium (Gibco#10010-023).
4. DNase I stock solution (13.33 mg/mL) (Roche#104159).
5. Trypsin-EDTA (0.25%) (Gibco#25300-062).
6. MEF culture medium with antibiotic
   - MEM Alpha (Gibco#12561-056).
   - Fetal bovine serum (FBS) (Gibco#10437-028).
   - Penicillin-streptomycin (100 ×) (Gibco#15070-063).
7. 50 mL tubes.
8. 15 mL tubes.
9. 100 mm tissue culture dishes.
10. 150 mm tissue culture dishes.
11. Sterile stir bar.
12. Sterile glycerol (80%).
13. Sterile sodium chloride solution (5 M).

2.1.2 Cryopreserving non-irradiated MEFs

1. Cryopreservation medium: Recovery Cell Culture Freezing Medium (Gibco#12648-010) (see Note 2).
2. Cryogenic vials (Corning#430659).
3. Cryogenic cooling container.

2.1.3 Irradiating of MEFs

1. Non-irradiated MEFs.
2. MEF culture medium.
   - MEM Alpha (Gibco#12561-056).
   - Fetal bovine serum (FBS) (Gibco#10437-028).
3. Trypsin-EDTA (0.25%) (Gibco#25300-062).
4. Phosphate-buffered saline (PBS) without calcium and magnesium (Gibco#10010-023).
5. 150 mm tissue culture dishes.

2.1.4 Cryopreserving irradiated MEFs

1. Cryopreservation medium: Recovery Cell Culture Freezing Medium (Gibco#12648-010).
2. Cryogenic vials (Corning#430659).
3. Cryo cooling container.

2.1.5 Thawing and culturing irradiated MEFs

1. MEF culture medium.
2. Different size tissue culture dishes.

2.2 Mouse embryonic stem cells (mESCs)

2.2.1 Derivation of mESCs

2.2.1.1 E3.5 blastocysts isolation

1. E3.5 pregnant mouse (see Note 1).
2. Irradiated mouse embryonic fibroblasts (MEFs) (see Note 3).
3. MEF culture medium.
4. ESC derivation medium.
   - KnockOut DMEM (Gibco#10829-018).
   - KnockOut serum replacement (KSR) (Invitrogen#10828-028).
   - L-glutamine (Gibco#25030).
   - MEM non-essentials amino acids (Gibco#11140-050).
   - β-Mercaptoethanol (Sigma#M7522) (see Note 4).
   - Penicillin-streptomycin (100 ×) (Gibco#15070-063) (see Note 5).
   - GSK3 inhibitor CHIR99021 (Stemgent#04-0004).
   - MEK inhibitor PD0325901 (Stemgent#04-0006).
   - LIF (10^7/mL) (Millipore#ESG1106).
5. Ethanol, 70% (v/v).
6. Sterile dissection tool: forceps and scissors (see Note 6).
7. 35 mm tissue culture dishes.

8. 3 mL syringe.
9. 23-Gage needles.
10. 4-Well tissue culture plates.
11. Phosphate-buffered saline (PBS) without calcium and magnesium (Gibco#10010-023).
12. Trypsin-EDTA (0.25%) (Gibco#25300-062).
13. DMSO (Sigma#D2650) (see Note 7).

2.2.1.2 Culturing ESCs

1. ESC culture medium.
   - KnockOut DMEM (Gibco#10829-018).
   - Fetal bovine serum embryonic stem cell qualified (ES-FBS) (Bio-Techne#S10250).
   - KnockOut serum replacement (KSR) (Invitrogen#10828-028).
   - L-Glutamine (Gibco#25030).
   - MEM non-essentials amino acids (Gibco#11140-050).
   - β-Mercaptoethanol (Sigma#M7522).
   - GSK3 inhibitor CHIR99021 (Stemgent#04-0004).
   - MEK inhibitor PD0325901 (Stemgent#04-0006).
   - LIF (10^7/mL) (Millipore#ESG1106).
2. MEF feeder tissue culture plates.
3. Trypsin-EDTA (0.25%) (Gibco#25300-062).
4. DMSO (Sigma#D2650).

2.2.2 Cryopreserving ESCs

1. Cryopreservation medium: Recovery Cell Culture Freezing Medium (Gibco#12648-010).
2. Cryogenic vials (Corning#430659).
3. Cryo cooling container.

2.2.3 Thawing of ESCs

1. MEF medium.
2. ESC culture medium.
3. MEF feeder plates.

2.2.4 Characterization of ESCs

1. TRIzol (Life Technologies#15596018) (see Note 8).
2. SuperScript III One-Step RT-PCR System (Invitrogen#18080-044) (see Note 9).

3. Primers for ESC-specific transcripts.
   Zfp42 (Rex1) forward: TGAAAGTGAGATTAGCCCCGAG
   Zfp42 (Rex1) reverse: GTCCCATCCCCTTCAATAGCAC
   Klf2 forward: TCGAGGCTAGATGCCTTGTGA
   Klf2 reverse: AAACGAAGCAGGCGGCAGA
   Klf4 forward: TGGTGCTTGGTGAGTTGTGG
   Klf4 reverse: GCTCCCCCGTTTGGTACCTT
   Stella forward: AGGCTCGAAGGAAATGAGTTTG
   Stella reverse: TCCTAATTCTTCCCGATTTTCG
   Stra8 forward: ACCCTGGTAGGGCTCTTCAA
   Stra8 reverse: GACCTCCTCTAAGCTGTTGGG
4. Materials for gel electrophoresis.
5. Reagents for immunofluorescence (IF) for detection of ESC-specific protein expression.
   - 1 × PBS.
   - 6-Well dish, or similar chamber to be used for dehydration and for washing coverslips.
   - Blocking buffer: 1 × PBS with 0.5 mg/mL BSA (NEB#B9000S), 50 μg/mL tRNA, and 0.2% Tween 20 (make and use a 10% Tween 20 stock). Prewarm to 37 °C.
   - Primary antibody of choice (see Note 10): NANOG antibody (ReproCELL#RCAB002P-F), OCT4 antibody (Santa Cruz SC-5279), REX1 antibody (Thermo Scientific PA5-27567).
   - Fluorescently conjugated secondary antibody: Alexa Fluor (Invitrogen) secondary antibodies work well with this protocol.
   - Small glass plate (see Note 11).
   - Parafilm.
   - Forceps.
   - IF chamber: A small humid chamber (see Note 12) for incubating slides, humidity provided by 1 × PBS.
   - Incubator set to 37 °C.
   - 1 × PBS with 0.2% Tween 20 (Sigma-Aldrich#P9416).
   - 4′,6-Diamidino-2-phenylindole, dihydrochloride (DAPI): dissolved in water at 5 mg/mL.
   - Mounting medium (see Note 13).
   - Microscope slides.
   - Coverslips (see Note 14).

2.3 Mouse epiblast stem cells (EpiSCs)

2.3.1 Derivation of EpiSCs from preimplantation embryos

2.3.1.1 Epiblast isolation and plating

1. E3.5 timed pregnant female mouse (see Note 15).
2. Sterile dissection tool: forceps and scissors.
3. Syringe and needle (23 gage).
4. K15F5 medium
   - KnockOut DMEM (Gibco#10829-018).
   - KnockOut serum replacement (KSR) (Invitrogen#10828-028).
   - Fetal bovine serum embryonic stem cell qualified (ES-FBS) (Bio-Techne#S10250).
   - L-Glutamine (Gibco#25030).
   - MEM non-essentials amino acids (Gibco#11140-050).
   - β-Mercaptoethanol (Sigma#M7522).
   - Penicillin-streptomycin (100 ×) (Gibco#15070-063).
5. EpiSC culture medium
   - KnockOut DMEM (Gibco#10829-018).
   - KnockOut serum replacement (KSR) (Invitrogen#10828-028).
   - L-Glutamine (Gibco#25030).
   - MEM non-essentials amino acids (Gibco#11140-050).
   - β-Mercaptoethanol (Sigma#M7522).
   - FGF2 stock solution (10 μg/mL) (R&D Systems 233-FB-025).
6. MEF feeder tissue culture plates.
7. Non-tissue culture dishes (e.g., 35 or 60 mm).

2.3.1.2 EpiSC culture

1. EpiSC passaging medium: Collagenase type IV (Invitrogen#17104-019) (1.5 mg/mL).
2. EpiSC culture medium.
3. MEF feeder plates.

2.3.2 Cryopreserving EpiSCs

1. Cryopreservation medium: Recovery cell culture freezing medium (Gibco#12648-010).
2. Cryogenic vials (Corning#430659).
3. Cryo cooling container.

2.3.3 Thawing of EpiSCs

1. MEF medium.
2. EpiSC culture medium.
3. MEF feeder plates.

2.3.4 EpiSC characterization

1. TRIzol (Life Technologies#15596018).
2. SuperScript III One-Step RT-PCR System (Invitrogen#18080-044).
3. Primers for EpiSC-specific transcripts.
   Pou5f1 forward: CGTTCTCTTTGGAAAGGTGTTC
   Pou5f1 reverse: GAACCATACTCGAACCACATCC
   Fgf5 forward: CTGTACTGCAGAGTGGCATCGG
   Fgf5 reverse: GACTTCTGCGAGGCTGCGACAGG
4. Materials for gel electrophoresis.
5. Reagents for immunofluorescence (IF) for detection of proteins
   - 1 × PBS.
   - 6-Well dish, or similar chamber to be used for dehydration and washing coverslips.
   - Blocking buffer: 1 × PBS with 0.5 mg/mL BSA (NEB#B9000S), 50 μg/mL tRNA, and 0.2% Tween 20 (make and use a 10% Tween 20 stock). Prewarm to 37 °C.
   - Primary antibody of choice (see Note 16):
     NANOG antibody (ReproCELL#RCAB002P-F), OCT4 antibody (Santa Cruz SC-5279), REX1 antibody (Thermo Scientific PA5-27567).
   - Fluorescently conjugated secondary antibody: Alexa Fluor (Invitrogen) secondary antibodies work well with this protocol.
   - Small glass plate.
   - Parafilm.
   - Forceps.
   - IF chamber: A small humid chamber for incubating slides, humidity provided by 1 × PBS.
   - Incubator set to 37 °C.
   - 1 × PBS with 0.2% Tween 20 (Sigma-Aldrich#P9416).
   - 4′,6-Diamidino-2-phenylindole, dihydrochloride (DAPI): dissolved in water at 5 mg/mL.
   - Mounting medium.
   - Microscope slides.
   - Coverslips.

2.4 Epiblast like stem cells (EpiLCs)

2.4.1 Culturing ESCs

1. ESC culture medium
   - KnockOut DMEM (Gibco#10829-018).
   - Fetal bovine serum embryonic stem cell qualified (ES-FBS) (Bio-Techne#S10250).
   - KnockOut serum replacement (KSR) (Invitrogen#10828-028).
   - L-Glutamine (Gibco#25030).
   - MEM non-essentials amino acids (Gibco#11140-050).
   - β-Mercaptoethanol (Sigma#M7522).
   - GSK3 inhibitor CHIR99021 (Stemgent#04-0004).
   - MEK inhibitor PD0325901 (Stemgent#04-0006).
   - LIF (10^7/mL) (Millipore#ESG1106).
2. MEF feeder tissue culture plates.
3. Trypsin-EDTA (0.25%) (Gibco#25300-062).
4. DMSO (Sigma#D2650).

2.4.2 Differentiating ESCs into EpiLCs

1. N2B27 medium
   - DMEM/F12 (Gibco#11330-032).
   - Neurobasal media (Gibco#21103-049).
   - L-Glutamine (Gibco#25030).
   - β-Mercaptoethanol (Sigma#M7522).
   - N2 supplement (Invitrogen#17502048).
   - B27 supplement (Invitrogen#17504-044).
   - GSK3 inhibitor CHIR99021 (Stemgent#04-0004).
   - MEK inhibitor PD0325901 (Stemgent#04-0006).
   - LIF (10^7/mL) (Millipore#ESG1106).
2. FGF2 stock solution (10 μg/mL) (R&D Systems 233-FB-025).
3. Activin A stock solution (10 μg/mL) (R&D Systems 338-AC-005).
4. Accutase (Sigma#A6964).
5. Fibronectin (Sigma#F1141).
6. BSA fraction V solution in PBS (0.1%, w/v) (Sigma-Aldrich A3311 or A1470).
7. Gelatin (0.2%) (Sigma#G2500).
8. DMSO (Sigma#D2650).

2.4.3 Characterization of EpiLCs

1. TRIzol (Life Technologies#15596018).
2. SuperScript III One-Step RT-PCR System (Invitrogen#18080-044).

3. Primers for ESC-specific transcripts.
   Zfp42 (*Rex1*) forward: TGGAAGCGAGTTCCCTTCTC
   Zfp42 (*Rex1*) reverse: GCCGCCTGCAAGTAATGAG
   Klf2 forward: TCGAGGCTAGATGCCTTGTGA
   Klf2 reverse: AAACGAAGCAGGCGGCAGA
   Klf4 forward: TGGTGCTTGGTGAGTTGTGG
   Klf4 reverse: GCTCCCCCGTTTGGTACCTT
   Stella forward: AGGCTCGAAGGAAATGAGTTTG
   Stella reverse: TCCTAATTCTTCCCGATTTTCG
   Stra8 forward: ACCCTGGTAGGGCTCTTCAA
   Stra8 reverse: GACCTCCTCTAAGCTGTTGGG
4. Primers for EpiLC-specific transcripts.
   Fgf5 forward: CTGTACTGCAGAGTGGCATCGG
   Fgf5 reverse: GACTTCTGCGAGGCTGCGACAGG
   Cer1 forward: CTCTGG GGAAGGCAGACCTAT
   Cer1 reverse: CCACAAACAGATCCGGCTT
5. Primers for differentiated EpiLCs.
   a. Mesoderm
      Brachyury forward: CTTCCAGGTCCAACGCCTAC
      Brachyury reverse: GTCGTCGCTCATGTTCTTCAA
   b. Ectoderm
      β-III tubulin forward: TAGACCCCAGCGGCAACTAT
      β-III tubulin reverse: GTTCCAGGTTCCAAGTCCACC
   c. Endoderm
      FoxA2 forward: TCCGACTGGAGCAGCTACTAC
      FoxA2 reverse: GCGCCCACATAGGATGACA
6. Materials for gel electrophoresis.
7. Reagents for immunofluorescence (IF).
   - 1 × PBS.
   - 6-Well dish, or similar chamber to be used for dehydration and washing coverslips.
   - Blocking buffer: 1 × PBS with 0.5 mg/mL BSA (NEB#B9000S), 50 μg/mL tRNA, and 0.2% Tween 20 (make and use a 10% Tween 20 stock). Prewarm to 37 °C.
   - Primary antibody of choice (see Note 17): NANOG antibody (ReproCELL#RCAB002P-F), OCT4 antibody (Santa Cruz SC-5279), REX1 antibody (Thermo Scientific PA5-27567).
   - Fluorescently conjugated secondary antibody: Alexa Fluor (Invitrogen) secondary antibodies work well with this protocol.

- Small glass plate.
- Parafilm.
- Forceps.
- IF chamber: A small humid chamber for incubating slides, humidity provided by 1 × PBS.
- Incubator set to 37 °C.
- 1 × PBS with 0.2% Tween 20 (Sigma-Aldrich#P9416).
- 4′,6-Diamidino-2-phenylindole, dihydrochloride (DAPI): dissolved in water at 5 mg/mL.
- Mounting medium.

8. Microscope slides.
9. Coverslips.

2.4.4 Cryopreservation of EpiLCs

1. Cryopreservation medium: Recovery cell culture freezing medium (Gibco#12648-010).
2. Cryogenic vials (Corning#430659).
3. Cryo cooling container.

2.5 Equipment

1. Incubator, humidified, 37 °C; 5% CO_2, 95% air.
2. Class II biosafety cabinet (tissue culture hood).
3. Automated cell counter.
4. Irradiation instrument.
5. Microscopes, inverted and stereomicroscope.
6. Pipettes.
7. Pipettors.
8. Vacuum Pump (Fisher Scientific, cat. no. 13-878-40 or equivalent).
9. Water bath (Thermo Scientific, model 2864 or equivalent).

3. Methods

Cell culture work should be performed under sterile conditions. All manipulation of cells and preparation of solutions should be done inside a certified tissue culture hood (class II biological safety cabinet). MEFs, ESCs, EpiSCs, and EpiLCs should be cultured at 37 °C in a humid atmosphere with 5% CO_2. MEFs can be derived from E(12.5–13.5) and E(15–17) mouse embryos. ESCs are derived from embryonic day (E) 3.5 mouse embryos. EpiSCs can be derived from E3.5, E4.5, and E5.5 mouse embryos.

3.1 Mouse embryonic fibroblasts (MEFs)

Mitotically inactive MEFs serve as feeder cells for the self-renewal of ESCs and EpiSCs.

3.1.1 MEF isolation

1. Warm MEF media (with penicillin-streptomycin) and PBS. Thaw trypsin and keep on ice. Prepare five 100 mm dishes containing sterile PBS.
2. Euthanize mouse at E(15–17) according to local animal care requirements. Remove uterine horns with embryos, cutting away the mesentery.
3. Cut along the entire length of the uterus with scissors, cut the placenta to remove each embryo, and place each embryo into a new 100 mm dish with PBS.
4. Remove the amniotic sac from each embryo and place each embryo in a new 100 mm dish.
5. Switch to a new set of sterile instruments (sterilize instruments by heating them in the "Germinator"); individually wash each embryo with PBS and place it in a new 100 mm dish.
6. Eviscerate and decapitate: Holding embryo by the head with abdomen facing up, use fine blunt forceps to puncture body wall and remove internal organs. Pinch off the head at the neck.
7. Place each carcass in a new 100 mm dish. Dispose of head and organs.
8. Take dish with embryos into tissue culture hood. Transfer embryos in a new 100 mm dish with 10 mL trypsin.
9. Mince carcasses with a sterile scalpel or oil-free razor blade in the hood. This should be thorough and take >5 min.
10. Use a 5 mL pipet to further homogenize the embryos. Do this quite thoroughly.
11. Use a 2 mL pipet to further break up the slurry. Do this quite thoroughly as well.
12. Use a 1 mL pipet and repeat step 10.
13. Use a 5 mL pipet to transfer entire slurry into a sterile 250 mL flask with a sterile stir bar.
14. Rinse plate with 10 mL trypsin to remove all remaining tissue and pipet into the 250 mL flask with the rest of the slurry.
15. Wash the 100 mm dish once more with 10 mL aliquot of trypsin.
16. The total volume in the flask should now exceed 30 mL.

17. OPTIONAL: Force tissue through collector screen in the cup with a sterile glass pestle. Continue to press tissue through the screen until as much as possible has gone through. Pipet 10 mL trypsin through screen, and also onto the bottom of the screen to remove all adhering tissue. Repeat with one more 10 mL aliquot of trypsin.
18. Add 0.3 mL DNase I stock solution to flask to reduce viscosity.
19. Place flask in 37 °C incubator. Every 5 min, place on a stir plate and stir gently for 1 min. Total incubation time at 37 °C should be approximately 30 min.
20. Pipet the entire solution into a 50 mL conical tube.
21. Pellet cells by centrifugation at 1500 rpm for 5 min.
22. Remove media, leaving a small amount in the tube to resuspend the pellet.
23. Wash pellet twice with 30 mL MEF media.
24. Resuspend cells in 15 mL MEF media.
25. Dilute sample 1:10 (4.5 mL PBS + 0.5 mL cell suspension in 15 mL tube). Count viable cells (not debris). Expect 5×10^7–10^8 cells from 10 E15 fetuses.
26. Plate cells at 6×10^6 per 150 mm dish.
27. Change the media after 24 h.
28. When confluent (3–5 days), split each 100 mm dish onto 5, 100 mm dishes. Freeze cells when the five dishes become confluent (freeze 6×10^6 cells per cryogenic vial with 1 mL freezing medium) or proceed with splitting once more 1:5 and follow the irradiation protocol (see Section 3.1.3).

3.1.2 Cryopreserving MEFs

1. Label cryogenic vials in the hood.
2. Aspirate media from the 150 mm dishes.
3. Wash cells in each dish with 15 mL room temperature PBS.
4. Trypsinize cells (10 mL 0.25% Trypsin-EDTA/1, 150 mm dish) at 37 °C for 6 min.
5. Rinse vigorously after removal from the incubator to dislodge cells from the bottom of the plate.
6. Transfer cell suspension from two 150 mm dishes to one 50 mL tube containing 10 mL MEF media.
7. Spin down cells in the 50 mL tubes by centrifugation at 1500 rpm for 5 min.
8. Remove the caps from labeled cryovials in the hood.

9. Aspirate trypsin solution from centrifuge tubes and tap each tube to loosen the pellet.
10. Combine the cells from all tubes into one common 50 mL tube using 10 mL MEF media.
11. Count the viable cells in the common 50 mL tube by automated cell counter and centrifuge (1500 rpm for 5 min) to form a pellet.
12. Aspirate supernatant tap the tube to loosen the pellet and resuspend in recovery cell culture freezing medium. The amount of freezing media depends on the number of viable cells (6×10^6 cells/mL of cell suspension).
13. Working quickly, add 1 mL cell suspension to each cryovial, screw the cap on tightly, and place the vials in a freezing container in a −80 °C freezer overnight.
14. Transfer vials from −80 °C freezer to the liquid N_2 freezer the next day. Keep vials on dry ice during the transfer.
15. Test thaw one vial (onto a gelatin-coated 150 mm dish) the day after transferring cells to the liquid N_2 freezer to determine viability. Culture for 2–3 days in an antibiotic-free MEF medium to check for contamination.

3.1.3 Irradiating and cryopreserving MEFs

The volumes listed are for 150 mm dishes.

1. Thaw one vial of non-irradiated MEFs into one 150 mm dish. Use MEF media.
2. When confluent, aspirate media and wash the cells with 6 mL of sterile 1 × PBS and aspirate PBS. Add 9 mL of Trypsin-EDTA (0.25%) to the cells and incubate for ~5 min at 37 °C. Then, pipette vigorously to break into single cells. Neutralize trypsin with 9 mL MEF media and pellet down the cells by centrifugation (1500 rpm for 5 min at room temperature). Resuspend the cells in MEF media and split 1:6.
3. When cells are confluent, repeat step 2 above.
4. After two rounds of splitting and growth, the 36 plates of MEFs are ready for irradiation.
5. Harvest the cells by trypsinizing 5–6 plates at a time in 37 °C incubator for ~5 min.
6. Pipet up and down in trypsin to break clumps of cells apart. If they are sticking together, use a glass pipet to pipet cells up and down.
7. Neutralize trypsin with MEF media.
8. Combine dishes of trypsinized cells into 50 mL conical tubes.

9. Collect cells by centrifuging the tubes at 1500 rpm for 5 min. Aspirate trypsin and resuspend the cells in ~25 mL of MEF media.
10. Irradiate the cells with 6000 rads.
11. Count the cells after irradiation with the automated counter.
12. Plate 2.5×10^6 cells/100 mm dish of irradiated MEFs for use within 1 week.
13. Collect the remaining irradiated MEFs via centrifugation.
14. Resuspend the cells in the freezing medium. Count the total number of viable cells and aliquot ~5×10^6 cells in 1 mL of the freezing medium per cryopreservation vial (see Note 18).
15. Tightly close the vials. Store vials in −80 °C freezer for 1–3 days, then transfer to longer term storage in a liquid nitrogen tank.

3.1.4 Preparation of MEF feeder tissue culture dishes

1. Remove a vial from liquid nitrogen storage and thaw quickly in 37 °C water bath until almost completely melted.
2. Gently add cell suspension to 10 mL of room temperature MEF medium in a 15 mL tube.
3. Spin tube down at 1500 rpm for 5 min, remove supernatant and gently resuspend pellet in prewarmed MEF medium.
4. Plate irradiated MEFs on tissue culture dishes in MEF media (use 750 μL MEF medium/4-well well; 1 mL/12-well well; 2 mL/6-well well; 10 mL/100 mm dish) (see Note 18).
5. After overnight plating, MEF feeders will be ready for culturing other cells.

3.2 Mouse embryonic stem cells (mESCs)

3.2.1 Derivation of ESCs

3.2.1.1 Preparation of MEF feeder plates

1. Thaw and plate irradiated MEFs onto 4-well tissue culture plates before the day of isolating embryos. The number of wells needed is equal to the number of embryos to be harvested and is thus dependent on the number of pregnant females used. The number of embryos received per female is highly variable but generally averages between 6 and 8.
2. On the day of embryo harvesting, remove the MEF media from 4-well plates and rinse the 4-well feeder plates with 1 × PBS. Replace the media with ES-derivation media.

3.2.1.2 Collecting mouse embryonic (E)3.5 mouse embryos

1. Dissect pregnant mouse in sterile conditions. Wipe the surrounding areas with 70% ethanol before beginning. Sterilize the scissors and forceps used for dissection with the dry-heat sterilization apparatus.
2. Remove the uteri of an E3.5 pregnant mouse. Trim away excess fat.
3. With a 3 mL syringe, collect 3 mL of ESC derivation media and attach a 23-gage needle to the syringe. While securing the uterine horn with forceps, insert the needle and flush the uterine horn with 1 mL ESC derivation media into a 35 mm dish. Repeat this process with the second uterine horn.

3.2.1.3 Plating and early culture

1. Using a P20 pipette set at 4 μL, collect the flushed embryos and wash through several drops of ESC derivation media in a sterile 35 mm dish before placing them into a final drop of ESC derivation media.
2. Plate blastocysts in MEF-plated 4-well dishes containing 750 μL ESC derivation media per well. Plate one blastocyst per well.
3. Incubate the plates at 37 °C, 5% CO_2. Do not disturb the plates for 48 h to allow for the blastocysts to hatch and attach to the feeder layer. The majority (80–90%) of blastocysts should hatch and attach to the feeder layer.
4. On day 3 post-plating, replace the media with fresh ESC derivation media. Continue to feed in this manner every other day until disaggregation occurs.
5. After the initial 48 h of incubation, monitor growth under a microscope daily. If outgrowths are prominent 4–5 days post-plating, proceed to the next step. If not, wait one or two more days before proceeding to the next step.

3.2.1.4 Disaggregation of blastocysts outgrowth

1. Aliquot 20 μL trypsin into each well of a 96-well plate. Incubate at 37 °C for at least 20 min.
2. Carefully wash each well with 500 μL of 1 × PBS and then add 500 μL of 1 × PBS.
3. Use 1 or 2 sterile needles to scrape the blastocyst outgrowth away from the surrounding MEF cells. Do this under the microscope. Use a P20 pipet set at 3 μL to place the outgrowth in one well of a 96-well plate containing prewarmed trypsin.

4. Once all the blastocyst outgrowths are in the 96-well dish, incubate in the 37 °C incubator for 5–10 min.
5. Now dissociate each outgrowth individually using a P20 pipet set at 10 μL. Vigorously pipet up and down to create a single cell suspension. Neutralize with 20 μL MEF media/well.
6. Plate embryos individually into wells of a MEF-plated 96-well plate with ESC derivation media. Change media the next day as early as possible.
7. Mouse ES colonies will be evident over the next 2–3 days (see Note 19).
8. Passage and culture ES cells every alternative day (1:3–1:4) with ES culture media.
9. Change media daily.

3.2.2 Passaging ESCs

1. Aspirate media and wash the cells with room temperature 1 × PBS.
2. Add 0.25% Trypsin-EDTA to cover the cell layer, usually 1/3 the amount used to culture the cells.
3. Return the cells to the incubator for 4–6 min. Observe cell dissociation with an inverted microscope.
4. As soon as colonies begin to dissociate, inactivate the trypsin by adding an equal amount of MEF medium.
5. Pipet up and down vigorously to dissociate into a single cell suspension.
6. Collect the cells in one tube and centrifuge for 5 min at 1500 rpm. Resuspend the pellet in 1 mL of ESC medium, creating a single cell suspension through gentle trituration.
7. Plate the cells on MEF feeder plates (with 1:3 splitting).

3.2.3 Culturing ESCs

1. Culture ESCs on MEF feeders in ESC culture medium supplemented with 3 μM GSK3 inhibitor CHIR99021 (Stemgent#04-0004), 1 μM MEK inhibitor PD0325901 (Stemgent#04-0006), and 1000 U/mL LIF (Millipore#ESG1106) (see Note 20).
2. When ~70–80% confluent, split ESCs with Trypsin-EDTA (0.25%) (see Section 3.2.2) into 1:3–1:4. ESCs. Change media daily (see Note 21).

3.2.4 Cryopreserving ESCs

1. When cells are ~70–80% confluent, aspirate the media and wash the cells with 1 × PBS.
2. Trypsinize the cells with 0.25% Trypsin-EDTA as mentioned above.

3. Pellet down the cells and resuspend in recovery cell culture freezing medium (Gibco#12648-010). Use 2mL freezing medium per one ~70–80% confluent 6-well well and place into two cryogenic vials, each containing 1mL cell suspension.
4. Put the vials in cryo cooling container and maintain at −80°C for 1 day. Then transfer vials to liquid nitrogen.

3.2.5 Thawing of ESCs

1. Remove a vial from liquid nitrogen storage and thaw quickly in 37°C water bath until almost completely melted.
2. Gently add cell suspension to 10mL of room temperature MEF medium in a 15mL tube.
3. Spin tube down at 1500rpm for 5min, remove supernatant and gently resuspend pellet in 2mL of prewarmed ESC culture medium.
4. Plate into one well of a 6-well plate. Place in a tissue culture incubator. Change media daily.
5. If recovery is adequate, colonies should be ready to passage in 2–3 days.

3.2.6 Characterization of ESCs

3.2.6.1 Morphological and molecular characterization of ESCs

The ESCs can be morphologically examined (Fig. 1) followed by profiling of the cells by RT-PCR and IF.

1. For RT-PCR, prepare RNAs by aspirating media from one confluent 6-well well and resuspending the ESCs in 1mL TRIzol reagent after washing with 1× PBS. These samples can be stored in −80°C for later use.

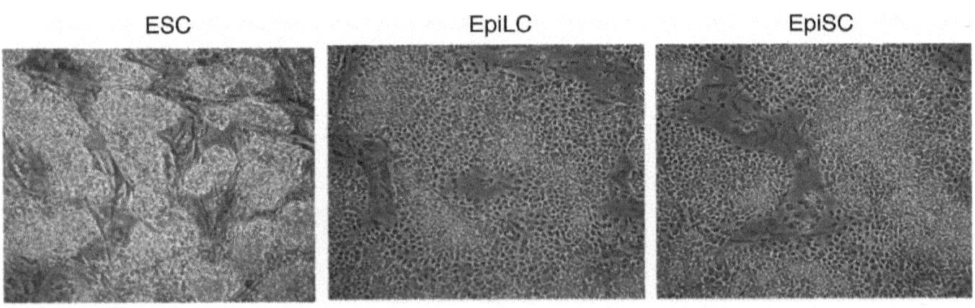

Fig. 1 Stereo micrographs of ESCs, EpilLCs, and EpiSCs highlighting differences in cellular morphology. *The figure was adopted from Gayen, S., Maclary, E., Buttigieg, E., Hinten, M., & Kalantry, S. (2015). A primary role for the Tsix lncRNA in maintaining random X-chromosome inactivation.* Cell Reports, *11(8), 1251–1265. doi: 10.1016/j.celrep.2015.04.039.*

2. Isolate RNA according to a standard TRIzol extraction protocol.
3. Analyze the expression of cell-specific markers by RT-PCR.
4. In addition to via RT-PCR, ESCs can be characterized through IF for proteins specific to ESCs.

3.2.6.2 IF-based detection of marker proteins in individual ESCs

1. Begin with fixed, permeabilized cells, plated on gelatinized glass coverslips and stored in 70% ethanol.
2. Make blocking buffer and warm to 37 °C.
3. Place sample coverslip in a 6-well dish that contains 2 mL of 1 × PBS in each well.
4. Wash briefly with three changes of 1 × PBS to remove ethanol.
5. Wash with 1 × PBS three times, 3 min each on a rocker.
6. Wrap a glass plate tightly with Parafilm for incubating coverslips for subsequent steps.
7. Block slides for 30 min at 37 °C in 50 μL prewarmed blocking buffer in a humid chamber: Place a 50 μL drop of blocking buffer on the parafilm-wrapped glass plate and invert the coverslip, sample-side down, into the blocking buffer. Place the parafilm-wrapped plate in the humid chamber, and incubate for 30 min at 37 °C. All incubations in blocking buffer, primary antibody, or secondary antibody should be set up in this manner.
8. Carefully lift coverslip from blocking buffer with forceps and place it into a 50 μL droplet of diluted primary antibody on a parafilm-wrapped plate. Incubate with 50 μL primary antibody diluted in prewarmed blocking buffer (dilution based on primary antibody you are using) in a humid chamber at 37 °C for 1 h (see Note 22).
9. Remove coverslip from primary antibody solution and place, sample-side up, in a 6-well dish. Wash three times with 1 × PBS/0.2% Tween 20 for 3 min on a rocker.
10. Incubate coverslip in 50 μL prewarmed blocking buffer on a parafilm-wrapped plate in a humid chamber for 5 min at 37 °C.
11. Incubate coverslip with 50 μL secondary antibody diluted in prewarmed blocking buffer in a humid chamber at 37 °C for 30 min. Antibody dilution depends on the secondary antibody used; Alexa Fluor-conjugated secondary antibodies should be used at a 1:300 dilution.
12. Remove coverslip from secondary antibody and wash three times with 1 × PBS/0.2% Tween 20 for 3 min each on a rocker. The first wash should contain a 1:100,000–1:200,000 dilution of DAPI (5 mg/mL)

(see Note 23). Then, rinse once briefly with PBS/0.2% Tween 20 and wash two more times for 5–7 min each while rocking to remove excess DAPI.

13. Remove coverslip from the dish, tap off excess liquid, and then mount on a slide, sample-side down, in mounting medium. Image samples or store at −20 °C for later imaging.

3.3 Mouse epiblast stem cells (mEpiSCs)

3.3.1 Derivation of EpiSCs from E3.5 preimplantation embryos

3.3.1.1 Preparation of MEF feeder plates

1. Thaw and plate inactivated MEFs onto 4-well plates before the day of isolating embryos. The number of wells needed is dependent on the number of pregnant females. The number of embryos recovered per female can be variable but generally averages between 6 and 8 for most mouse strains. Blastocysts are plated in individual wells.
2. On the day of the embryo harvest, remove the MEF media and rinse the 4-well feeder plates with PBS. Replace the media with K15F5 media.

3.3.1.2 Collecting E3.5 blastocysts

1. Dissect the mice in a sterile condition. Wipe the surroundings with 70% ethanol before beginning. Sterilize the scissors and forceps used for dissection with the dry-heat sterilization apparatus.
2. Remove the uteri of a E3.5 pregnant mouse. Trim away excess fat.
3. With a 3 mL syringe, collect 3 mL of K15F5 medium and attach a 23-gage needle. While securing the uterine horn with forceps, insert the needle and flush the uterine horn with 1 mL of K15F5 medium into a 35 mm dish. Repeat this process with the second uterine horn.

3.3.1.3 Plating and early culture

1. Using a P20 pipette set at 4 μL, collect the flushed embryos and wash through several drops of K15F5 medium before placing them into a final drop of K15F5 medium.
2. Plate E3.5 blastocysts in MEF-plated 4-well dishes containing 750 μL K15F5 medium. Only plate one blastocyst per well in a 4-well MEF feeder plate. Always use fresh tips for plating each blastocyst.
3. Incubate the plates at 37 °C, 5% CO_2. Do not disturb the plates for 48 h to allow for the blastocysts to hatch and attach to the feeder layer. The majority (80–90%) of blastocysts should hatch and attach to the feeder layer.

4. On day 3 post-plating, replace the medium with fresh medium. Continue to feed in this manner every day until disaggregation.
5. On day 5, begin to monitor the growth of the outgrowth under a microscope daily. If outgrowths are prominent at day 5 of plating, proceed to the next steps below. If not, wait one or two more days before proceeding to the next step.

3.3.1.4 Disaggregation of blastocysts outgrowth

1. Aliquot 20 μL trypsin in each well of a 96-well plate. Incubate at 37 °C for at least 20 min.
2. When the trypsin is warmed up, change media in each well of 4-well plates with sterile 1 × PBS. Repeat for a total of 2 ×.
3. Wipe down the surface around the microscope and the microscope itself with 70% ethanol.
4. Using a P20 pipet set at 3 μL, scrape up the blastocyst outgrowth under the microscope and place in a 96-well well with prewarmed trypsin. Exercise as sterile technique as possible.
5. Process all of the blastocyst outgrowths as above.
6. Incubate in the 37 °C incubator for 10 min.
7. Once all the blastocyst outgrowths are in the 96-well dish, pipet up and down very gently in each well using the P20 pipet set at 10 μL.
8. Watch intermittently to make sure the blastocyst outgrowths are dissociating.
9. Check each of the 96-well wells under the microscope—the outgrowths should now be disaggregated in small clumps. Partial dissociation is important for EpiSC derivation (see Note 24).
10. Plate each of the partial dissociated embryos into a MEF-plated well of 24-well plate for an additional 6 days in K15F5 medium (750 μL/well).
11. At this point, passage each culture by a brief exposure (2–3 min) to 0.25% trypsin/EDTA, inactivate trypsin with FBS-containing MEF medium, gently tritrate to prevent complete single cell dissociation of any pluripotent clusters, and plate into a 6-well plate containing feeders in K15F5 medium (2 mL/well).
12. Morphologically distinct mouse ES cell and/or EpiSC colonies will be evident over the next 4–8 days (16–20 total days from initial blastocyst explants). Separately expand these colony types in a medium designed to support either mouse ES cells or EpiSCs.
13. Pick individual EpiSC colonies and dissociate the EpiSC colonies manually into small clusters using a glass needle and plate into 24-well plate

containing feeders in EpiSC culture medium. Passage EpiSCs every third day using 1.5 mg/mL collagenase type IV (Invitrogen) and triturate into small clumps of 10–100 cells.

14. Continue growing EpiSC cells by passaging into 12-well (use 1.5 mL media) and 6-well (use 2 mL media) plates (MEF plated), respectively, with EpiSC culture medium.

3.3.2 Culturing EpiSCs

1. To passage EpiSCs, aspirate EpiSC culture medium from each well of 6-well plate, add 1 mL of EpiSC passaging medium (1.5 mg/mL collagenase type IV) (for one 6-well well), and incubate at 37 °C for 8–12 min (see Note 25).
2. Add 1 mL of EpiSC culture medium to each well and dislodge colonies by gentle pipetting to create small clumps (not to single cells). Combine colony suspensions from each plate into one 15-mL conical tube.
3. Centrifuge at 1500 rpm for 5 min. Gently resuspend the cell pellet in 2 mL EpiSC culture medium and perform another spin. Repeat the rinse/spin cycle one more time for a total of three. FGF2 is omitted from the EpiSC culture medium when used for rinsing.
4. Resuspend colonies in 2 mL of EpiSC culture medium and transfer to an individual well of a 6-well feeder plate.
5. Keep the plate in the incubator.
6. EpiSCs are typically split 1:4–1:6 every 2–3 days. The density at which each EpiSC line is grown is very important and must be determined empirically.
7. EpiSC culture medium should be changed daily.

3.3.3 Cryopreservation of EpiSCs

1. Isolate EpiSCs into small clumps after collagenase treatment as mentioned above.
2. Resuspend pellet from one 6-well well in 1 mL of recovery cell culture freezing medium (Gibco#12648-010) and place into two cryogenic vials each containing 1 mL cell suspension with freezing medium.
3. Place tubes in cryo cooling container and store at −80 °C. The following day, transfer vials to a liquid nitrogen freezer for permanent storage.

3.3.4 Thawing of EpiSCs

1. Remove 1 vial from liquid nitrogen storage and thaw in 37 °C water bath until almost completely melted.

2. Gently add cell suspension to 10 mL of room temperature MEF medium.
3. Spin down at 1500 rpm for 5 min, remove supernatant and gently resuspend pellet in 2 mL of prewarmed EpiSC culture medium.
4. Remove MEF medium from MEF feeders and plate suspension into one well of 6-well plate. Place in a tissue culture 37 °C incubator. Change media daily.
5. If recovery is adequate, colonies should be ready to passage in 2–3 days.

3.3.5 Characterization of EpiSCs

3.3.5.1 Morphological and molecular characterization of EpiSCs

The EpiSCs can be morphologically examined (Fig. 1) followed by profiling of the cells by RT-PCR and IF.

1. For RT-PCR, prepare RNAs by aspirating media from ~1 confluent 6-well well of EpiSCs and by adding 1 mL of TRIzol reagent after washing with 1 × PBS. These samples can be stored in 1.5 mL tubes at −80 °C for later use.
2. Isolate RNA according to a standard TRIzol extraction protocol.
3. Analyze the expression of cell-specific markers by RT-PCR.
4. In addition to via RT-PCR, the EpiLCs can be characterized through IF for proteins specific to EpiSCs.
5. In addition to RT-PCR and IF, female EpiSCs can also be characterized through RNA FISH to visualize expression of *Xist* RNA, a marker of X-chromosome inactivation, which should be expressed in EpiSCs.

3.3.5.2 IF-based detection of marker proteins in individual EpiSCs

1. Begin with fixed, permeabilized cells, plated on gelatinized glass coverslips and stored in 70% ethanol.
2. Make blocking buffer and warm to 37 °C.
3. Place sample coverslip in a 6-well dish that contains 2 mL of 1 × PBS in each well.
4. Wash briefly with three changes of 1 × PBS to remove ethanol.
5. Wash again with 1 × PBS three times, 3 min each on a rocker.
6. Wrap a glass plate tightly with Parafilm for incubating coverslips in subsequent steps.
7. Block slides for 30 min at 37 °C in 50 μL prewarmed blocking buffer in a humid chamber: Place a 50 μL drop of blocking buffer on the parafilm-wrapped glass plate and invert the coverslip, sample-side down, into the blocking buffer. Place the parafilm-wrapped plate in the humid chamber and incubate for 30 min at 37 °C. All incubations

in blocking buffer, primary antibody, or secondary antibody should be set up in this manner.

8. Carefully lift coverslip from blocking buffer with forceps and place into a 50 μL droplet of diluted primary antibody on a parafilm-wrapped plate. Incubate with 50 μL primary antibody diluted in prewarmed blocking buffer (dilution based on primary antibody you are using) in a humid chamber at 37 °C for 1 h (see Note 26).
9. Remove coverslip from primary antibody solution and place, sample-side up, in a 6-well dish. Wash three times with 1 × PBS/0.2% Tween 20 for 3 min on a rocker.
10. Incubate coverslip in 50 μL prewarmed blocking buffer on a parafilm-wrapped plate in a humid chamber for 5 min at 37 °C.
11. Incubate coverslip with 50 μL secondary antibody diluted in prewarmed blocking buffer in a humid chamber at 37 °C for 30 min. Antibody dilution depends on the secondary antibody used; Alexa Fluor-conjugated secondary antibodies should be used at a 1:300 dilution.
12. Remove coverslip from secondary antibody and wash three times with 1 × PBS/0.2% Tween 20 for 3 min each on a rocker. The first wash should contain a 1:100,000–1:200,000 dilution of DAPI (5 mg/mL) (see Note 23). Then, rinse once briefly with PBS/0.2% Tween 20 and wash two more times for 5–7 min each while rocking to remove excess DAPI.
13. Remove coverslip from the dish, tap off excess liquid, and then mount on a slide, sample-side down, in mounting medium. Image samples or store at −20 °C for later imaging.

3.4 EpiLCs

3.4.1 Generating EpiLCs from ESC

3.4.1.1 Preparation of gelatin-coated tissue culture dishes

1. Prepare 0.2% gelatin with distilled water and autoclave. To prepare gelatin-coated tissue culture dishes, apply 0.2% gelatin for 10–15 min. Then, aspirate gelatin and dry gelatin by keeping the lid open inside tissue culture hood.

3.4.1.2 Preparation of fibronectin-coated tissue culture dishes

1. Prepare 10 μg/mL fibronectin solution and apply to tissue culture dishes for 20 min. Aspirate the solution and let the dishes dry for 30 min by keeping the lid open inside the tissue culture hood.

3.4.1.3 Differentiating ESCs into EpiLCs

1. Culture ES cells on MEF feeders (6-well well) in 2i/LIF or serum/LIF conditions (see Note 27). When ESCs become confluent, add Trypsin-EDTA (0.25%) to the cells (700 μL to 1 mL Trypsin/1, 6-well well) and incubate at 37 °C for 5 min in the incubator. Pipet vigorously and neutralize trypsin with any media containing FBS. Then, pellet down the cells with centrifugation at 1500 rpm for 5 min.
2. Resuspend the cells with 1 mL N2B27 media and again spin down the cells and aspirate media to remove any leftover trypsin with the cells.
3. Resuspend the cells in N2B27 medium supplemented with 3 μM GSK3 inhibitor CHIR99021 (Stemgent#04-0004), 1 μM MEK inhibitor PD0325901 (Stemgent#04-0006), 1000 U/mL LIF (Millipore# ESG1106) and plate into a gelatin-coated 6-well well (if necessary all the cells, plate onto 3, 6-well well from 1 confluent 6-well well; or maintain 1, 6-well well and discard the remaining cells). Change media daily.
4. After 2 days, add Accutase (see Note 28) and incubate for 5–10 min at room temperature. Check the flask frequently (in every 1–2 min) to see rounded cells while remaining attached to the bottom of the flask. Then, smack the flask against the palm of your hand to dislodge any adhered cells. Gently disperse the cells and take a sample of the cell suspension to determine the viable cell density.
5. Add an aliquot of the detached cells to fresh N2B27 media in new gelatin-coated dishes. Place the dishes into the incubator.
6. Culture the cells in gelatin-coated dishes for four passages. Each passage should be ~2 days and change media daily. At each passage, the cells should be split from 1:2 to 1:3. During culture at passages 1–4, colonies should look round and shiny.
7. To differentiate ESCs into EpiLCs, culture ESCs for 48 h in N2B27 media supplemented with 10 ng/mL FGF2 (R&D Systems#233-FB) and 20 ng/mL Activin A (R&D Systems#338-AC) in Fibronectin (10 μg/mL) (Sigma#F1141)-coated tissue culture dishes.
8. For further differentiation, culture the cells in N2B27 medium without FGF2 and Activin A for additional days.

3.4.2 Cryopreserving EpiLCs

As ESCs begin to differentiate into EpiLCs, the cells can be frozen at passages 1–4 before the generation of EpiLCs (see Note 29).

1. Dissociate cells using Accutase or 0.05% Trypsin-EDTA at passage 1–4 and pellet down the cells by centrifugation with 1500 rpm at 37 °C.

2. Resuspend pellet of one 6-well well in 1 mL of recovery cell culture freezing medium (Gibco#12648-010) and place into two cryogenic vials each containing 1 mL cell suspension with freezing medium.
3. Place tubes in cryo cooling container and store at −80 °C. The following day, transfer vials to a liquid nitrogen freezer for permanent storage.
4. Use one cryogenic vial for thawing cells into one well of gelatin-coated 6-well plate.

3.4.3 Thawing cells for EpiLC generation

1. Prepare gelatin-coated tissue culture plate.
2. Remove one vial from liquid nitrogen storage and thaw in 37 °C water bath until almost completely melted.
3. Gently add cell suspension to 10 mL of room temperature MEF medium.
4. Spin down at 1500 rpm for 5 min, remove supernatant and gently resuspend pellet in 1 mL of prewarmed N2B27 medium.
5. Spin down cell again at 1500 rpm for 5 min, remove supernatant and gently resuspend pellet in 2 mL of prewarmed N2B27 medium supplemented with 3 μM GSK3 inhibitor CHIR99021 (Stemgent#04-0004), 1 μM MEK inhibitor PD0325901 (Stemgent#04-0006), 1000 U/mL LIF (Millipore#ESG1106).
6. Plate suspension into one well of 6-well gelatin-coated plate. Place in tissue culture incubator. Change media daily.
7. If recovery is adequate, colonies should be ready to passage in 2 days.
8. Follow the protocol for the conversion of ESCs into EpiLCs as mentioned in Section 3.4.1.3.

3.4.4 Characterization of EpiLCs

3.4.4.1 Morphological and molecular characterization of EpiLCs

The differentiation of ESCs into EpiLCs can be morphologically examined (Fig. 1) followed by profiling of the cells by RT-PCR and IF.

1. For RT-PCR, prepare RNAs by aspirating media from one confluent 6-well well and resuspending the ESCs in 1 mL TRIzol reagent after washing with 1 × PBS. These samples can be stored in − 80 °C for later use.
2. Isolate RNA according to a standard TRIzol extraction protocol.
3. Analyze the expression of cell-specific markers by RT-PCR.
4. In addition to via RT-PCR, the EpiLCs can be characterized through IF for proteins specific for ESCs, EpiLCs, and differentiated EpiLCs.
5. In addition to RT-PCR and IF, female EpiLCs can also be characterized through RNA FISH for Xist RNA, a marker of X-chromosome inactivation, which is only expressed in EpiLCs and differentiated EpiLCs.

3.4.4.2 IF-based detection of marker proteins in individual EpiLCs

1. Begin with fixed, permeabilized cells, plated on gelatinized glass coverslips and stored in 70% ethanol.
2. Make blocking buffer and warm to 37 °C.
3. Place sample coverslip in a 6-well dish that contains 2 mL of 1 × PBS in each well.
4. Wash briefly with three changes of 1 × PBS to remove ethanol.
5. Wash with 1 × PBS three times, 3 min each on a rocker.
6. Wrap a glass plate tightly with Parafilm for incubating coverslips in subsequent steps.
7. Block slides for 30 min at 37 °C in 50 μL prewarmed blocking buffer in a humid chamber: Place a 50 μL drop of blocking buffer on the parafilm-wrapped glass plate and invert the coverslip, sample-side down, onto the blocking buffer. Place the parafilm-wrapped plate in the humid chamber, and incubate for 30 min at 37 °C. All incubations in blocking buffer, primary antibody, or secondary antibody should be set up in this manner.
8. Carefully lift coverslip from blocking buffer with forceps and place it into a 50 μL droplet of diluted primary antibody on a parafilm-wrapped plate. Incubate with 50 μL primary antibody diluted in prewarmed blocking buffer (dilution based on primary antibody you are using) in a humid chamber at 37 °C for 1 h (see Note 30).
9. Remove coverslip from primary antibody solution and place, sample-side up, in a 6-well dish. Wash three times with 1 × PBS/0.2% Tween 20 for 3 min on a rocker.
10. Incubate coverslip in 50 μL prewarmed blocking buffer on a parafilm-wrapped plate in a humid chamber for 5 min at 37 °C.
11. Incubate coverslip with 50 μL secondary antibody diluted in prewarmed blocking buffer in a humid chamber at 37 °C for 30 min. Antibody dilution depends on the secondary antibody used; Alexa Fluor-conjugated secondary antibodies should be used at a 1:300 dilution.
12. Remove coverslip from secondary antibody and wash three times with 1 × PBS/0.2% Tween 20 for 3 min each on a rocker. The first wash should contain a 1:100,000–1:200,000 dilution of DAPI (5 mg/mL). Then, rinse once briefly with PBS/0.2% Tween 20 and wash two more times for 5–7 min each while rocking to remove excess DAPI.
13. Remove coverslip from the dish, tap off excess liquid, and then mount on a slide, sample-side down, in mounting medium. Image samples or store at −20 °C for later imaging.

4. Discussion

The slow and stepwise conversion of ESCs uniformly into EpiLCs permits the molecular dissection of transcriptional and epigenetic processes that tip the balance between self-renewal of naïve pluripotent stem cells vs differentiation of the naïve pluripotent stem cells into primed pluripotent stem cells. EpiLCs, however, are only transiently present in culture as ESCs differentiate. EpiSCs, on the other hand, are self-renewing primed pluripotent cells that can be stably propagated in culture.

Although EpiLCs morphologically and transcriptionally mimic EpiSCs, the two cell types are not identical. Both EpiLCs and EpiSCs share common expression patterns for some genes. For example, the pluripotency gene *Oct3/4* (*Pou5f1*) is expressed in EpiSCs as well as in EpiLCs (Hayashi et al., 2011). Other pluripotency genes such as *Sox2*, *Nanog*, *Klf5*, however, are expressed in EpiSCs but downregulated in EpiLCs (Hayashi et al., 2011). Genes that characterize the postimplantation epiblast, e.g., *Wnt3*, *Fgf5*, and *Dnmt3b*, are upregulated in EpiSCs as well as EpiLCs (Hayashi et al., 2011). Endoderm marker genes such as *Gata4*, *Gata6*, and *Sox17* are expressed highly in EpiSCs and a very low level in EpiLCs (Hayashi et al., 2011). *Blimp1*, a marker of primordial germ cells, is also expressed in a subset of EpiSCs but downregulated in EpiLCs (Hayashi et al., 2011). The differentiation of ESCs into EpiLCs may be a useful as a model system to investigate the transcriptional and epigenetic mechanisms underlying the ICM to epiblast differentiation. EpiLCs as well as EpiSCs may also serve as starting materials for the induction of other lineages derived from the epiblast.

5. Recipes

DNase I solution (13.33 mg/mL).

Reagent	**Amount**
DNase I (Roche#104159)	100 mg
Sterile sodium chloride (NaCl) (5 M)	0.45 mL
Sterile 80% glycerol	4.69 mL
Sterile distilled water	2.36 mL

Aliquots store at −20 °C.

0.2% gelatin solution.

Reagent	Amount
Gelatin	200 mg
Distilled water	100 mL

Dissolve gelatin to distilled water and autoclave (final volume 100 mL). Keep it at 4 °C.

BSA fraction V solution in PBS (0.1%, w/v).

Reagent	Amount
BSA fraction V (Sigma-Aldrich A3311 or A1470)	10 mg
Phosphate-buffered saline (PBS) without Ca^{2+}/Mg^{2+}	10 mL

Resuspend BSA in PBS. Filter through a 0.45-μm filter, aliquot, and store at −80 °C.

FGF2 stock solution (10 μg/mL).

Reagent	Amount
Recombinant fibroblast growth factor-2 (FGF2; R&D Systems 233-FB-025)	25 μg
BSA fraction V solution in PBS (0.1%, w/v)	2.5 mL

Resuspend lyophilized FGF2 in BSA/PBS, mix well, aliquot 50 μL, and freeze at −80 °C. Thaw each aliquot as needed and store it at 4 °C. Do not refreeze the thawed aliquot.

Activin A stock solution (10 μg/mL).

Reagent	Amount
Recombinant Activin A (ActA; R&D Systems 338-AC-005)	5 μg
BSA fraction V solution in PBS (0.1%, w/v)	0.5 mL

Resuspend lyophilized Activin A in its vial, mix well, aliquot 50 μL, and freeze at −80 °C. Thaw each aliquot as needed and store it at 4 °C. Do not refreeze.

CHIR99021 stock solution (10 mM).

Reagent	Amount
GSK3 inhibitor CHIR99021 (Stemgent#04-0004)	5 mg
DMSO	1.07 mL

Protect from light, aliquots store at −20 °C. The above data is based on the product molecular weight 465.34. Batch specific molecular weights may vary from batch to batch due to the degree of hydration, which will affect the solvent volumes required to prepare stock solutions. Thaw each aliquot as needed and store it at 4 °C. Do not refreeze.

MEK inhibitor PD0325901 (10 mM).

Reagent	Amount
PD0325901 (Stemgent#04-0006)	2 mg
DMSO	414.8 μL

Protect from light, aliquots store at −20 °C. Thaw each aliquot as needed and store it at 4 °C. Do not refreeze.

MEF medium.

Reagent	Amount
MEM Alpha (1 ×) (Gibco#12561-056)	90 mL
FBS (Gibco#10437-028)	10 mL

Filter and store the medium (final volume 100 mL) at 4 °C and use within 2 weeks.

EpiSC dissociation medium: Collagenase type IV (1.5 mg/mL).

Reagent	Amount	Final concentration
Collagenase type IV (Invitrogen#17104-019)	0.5 mg	1.5 mg/mL
EpiSC culture medium (without FGF2)	333.33 mL	

Dissolve 0.5 mg of collagenase type IV in 333.33 mL of EpiSC medium. Filter, aliquot in 10 mL, and freeze at −80 °C. Thaw each aliquot as needed and store it at 4 °C. Do not refreeze.

ESC derivation medium.

Reagent	Amount	Final concentration
KnockOut DMEM (Gibco#10829-018)	76 mL	
KnockOut serum replacement (KSR) (Invitrogen#10828-028)	20 mL	20%
Penicillin-streptomycin (100 ×) (Gibco#15070-063)	1 mL	1 ×
L-Glutamine (200 mM) (Gibco#25030)	1 mL	2 mM
MEM non-essentials amino acids (100 ×) (Gibco#11140-050)	1 mL	1 ×
β-Mercaptoethanol (10 mM) (Sigma#M7522)	1 mL	0.1 mM
Supplements		
GSK3 inhibitor CHIR99021 (10 mM) (Stemgent#04-0004)	30 μL	3 μM
MEK inhibitor PD0325901 (10 mM) (Stemgent#04-0006)	10 μL	1 μM
LIF (10^7/mL) (Millipore#ESG1106)	10 μL	10^3/mL

Filter and store the medium (final volume 100 mL) at 4 °C and use within 2 weeks. Add supplements (GSK3 inhibitor CHIR99021, MEK inhibitor PD0325901, and LIF) to the media during freshly during use of the media. Prepare β-mercaptoethanol 10 mM stock from 14.3 M stock (Sigma-Aldrich M7522) by adding 6.25 μL to 9 mL of sterile distilled water.

ESC culture medium.

Reagent	Amount	Final concentration
KnockOut DMEM (Gibco#10829-018)	76 mL	
Fetal bovine serum embryonic stem cell qualified (ES-FBS) (Bio-Techne#S10250)	5 mL	5%
KnockOut serum replacement (KSR) (Invitrogen#10828-028)	15 mL	15%
L-glutamine (200 mM) (Gibco#25030)	1 mL	2 mM
MEM non-essentials amino acids (100 ×) (Gibco#11140-050)	1 mL	1 ×
β-Mercaptoethanol (10 mM) (Sigma#M7522)	1 mL	0.1 mM
Supplements		
GSK3 inhibitor CHIR99021 (10 mM) (Stemgent#04-0004)	30 μL	3 μM
MEK inhibitor PD0325901 (10 mM) (Stemgent#04-0006)	10 μL	1 μM
LIF (10^7/mL) (Millipore#ESG1106)	10 μL	10^3/mL

Filter and store the medium (final volume 100 mL) at 4 °C and use within 2 weeks. Add supplements (GSK3 inhibitor CHIR99021, MEK inhibitor PD0325901, and LIF) to the media during freshly during use of the media. ESC can be cultured in ESC media supplemented with LIF only. Prepare β-mercaptoethanol 10 mM stock from 14.3 M stock (Sigma-Aldrich M7522) by adding 6.25 μL to 9 mL of sterile distilled water.

K15F5 medium.

Reagent	Amount	Final concentration
KnockOut DMEM (Gibco#10829-018)	76 mL	
KnockOut serum replacement (KSR) (Invitrogen#10828-028)	15 mL	15%
Fetal bovine serum embryonic stem cell qualified (ES-FBS) (Bio-Techne#S10250)	5 mL	5%
Penicillin-streptomycin (100 ×) (Gibco#15070-063)	1 mL	1 ×
L-Glutamine (200 mM) (Gibco#25030)	1 mL	2 mM
MEM non-essentials amino acids (100 ×) (Gibco#11140-050)	1 mL	1 ×
β-Mercaptoethanol (10 mM) (Sigma#M7522)	1 mL	0.1 mM

Filter and store the medium (final volume 100 mL) at 4 °C and use within 2 weeks. Prepare β-mercaptoethanol 10 mM stock from 14.3 M stock (Sigma-Aldrich M7522) by adding 6.25 μL to 9 mL of sterile distilled water.

EpiSC culture medium.

Reagent	Amount	Final concentration
KnockOut DMEM (Gibco#10829-018)	76 mL	
KnockOut serum replacement (KSR) (Invitrogen#10828-028)	20 mL	20%
L-Glutamine (200 mM) (Gibco#25030)	1 mL	2 mM
MEM non-essentials amino acids (100 ×) (Gibco#11140-050)	1 mL	1 ×
β-Mercaptoethanol (10 mM) (Sigma#M7522)	1 mL	0.1 mM
FGF2 (R&D Systems) (10 μg/mL)	100 μL	10 ng/μl

Filter and store the medium (final volume 100 mL) at 4 °C and use it within 2 weeks. Add FGF2 supplement freshly.

N2B27 medium.

Reagent	Amount	Final concentration
DMEM/F12 (Gibco#11330-032)	47.5 mL	
Neurobasal media (Gibco#21103-049)	47.5 mL	
L-Glutamine (200 mM) (GIBCO#25030)	1 mL	2 mM
β-Nercaptoethanol (10 mM) (Sigma#M7522)	1 mL	0.1 mM
N2 supplement (100 ×) (Invitrogen#17502048)	1 mL	1 ×
B27 supplement (50 ×) (Invitrogen#17504-044)	2 mL	1 ×

Filter and store the medium (final volume 100 mL) at 4 °C and use it within 2 weeks. Prepare β-mercaptoethanol 10 mM stock from 14.3 M stock (Sigma-Aldrich M7522) by adding 6.25 μL to 9 mL of sterile distilled water.

6. Notes

1. All relevant governmental and institutional standard rules and regulations must be maintained during the experiments involving rodents. Timed natural matings should be used and 12h from the middle of the dark period indicate day 0.5 (E0.5).
2. Other freezing media can be used to freeze the cells.
3. Mitotically inactive MEFs served as feeders for ESC or EpiSC culture. Non-irradiated MEFs (Mitomycin-C treated) can be used as feeder cells.
4. This reagent is a biohazard; adequate safety instructions should be taken when handling.
5. We use penicillin-streptomycin in media only for the cell derivation and not for the cell culture.
6. Properly autoclaved and sharpened forceps and scissors should be used for any kind of dissection.
7. This reagent is a biohazard; adequate safety instructions should be taken when handling.
8. RNA purification methods other than TRIzol can be used in place of TRIzol.
9. Any robust RT-PCR system should work. For RT-PCR, random primers or gene specific reverse primer can be used for cDNA synthesis. Instead of RT-PCR, qPCR may be useful to detect the expression of specific transcripts in different cell lines of ESCs, EpiSCs, and EpiLCs.
10. Other primary antibodies specific to ESCs can be used if good antibodies exist.
11. Short glass plates designed for casting protein gels are a good size for hybridization. For example, Bio-Rad's MiniPROTEAN Short Plates fit well in the humid chamber described in Note 12.
12. For humid chambers, a microscope slide box (i.e., one that holds 100 slides) works well. Place paper towels soaked in 1 × PBS at the bottom of the box to create a humid chamber for immunofluorescence.
13. For mounting medium, use Vectashield (Vector Labs) or similar anti-fade mounting medium and seal coverslips with clear nail polish after mounting on slides.
14. Use 22 × 22 mm coverslips, which fit well within a single well of a 6-well dish. All the dehydration and washing steps can be performed in the wells of this dish.

15. EpiSC can be derived from E3.5, E4.5, and E5.5 mouse embryos. E3.5 are preferable for EpiSC derivation to avoid microdissection of the embryo.
16. Other primary antibodies specific to EpiSCs can be used if good antibodies exist.
17. Other primary antibodies specific to EpiLCs can be used if good antibodies exist.
18. The concentration of MEFs for plating should approximately be:
    - 4-well plate: 2×10^5 cells/well; 8×10^5 cells/plate.
    - 6-well plate: 8×10^5 cells/well; 5×10^6/plate.
    - 10 cm dish: 5×10^6/dish
19. The ES colonies are now in passage 1. Increase passage number by 1 with each subsequent exposure of trypsin and carefully track passage number by labeling plates and tubes.
20. ESCs can be cultured in 2i/LIF or in serum/LIF conditions.
21. Generally, ESCs become confluent in every 2–3 days. So, split ESCs after every 2–3 days.
22. In ESC, NANOG, OCT4, and REX1; all three proteins should be expressed.
23. Dilute 5 mg/mL DAPI 1:500 in RNase/DNase-free ultrapure water, and store in the dark at −20 °C. Add 6–10 μL of this dilution into each well while washing samples.
24. Caution should be taken not to produce a single cell suspension.
25. The time required for the collagenase to free EpiSC colonies from the plate can vary. Check the colonies in every few minutes. They are ready when the edges begin to retract from the plate.
26. NANOG and OCT4 should be expressed in EpiSCs but REX1 should not be expressed.
27. The ESCs can be cultured in serum/LIF conditions. The conversion of ESCs to EpiLCs can be performed starting from ESCs either cultured in 2i/LIF or in serum/LIF conditions. However, ESC to EpiLC conversion seems to be more robust and with lower levels of cell death starting with ESCs cultured in 2i/LIF.
28. Trypsin-EDTA (0.05%) can be used for splitting instead of Accutase. But, incubate the cells for a maximum of 2–3 min at 37 °C after adding trypsin to the cells. Then, neutralize the trypsin by adding at least an equal volume of FBS-containing media. Pellet the cells by centrifugation at 1500 rpm for 5 min at room temperature.

29. EpiLC cannot be preserved but it is possible to preserve the differentiating ESCs at passage numbers 1, 2, 3 or 4 before EpiLC conversion. However, it is recommended to continue the ESC to EpiLC conversion.

30. NANOG and OCT4 should be expressed in ESCs, EpiLCs and differentiated EpiLCs but REX1 will not.

Acknowledgments

We thank members of the Kalantry laboratory for discussions and help with the manuscript. We especially acknowledge Marissa Cloutier for proofreading and editing the manuscript. We also thank to Dr. Thomas L. Saunders for providing the protocol for derivation of MEF cells. Work in the Kalantry lab is funded by the NIH (R01GM124571 and R01HD095463) and March of Dimes (1-FY18-526).

References

Bao, S., Tang, F., Li, X., Hayashi, K., Gillich, A., Lao, K., et al. (2009). Epigenetic reversion of post-implantation epiblast to pluripotent embryonic stem cells. *Nature, 461*(7268), 1292–1295. https://doi.org/10.1038/nature08534.

Bradley, A., Evans, M., Kaufman, M. H., & Robertson, E. (1984). Formation of germ-line chimaeras from embryo-derived teratocarcinoma cell lines. *Nature, 309*(5965), 255–256. https://doi.org/10.1038/309255a0.

Brons, I. G., Smithers, L. E., Trotter, M. W., Rugg-Gunn, P., Sun, B., Chuva de Sousa Lopes, S. M., et al. (2007). Derivation of pluripotent epiblast stem cells from mammalian embryos. *Nature, 448*(7150), 191–195. https://doi.org/10.1038/nature05950.

Brook, F. A., & Gardner, R. L. (1997). The origin and efficient derivation of embryonic stem cells in the mouse. *Proceedings of the National Academy of Sciences of the United States of America, 94*(11), 5709–5712. https://doi.org/10.1073/pnas.94.11.5709.

Evans, M. J., & Kaufman, M. H. (1981). Establishment in culture of pluripotential cells from mouse embryos. *Nature, 292*(5819), 154–156. https://doi.org/10.1038/292154a0.

Gardner, R. L., & Lyon, M. F. (1971). X chromosome inactivation studied by injection of a single cell into the mouse blastocyst. *Nature, 231*(5302), 385–386. https://doi.org/10.1038/231385a0.

Gayen, S., Maclary, E., Buttigieg, E., Hinten, M., & Kalantry, S. (2015). A primary role for the Tsix lncRNA in maintaining random X-chromosome inactivation. *Cell Reports, 11*(8), 1251–1265. https://doi.org/10.1016/j.celrep.2015.04.039.

Guo, G., Yang, J., Nichols, J., Hall, J. S., Eyres, I., Mansfield, W., et al. (2009). Klf4 reverts developmentally programmed restriction of ground state pluripotency. *Development, 136*(7), 1063–1069. https://doi.org/10.1242/dev.030957.

Hackett, J. A., & Surani, M. A. (2014). Regulatory principles of pluripotency: From the ground state up. *Cell Stem Cell, 15*(4), 416–430. https://doi.org/10.1016/j.stem.2014.09.015.

Hayashi, K., Ohta, H., Kurimoto, K., Aramaki, S., & Saitou, M. (2011). Reconstitution of the mouse germ cell specification pathway in culture by pluripotent stem cells. *Cell, 146*(4), 519–532. https://doi.org/10.1016/j.cell.2011.06.052.

Mak, W., Nesterova, T. B., de Napoles, M., Appanah, R., Yamanaka, S., Otte, A. P., et al. (2004). Reactivation of the paternal X chromosome in early mouse embryos. *Science, 303*(5658), 666–669. https://doi.org/10.1126/science.1092674.

Martin, G. R. (1981). Isolation of a pluripotent cell line from early mouse embryos cultured in medium conditioned by teratocarcinoma stem cells. *Proceedings of the National Academy of Sciences of the United States of America, 78*(12), 7634–7638. https://doi.org/10.1073/pnas.78.12.7634.

McMahon, A., Fosten, M., & Monk, M. (1983). X-chromosome inactivation mosaicism in the three germ layers and the germ line of the mouse embryo. *Journal of Embryology and Experimental Morphology, 74*, 207–220. Retrieved from https://www.ncbi.nlm.nih.gov/pubmed/6886595.

Monk, M., & Harper, M. I. (1979). Sequential X chromosome inactivation coupled with cellular differentiation in early mouse embryos. *Nature, 281*(5729), 311–313. https://doi.org/10.1038/281311a0.

Nakamura, T., Okamoto, I., Sasaki, K., Yabuta, Y., Iwatani, C., Tsuchiya, H., et al. (2016). A developmental coordinate of pluripotency among mice, monkeys and humans. *Nature, 537*(7618), 57–62. https://doi.org/10.1038/nature19096.

Rastan, S. (1983). Non-random X-chromosome inactivation in mouse X-autosome translocation embryos—Location of the inactivation centre. *Journal of Embryology and Experimental Morphology, 78*, 1–22. Retrieved from https://www.ncbi.nlm.nih.gov/pubmed/6198418.

Rastan, S., & Robertson, E. J. (1985). X-chromosome deletions in embryo-derived (EK) cell lines associated with lack of X-chromosome inactivation. *Journal of Embryology and Experimental Morphology, 90*, 379–388. Retrieved from https://www.ncbi.nlm.nih.gov/pubmed/3834036.

Rossant, J. (1976). Postimplantation development of blastomeres isolated from 4- and 8-cell mouse eggs. *Journal of Embryology and Experimental Morphology, 36*(2), 283–290. Retrieved from https://www.ncbi.nlm.nih.gov/pubmed/1033982.

Tarkowski, A. K. (1959). Experiments on the development of isolated blastomeres of mouse eggs. *Nature, 184*, 1286–1287. https://doi.org/10.1038/1841286a0.

Tarkowski, A. K., & Wroblewska, J. (1967). Development of blastomeres of mouse eggs isolated at the 4- and 8-cell stage. *Journal of Embryology and Experimental Morphology, 18*(1), 155–180. Retrieved from https://www.ncbi.nlm.nih.gov/pubmed/6048976.

Tesar, P. J., Chenoweth, J. G., Brook, F. A., Davies, T. J., Evans, E. P., Mack, D. L., et al. (2007). New cell lines from mouse epiblast share defining features with human embryonic stem cells. *Nature, 448*(7150), 196–199. https://doi.org/10.1038/nature05972.

CHAPTER FIVE

Differentiation of human pluripotent stem cells toward pharyngeal endoderm derivatives: Current status and potential

Margaret E. Magaletta, Richard Siller, René Maehr*
Program in Molecular Medicine, Diabetes Center of Excellence, University of Massachusetts Medical School, Worcester, MA, United States
*Corresponding author: e-mail address: rene.maehr@umassmed.edu

Contents

1. Introduction 176
2. Overview of the pharyngeal apparatus formation within the gut tube 178
3. Pharyngeal endoderm development and lineage specification within the pharyngeal pouches 180
4. Pharynx derivative pluripotent stem cell differentiation protocols: Current status 182
   4.1 Parathyroid 185
   4.2 Thyroid 190
   4.3 Thymus 191
5. Applications of hPSCs for studying pharyngeal endoderm development and disease 195
6. Future directions for hPSC differentiation approaches toward pharyngeal derivatives 196
   6.1 Reprogramming hPSCs toward pharyngeal derivatives 196
   6.2 Single cell-omics for informing and assessing hPSC differentiation 197
7. Concluding remarks 199
Acknowledgments 199
References 199

Abstract

The pharyngeal apparatus, a transient embryological structure, includes diverse cells from all three germ layers that ultimately contribute to a variety of adult tissues. In particular, pharyngeal endoderm produces cells of the inner ear, palatine tonsils, the thymus, parathyroid and thyroid glands, and ultimobranchial bodies. Each of these structures and organs contribute to vital human physiological processes, including central immune tolerance (thymus) and metabolic homeostasis (parathyroid and thyroid glands, and ultimobranchial bodies). Thus, improper development or damage to pharyngeal endoderm derivatives leads to complicated and severe human maladies, such as autoimmunity,

Current Topics in Developmental Biology, Volume 138
ISSN 0070-2153
https://doi.org/10.1016/bs.ctdb.2020.01.004

immunodeficiency, hypothyroidism, and/or hypoparathyroidism. To study and treat such diseases, we can utilize human pluripotent stem cells (hPSCs), which differentiate into functionally mature cells *in vitro* given the proper developmental signals. Here, we discuss current efforts regarding the directed differentiation of hPSCs toward pharyngeal endoderm derivatives. We further discuss model system and therapeutic applications of pharyngeal endoderm cell types produced from hPSCs. Finally, we provide suggestions for improving hPSC differentiation approaches to pharyngeal endoderm derivatives with emphasis on current single cell-omics and 3D culture system technologies.

1. Introduction

Human pluripotent stem cells (hPSCs) have greatly improved our ability to interrogate human development and disease in the laboratory. hPSCs, either derived from the inner cell mass of a zygote (human embryonic stem cells; hESCs) or generated through factor-induced reprogramming of somatic cells (human induced pluripotent stem cells; hiPSCs) have the capacity for self-renewal and provide a replenishable source of genetically identical material, while also maintaining the potential to differentiate into all cell types of the human body (Takahashi et al., 2007; Thomson et al., 1998). Due to these qualities, the application of hPSCs enables precise investigation into the key signaling pathways that coordinate normal development, as well as the genetic predispositions linked to human disease.

To unlock the potential of hPSCs for studying human development and disease, efforts have been made to devise step-wise differentiation protocols designed to recapitulate *in vivo* developmental cell fate decisions *in vitro*. Indeed, much progress has been made toward generating various neuronal lineages (Tao & Zhang, 2016) as well as mesoderm-derived cardiac and kidney lineages (Burridge, Keller, Gold, & Wu, 2012; Morizane, Miyoshi, & Bonventre, 2017). Moreover, as the endoderm lineage produces various disease-relevant organs, much work has also focused on deriving endodermal cell types from hPSCs (Ikonomou & Kotton, 2015). However, despite over 20 years of hPSC differentiation studies, not all endoderm cell types have been derived *in vitro*. Current research efforts within the endoderm lineage field focus mainly on the pancreatic, lung, intestinal, and liver lineages (Jacob et al., 2017; Mun et al., 2019; Munera et al., 2017; Nair et al., 2019; Yiangou, Ross, Goh, & Vallier, 2018). One type of endoderm that has received comparably less attention concerning hPSC differentiation protocol development and functional analyses is pharyngeal endoderm and its derivatives, including thymus, parathyroid, and thyroid. The epithelial cells within these pharyngeal organs perform specialized, essential biological functions ranging from hormone production and metabolic regulation to immune cell education *via* tissue-specific antigen presentation (Abramson & Anderson, 2017; Mannstadt et al., 2017; Nilsson & Fagman, 2017).

Importantly, errors during pharyngeal apparatus development precipitate severe health consequences, contributing to one-third of congenital disorders (Jones & Trainor, 2004). In particular, 22q11.2 deletion syndrome (22q11.2DS) has been linked to defective pharyngeal endoderm development, causing hypoplasia or aplasia of the thymus and parathyroid resulting in immunodeficiency, autoimmunity, and hypocalcemia in addition to general craniofacial dysmorphia and cardiac defects (McDonald-McGinn et al., 2015). As the most common microdeletion disorder, 22q11.2DS affects approximately 1 in every 4000 live births and as many as 1 in 1000 fetuses (Grati et al., 2015). Many of the phenotypes observed in 22q11.2DS have been traced to loss of *TBX1*, and mutations in *TBX1* alone can also cause the same 22q11.2DS phenotypes (Gong et al., 2001; Ogata et al., 2014; Yagi et al., 2003). Gain of function mutations in *TBX1* also causes 22q11.2DS defects (Zweier, Sticht, Aydin-Yaylagul, Campbell, & Rauch, 2007). Similarly, *PAX1* homozygous null mutations cause otofaciocervical syndrome (OFCS), characterized by various craniofacial defects as well as thymus development defects leading to severe combined immunodeficiency (SCID) (Paganini et al., 2017).

Apart from disorders affecting pharyngeal apparatus development, certain genetic lesions induce organ-specific health consequences. For example, FOXN1 deficiency causes thymic hypo-/aplasia leading to immunodeficiency (Du et al., 2019; Rota & Dhalla, 2017). *GATA3* or *GCM2* mutations lead to parathyroid hypoplasia-induced hypoparathyroidism resulting in hypocalcemia (Doyle, Kirwin, Sol-Church, & Levine, 2012; Van Esch et al., 2000). Similarly, mutations in *PAX8* and *NKX2-1* induce hypothyroidism as a result of improper thyroid development (Krude et al., 2002; Macchia et al., 1998). Given the vital functions of pharyngeal endoderm derivatives in processes ranging from immune competence to hormone-mediated physiological homeostasis, the ability to coordinate highly efficient differentiation of hPSCs toward pharyngeal endoderm derivatives would significantly enhance our ability to interrogate disorders associated with these organs, produce cellular replacement therapies, and discover novel drug treatments.

In this review, we discuss the current understanding of pharyngeal endoderm development *in vivo* and how that knowledge translates to *in vitro* studies for differentiating hPSCs. Additionally, we will highlight the current gaps in our understanding of differentiation toward pharyngeal endoderm and the limitations we face in generating these cells *in vitro*.

2. Overview of the pharyngeal apparatus formation within the gut tube

The developing gut tube, which derives from definitive endoderm (DE), broadly segregates into four regions known as the anterior and posterior foreguts, midgut, and hindgut. The midgut and hindgut form the small and large intestines, respectively, and the foregut regions produce a series of organs and structures (reviewed in Zorn & Wells, 2009). The anterior-most region of the foregut is known as the pharyngeal endoderm (Fig. 1).

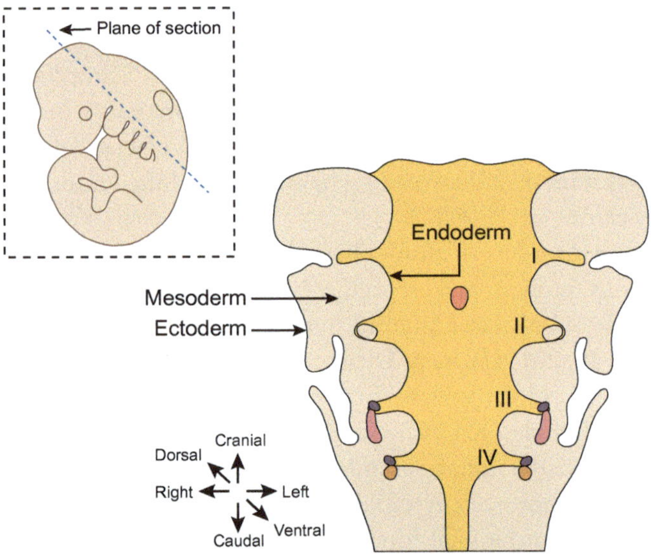

Fig. 1 Diagram of the developing pharyngeal arches during embryonic development. Coronal section of a mouse embryo shows the pouches labeled I–IV. Each pharyngeal pouch differentiates toward different lineages. Pouch I gives rise to the eustachian tube of the middle ear, pouch II gives rise to the palatine tonsils, pouch III gives rise to the thymus and inferior parathyroid glands, and pouch IV gives rise to the ultimobranchial body and superior parathyroid glands. The thyroid develops on the pharyngeal endoderm floor.

Pharyngeal endoderm and the surrounding tissue together comprise the pharyngeal apparatus, an evolutionarily conserved embryological structure that gives rise to various anatomical features and organs within the head and neck region. Precise anterior-posterior patterning of the pharyngeal apparatus produces five bilateral, segmented outgrowths called pharyngeal arches (Frisdal & Trainor, 2014). Each pharyngeal arch comprises cell types from all three germ layers organized as a neural crest/mesodermal core lined with endoderm and surrounded by an ectodermal surface. Between each arch, endoderm forms four sets of paired outpockets called pharyngeal pouches, each of which gives rise to distinct lineages (Fig. 1).

Pharyngeal endoderm functions both to supply morphogenic signals to surrounding tissue and to produce epithelial cells with a range of specialized functions. We refer to another review for a more detailed explanation of the functions of pharyngeal endoderm during pharyngeal arch formation (Graham, Okabe, & Quinlan, 2005). In brief, many recent studies have described critical roles of pharyngeal endoderm in signaling to form and pattern the pharyngeal arches. Such functions have been characterized in several model organisms including zebrafish, chicks, and mice (Arnold et al., 2006; Begbie, Brunet, Rubenstein, & Graham, 1999; Piotrowski et al., 2003; Piotrowski & Nusslein-Volhard, 2000; Trokovic, Trokovic, Mai, & Partanen, 2003; Yoshida et al., 2017).

In addition to the various morphogenic functions of the pharyngeal endoderm, this tissue also exhibits the intrinsic competence to produce several unique organs including the thyroid, thymus, parathyroid, and ultimobranchial body (UBB). Cells within each pharyngeal pouch, labeled I–IV, harbor specific lineage fates: the first pharyngeal pouch produces the eustachian tube of the middle ear, the second pouch produces the palatine tonsils, the third pouch produces the thymus and inferior parathyroid, and the fourth pouch produces the ultimobranchial body and superior parathyroid (Fig. 1). Additionally, the thyroid body develops on the pharyngeal floor. Ectopic transplant assays have demonstrated the endoderm-intrinsic capacity for thymus formation (Gordon et al., 2004; Le Douarin & Jotereau, 1975). Similarly, lineage trace assays indicate that thyroid-resident C cells (derived from the UBB) are exclusively derived from pharyngeal endoderm, rather than neural crest (Johansson et al., 2015). Pouch-derived endodermal cells further differentiate into many specialized, clinically-relevant cell types in mature organs such as: medullary and cortical thymic epithelial cells, which function to facilitate maturation and selection of thymocytes (Gordon et al., 2004); calcitonin-secreting parafollicular cells of the thyroid (Johansson et al., 2015); and parathyroid hormone-producing cells within the parathyroid gland (Boyd, 1950). Ultimately, the highly organized anatomy of the pharyngeal pouches complements underlying molecular mechanisms to coordinate the specification and maturation of several unique lineages simultaneously.

3. Pharyngeal endoderm development and lineage specification within the pharyngeal pouches

During morphogenesis of the pharyngeal endoderm, pharyngeal pouch formation and subsequent lineage-specific differentiation requires both signaling molecule patterning and genetic mechanisms. Knowledge of the processes whereby pharyngeal endoderm differentiates into various lineages enables the design of hPSC differentiation approaches. Precise coordination of signaling pathways and transcription factor networks orchestrates the differentiation of specific cell types within the pharyngeal endoderm. In this section, we will highlight the some of the key studies responsible for elucidating critical pharyngeal endoderm differentiation mechanisms.

Various signaling molecules function to pattern the pharyngeal endoderm and direct subsequent organ development. For example, retinoic acid (RA) expression induces a posterior fate and promotes proper caudal pouch segmentation (Kopinke, Sasine, Swift, Stephens, & Piotrowski, 2006). Mice mutant for Aldh1a2 or Rdh10, leading to RA deficiency, have defects in posterior arch development (Sandell et al., 2007; Vermot, Niederreither, Garnier, Chambon, & Dolle, 2003). Importantly, loss of retinoic acid synthesis in the developing embryo recapitulated phenotypes of DiGeorge syndrome, which represents a subset of disorders within 22q11.2DS (Vermot et al., 2003). Similarly, Wnt11r and Wnt4a signals from surrounding mesoderm and ectoderm both promote proper pouch formation (Choe et al., 2013). Later in development, Wnt glycoproteins regulate the expression of Foxn1 expression in the thymus (Balciunaite et al., 2002). However, enhanced activation of the β-catenin-dependent Wnt pathway in thymic epithelial cells (TECs) inhibits proper thymus morphogenesis (Swann, Happe, & Boehm, 2017). Likewise, overexpression of secreted Wnt ligand also disrupts thymus, thyroid, and parathyroid morphogenesis (Swann, Happe, & Boehm, 2017). Fibroblast growth factor (Fgf) signaling also plays a key role during pouch development. Specifically, *Fgf3* and *Fgf8* mutant mice fail to from pharyngeal pouches leading to arch disorganization and fusion (Crump, Maves, Lawson, Weinstein, & Kimmel, 2004). *Fgf8* mutant mice phenocopy many symptoms of 22q11.2 deletion syndrome, underscoring the importance of this signaling molecule during pharyngeal endoderm development (Frank et al., 2002). Other *Fgf* signaling molecules are required for later organogenesis events. Expression of Fgfr2-IIIB, which responds to mesenchymal signals of Fgf7 and Fgf10, is required for thymus development (Revest, Suniara, Kerr, Owen, & Dickson, 2001). In addition to anterior-posterior patterning, signaling molecules also facilitate dorsoventral patterning of the developing pharyngeal pouches. For instance, sonic hedgehog (SHH) signals from the dorsal pharyngeal apparatus specifies the parathyroid domain and ventral BMP4 activity specifies the thymus domain of the third pharyngeal pouch (Figueiredo et al., 2016; Gordon, Patel, Mishina, & Manley, 2010; Grevellec, Graham, & Tucker, 2011; Moore-Scott & Manley, 2005; Patel, Gordon, Mahbub, Blackburn, & Manley, 2006; Swann, Krauth, Happe, & Boehm, 2017). Endoderm-specific cre-mediated activation and inhibition of SHH signaling confirmed the function of SHH in promoting parathyroid-specific expression of Tbx1 and suppressing Foxn1 expression (Bain et al., 2016).

Regionalized expression of transcription factors within the pharyngeal endoderm regulates pouch identity. Anterior-posterior patterning of the pharyngeal pouches results in part from nested expression of *Hox* genes (for a comprehensive review, we refer you to Gordon, 2018). Among the *Hox* genes expressed in the pharyngeal endoderm, *Hoxa3* performs a significant role through various functions (Gordon, 2018). *Pbx1* null mice present many of the same defects observed in *Hoxa3* null mice including hypoplastic thymus lobes, parathyroid glands, and UBBs, likely due to interactions between Pbx1 and Hox proteins (Knoepfler, Lu, & Kamps, 1996; Manley, Selleri, Brendolan, Gordon, & Cleary, 2004). Despite the athymia and aparathyroidism phenotypes observed in *Hoxa3* null mice, subsequent experiments demonstrated that *Hoxa3* functions to promote organ maturation rather than lineage specification (Chojnowski et al., 2014). Other transcription factors involved in pharyngeal pouch development and organ formation include *Pax1*, *Pax9*, *Eya1*, *Six1*, and *Six4* (Manley & Condie, 2010). *Hoxa3* probably functions upstream of these transcription factors, since *Hoxa3* expression remains unaffected in these mutants (Gordon, 2018). *Pax9* null embryos fail to develop the thymus, parathyroid, and UBBs, while *Pax1* null embryos show thymus hypoplasia (Peters, Neubuser, Kratochwil, & Balling, 1998; Wallin et al., 1996).

Other transcription factors important for pharyngeal endoderm survival and development include *Nkx2-5* and *Nkx2-6*. *Nkx2-6* is expressed in the ventrolateral region of all pharyngeal pouches from E8.5 to E11.5 in mouse development (Biben, Hatzistavrou, & Harvey, 1998). Mutant embryos deficient for both *Nkx2-5* and *Nkx2-6* showed impaired pharyngeal endoderm development, with apoptotic markers upregulated and proliferation markers reduced (Tanaka, Schinke, Liao, Yamasaki, & Izumo, 2001). The transcription factor *Tbx1* is expressed throughout the developing pharyngeal endoderm, and endoderm-specific deletion of *Tbx1* induces the 22q11.2DS phenotype in mice (Arnold et al., 2006).

In addition to the aforementioned transcription factors involved in early pharyngeal endoderm development, various transcription factors also promote organ-specific differentiation within the pharyngeal pouches. Within the third pharyngeal pouch, *Foxn1* and *Gcm2* are expressed in the thymus and parathyroid domains, respectively (Gordon, Bennett, Blackburn, & Manley, 2001). While neither transcription factor specifies the thymus or the parathyroid, both are required for organ maturation (Blackburn et al., 1996; Liu, Yu, & Manley, 2007). The transcription factors Nkx2-1, Hhex, Pax8, and Foxe1 function cooperatively to facilitate proper thyroid development (Fernandez, Lopez-Marquez, & Santisteban, 2015; Parlato et al., 2004). As expected, genetic ablation of these transcription factors leads to failed thyroid formation or hypoplasia (Clifton-Bligh et al., 1998; De Felice et al., 1998; Kimura et al., 1996; Mansouri, Chowdhury, & Gruss, 1998; Martinez Barbera et al., 2000). Interestingly, neither Nkx2-1 nor Hhex are required for thyroid specification; instead, Nkx2-1 promotes organ maturation and Hhex maintains the thyroid progenitor pool (Kimura, Ward, & Minoo, 1999; Parlato et al., 2004).

4. Pharynx derivative pluripotent stem cell differentiation protocols: Current status

The field of hPSC-derived endodermal lineages primarily relies upon knowledge of *in vivo* developmental animal studies to inform *in vitro* differentiation protocols. As a general approach for procuring hPSC-derived pharyngeal endoderm, differentiation strategies should broadly mimic three stages of *in vivo* development: specification of definitive endoderm (DE), patterning toward anterior foregut endoderm (AFE), and formation of pharyngeal endoderm. Each of these stages express a unique pattern of transcription factors that mimic expression patterns observed during *in vivo* development, as described above, and can be measured to assess differentiation efficiency (Fig. 2). Competency of each intermediate naturally decreases across differentiation, ultimately producing tissue-specific cells from pluripotent cells. Thus, to ensure the highest possible differentiation efficiency toward pharyngeal cell types, each differentiation intermediate requires careful validation. Differentiation intermediates not only require identity verification to prevent the formation of erroneous cells, but also to authenticate the biological relevance of the final differentiation products.

To induce DE commitment rather than ectoderm or mesoderm commitment, hPSC differentiation protocols generally rely on high levels of Activin A to activate SMAD-mediated transcriptional programs. This was first demonstrated in mESC cultures and shortly thereafter adapted to

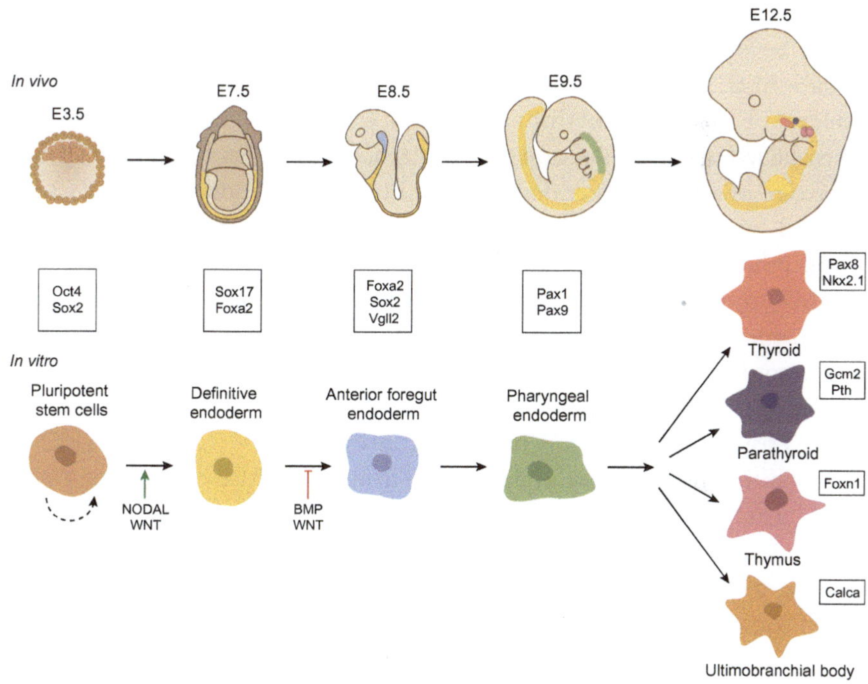

Fig. 2 Differentiation protocols aim to mimic normal embryonic development. The top panel provides an overview of the development *in vivo* of the pharynx and its derivatives from the pluripotent epiblast through definitive endoderm, anterior foregut endoderm, pharyngeal endoderm and finally specification to the pharyngeal-derived organs. Key transcription factors at each stage are highlighted in the boxes. The bottom panel demonstrates how this developmental knowledge can be translated to *in vitro* stem cell protocols using activation and inhibition of key signaling pathways temporally to direct the differentiation of the cells.

hPSC differentiation methods (D'Amour et al., 2005; Kubo et al., 2004). Moreover, since PI3K signaling can attenuate Nodal signaling, others determined that PI3K inhibition further improved hPSC differentiation efficiency to DE (McLean et al., 2007). Apart from TGFβ signaling activation, many hPSC differentiation protocols also induce WNT signaling through addition of WNT3a and inhibit GSK-3 signaling to better facilitate the transition to endoderm (Brolen et al., 2010; Ogawa et al., 2013; Teo, Valdez, Dirice, & Kulkarni, 2014). Notably, duration of treatment with Activin A or Wnt3a was found to affect competency toward the pancreatic and hepatic lineages (Toivonen et al., 2013).

Next, in favor of deriving pharyngeal cell types instead of other endoderm cell types like the pancreatic and hepatic lineages, DE must be patterned toward AFE. The anterior region of the gut tube expresses elevated levels of FOXA2 and SOX2 (Green et al., 2011; Zorn & Wells, 2009). Several groups have published protocols for differentiating hPSCs toward AFE. Initially, dual inhibition of BMP and TGFβ was shown to induce AFE commitment from DE (Green et al., 2011). Modification of this approach led to the production of organized, stratified epithelia that express a repertoire of pharyngeal endoderm-specific genes in addition to SOX2 and FOXA2 (Kearns et al., 2013). However, subsequent studies showed that TGFβ inhibition was disposable and instead implemented dual inhibition of BMP and WNT signaling to induce AFE commitment (Davenport, Diekmann, Budde, Detering, & Naujok, 2016; Loh et al., 2014). Select lung, esophagus, thyroid, and thymus differentiation protocols also pass through AFE-like intermediates (Longmire et al., 2012; Parent et al., 2013; Trisno et al., 2018; Wong et al., 2015).

To distinguish pharyngeal derivatives from other AFE derivatives such as the lung and esophagus, AFE can be further differentiated toward pharyngeal endoderm. To date, only a few publications describe formation of pharyngeal-like endoderm from AFE. Green et al. and Kearns et al. observe expression of *PAX9* and *TBX1* in addition to *SOX2* and *FOXA2* in their hPSC-derived AFE cell type, indicating some level of competency to produce pharyngeal endoderm (Green et al., 2011; Kearns et al., 2013). Furthermore, in an attempt to ventralize the AFE toward pharyngeal pouch identity using WNT3a, FGF10, KGF, BMP4, and EGF, Green et al. observed expression of *PAX1* and *NKX2-5*, as well as lung-associated genes *NKX2-1* and *P63* (Green et al., 2011). Kearns et al. produced purified Pax9+ cells with elevated expression of *Pax1*, *Tbx1*, *Eya1*, and *Six1* in comparison to PSCs and unpurified AFE (Kearns et al., 2013).

Another group reports to derive pharyngeal endoderm from mouse embryonic stem cells (mESCs) as a byproduct of an embryoid body cardiogenesis differentiation protocol (Hidaka et al., 2010). These *E*-cadherin/Nkx2-5 + cells express low levels of pharyngeal markers including *Pax1*, *Pax9*, *Six1*, *Eya1*, *Tbx1*, and *Vgll2*. However, the reported pharyngeal endoderm cells only comprise ~1% of the differentiated cells and do not successfully mature into functional pharyngeal derivatives (Hidaka et al., 2010). Importantly, reports often fall short with respect to detailed evaluation of pharyngeal endoderm as an intermediate and mainly assess the terminally differentiated cells (Parent et al., 2013; Sun, Luo, Li, & Zhao, 2013). However, given the general lack of protocols producing functional pharyngeal endoderm derivatives, the efficient production of competent pharyngeal endoderm likely requires more attention. For a compilation of protocols for differentiating PSCs toward pharyngeal derivatives, see Table 1.

4.1 Parathyroid

The parathyroid acts as the primary organ responsible for regulating calcium levels in the body through the production of parathyroid hormone (PTH) (Abate & Clarke, 2016). Impaired PTH secretion, a condition known as hypoparathyroidism, ultimately leads to chronic hypocalcemia and low-turnover bone disease. Known causes of hypoparathyroidism include hypoplasia, parathyroid damage, and parathyroid or thyroid surgery (Bilezikian et al., 2011; Clarke et al., 2016; Shoback et al., 2016). Although hormone replacement therapy through repeated injections of parathyroid hormone fragments can alleviate some disease burden, this treatment only partially substitutes for parathyroid function (Sikjaer, Rejnmark, Rolighed, Heickendorff, & Mosekilde, 2011; Winer, Sinaii, Peterson, Sainz, & Cutler, 2008). Thus, a reliable source of parathyroid cells would be a useful tool to study genetic lesions causing hypoparathyroidism and, in the future, hPSC-derived parathyroid cells could be used as a cellular therapy.

The first report of parathyroid differentiation from hPSCs, published in 2009, showed expression of key pharyngeal endoderm genes including *EYA1*, *SIX1*, *PAX1*, and parathyroid genes including *GCM2*, *CaSR* and *PTH* following co-culture with mouse embryonic fibroblasts (MEFs) in the presence of Activin A (Bingham et al., 2009). Immunostaining revealed protein expression of CaSR, although the cells also retained expression of definitive endoderm genes SOX17 and CXCR4. Importantly, ELISA measurements confirmed the presence of PTH in conditioned media, suggesting the cells

Table 1 Summary of hPSC differentiation protocols toward pharyngeal derivatives.

Target cell type	References	Starting hPSC type	Definitive endoderm induction	Anterior foregut endoderm induction	Pharyngeal endoderm induction	Terminal cell type induction	Functional characterization of endpoint cells
Parathyroid	Bingham, Cheng, Woods Ignatoski, and Doherty (2009)	• BG01 hESCs	Days 1–5: Activin A with increasing levels of FBS (0–2%) on MEFs	NA	NA	Day 5: Lift with TrypLE select, remove MEFs and culture in 5% FBS with Activin A until day 13 Day 13–26: Remove Activin A, then continue culture in 5% FBS	• PTH secretion detected *via* ELISA • CaSR, CXCR4, and SOX17 detected *via* immunofluorescence • Transcript expression of *EYA1*, *SIX1*, *PAX1*, *GCM2*, *CASR*, and *PTH*
	Woods Ignatoski, Bingham, Frome, and Doherty (2010)	• BG01 hESCs • H1 hESCs	Bingham protocol above, with the addition of SHH	NA	NA	Bingham protocol above, with addition of SHH until day 12	• PTH secretion detected *via* ELISA • *CaSR*, *PTH*, and *GCM2* transcript expression
Thyroid	Kurmann et al. (2015)	• Hypothyroid patient-derived hiPSCs (children with NKX2.1 coding sequence mutations) • BU3 hiPSCs • iPSC17 hiPSCs • RUES2 hESCs	cSFD medium 3-Day treatment with Activin A in embryoid body suspension culture	Green et al. (2011) protocol	NA	cSFD medium Treatment with BMP4 and FGF2 FGF2, FGF10, and heparin sodium salt Day 22, switch to DCI +K medium	• NKX2-1 and PAX8 protein expression *via* immunofluorescence • *NKX2-1*, *PAX8*, *TG*, *TPO*, and *TSH-R* transcript expression

	Ma, Morshed, Latif, and Davies (2017)	• H9 hESCs	RPMI1640 with 1% B27 4-Day treatment with Activin A	NA	NA	RPMI1640 with 1% B27 Treatment with ethacridine and TSH for up to 21 days. Cells then embedded in Matrigel supplemented with TSH	• Protein expression of NKX2-1, PAX8, TG *via* immunofluorescence • Transcript expression of *PAX8*, *NKX2-1*, *TG*, *TPO*, *TSH-R*, and *NIS* • Formation of thyroid follicles in 3D culture • Functional response to TSH *in vitro*
Thymus	Sun et al. (2013)	• H1 hESCs • H9 hESCs	X-VIVO 10 basal medium 5-Day treatment with Activin A	NA	X-VIVO 10 basal medium. 4 days treatment with retinoic acid and IWR1	X-VIVO 10 basal medium 5 days treatment with BMP4 and WNT3a	• Transcript expression of *TBX1*, *HOXA3*, *PAX9*, *EYA1*, *FOXN1* and *K5* • Protein expression of HOXA3, TBX1, SIX1, PBX1, FOXN1, KERATIN 5, and KERATIN 8 *via* immunofluorescence • *In vivo* transplantation enhanced maturation and supported mouse and human T cell development

Continued

Table 1 Summary of hPSC differentiation protocols toward pharyngeal derivatives.—cont'd

Target cell type	References	Starting hPSC type	Definitive endoderm induction	Anterior foregut endoderm induction	Pharyngeal endoderm induction	Terminal cell type induction	Functional characterization of endpoint cells
	Parent et al. (2013)	• CyT49 hESCs • HUES4 hESCs	Cells were seeded on MEFs. RPMI1640 with increasing concentrations of KSR (0% on day 1, 0.2% on days 2–3, and 2% on day 4). Supplements: 5-day treatment with Activin A, first day additionally treated with WNT3a	RPMI1640 with 0.5% B27 on days 5–7. Supplements: All-trans retinoic acid, BMP4, LY364947 through day 7	DMEM/F12 with 0.5% B27. Supplements: WNT3a, all-trans retinoic acid, BMP4, LY364947, FGF8b, KAAD-cyclopamine was used from days 7 to 9	DMEM/F12 with 0.5% B27. Supplements: WNT3a, all-trans retinoic acid, BMP4, FGF8b, KAAD-cyclopamine was used from days 9 to 11	• Transcript expression of *FOXN1*, *HOXA3*, *EYA1*, and *EPCAM* • HOXA3 and EPCAM protein level *via* immunofluorescence • Transplanted TEPs matured *in vivo* and expressed Keratin 5 and 8, MHCII, CCL25, CXCL12, SCF, and DLL4 • Transplanted TEPs support mouse T cell development • Transplanted TEPs induce allogenic skin graft rejection

Soh et al. (2014)	• HES3 hESCs • MEL1 hESCs • H9 hESCs • DF19-9-7T hiPSCs • FOXN1-GFP reporter hESCs	APEL or BPEL medium. 4-Day treatment with Activin A in spin aggregated 3D EB culture	APEL or BPEL medium 3D EB culture for 3 days followed by plating in 2D culture for 6 days	NA	AEL or BEL medium. 14-day treatment with KGF	• Transcript expression of *KRT5*, *KRT14*, *BCL11B*, *TP63*, *DLL1*, *NOTCH1*, etc. • Flow cytometry analysis of EPCAM expression • Identified Integrin-B4, HLA-DR, and EPCAM as surface markers of hPSC-derived TEPs
Su et al. (2015)	• H9 hESCs • CT2 hESCs	4-Day treatment with the Human Pluripotent cell-derived Endoderm Differentiation Kit (R&D systems)	NA	NA	10-Day treatment with human FGF7, FGF10, EGF, BMP4, all-trans retinoic acid, recombinant FOXN1 protein, and recombinant HOXA3 protein	• Transcript expression of *FOXN1*, *HOXA3*, *TBX1*, *PAX9*, *EYA1*, and *PAX1* • Protein expression (flow cytometry and immunofluorescence) of EPCAM, Keratin 5, and Keratin 8 • Maturation of TEPs to TECs *in vivo* with CCL25, HLA-DR, AIRE, and Delta protein expression (flow cytometry) • Transplanted TEPs support mouse and human T cell development

acquired a secretory function (Bingham et al., 2009). Subsequently, a modified protocol including the addition of SHH demonstrated enhanced levels of *PTH*, *GCM2*, and *CaSR*, as well as DE genes *SOX17* and *CXCR4* (Woods Ignatoski et al., 2010). Interestingly, Green et al. also reported expression of *GCM2* in hPSC-derived AFE cells, albeit at low levels, after treatment with SHH or FGF8 (Green et al., 2011).

These protocols demonstrated the ability to differentiate hPSCs into cells that express typical parathyroid-specific genes, and even produce PTH. However, the complete functional attributes of such cells have not been explored to date. For instance, functional tests are needed to show the capacity of cells to regulate intracellular calcium levels in response to activation of the calcium-sensing receptor. Ideally, these cells will also have the capacity to re-establish physiological homeostasis in a mouse model of hypoparathyroidism.

4.2 Thyroid

Through production of thyroid hormones, the thyroid functions as the primary endocrine gland that controls metabolic homeostasis in the body (Fisher & Klein, 1981). Hypothyroidism, a disorder characterized by insufficient production of thyroid hormones, emerges as a consequence of surgical thyroid removal, congenital mutations affecting thyroid development, or autoimmune destruction of the thyroid (Hollenberg, Choi, Serra, & Kotton, 2017; Macchia et al., 1998; Pyzik, Grywalska, Matyjaszek-Matuszek, & Rolinski, 2015). While hypothyroidism subsides with daily synthetic hormone treatment, clinical efficacy remains low. Thus, cell replacement therapy could supersede hormone replacement therapy as a more long-term and patient-friendly remedy.

Until recently, the majority of published protocols differentiate mESCs, rather than hPSCs, toward the thyroid lineage (Arufe et al., 2006; Arufe, Lu, & Lin, 2009; Dame et al., 2017; Jiang et al., 2010; Longmire et al., 2012; Ma, Latif, & Davies, 2009). Yet more current studies demonstrate that mESC differentiation protocols also achieve differentiation of hPSCs, suggesting an evolutionarily conserved thyroid differentiation pathway. After establishing differentiation of mESCs into bipotent Nkx2-1 positive lung and thyroid progenitors (Longmire et al., 2012), Kurmann et al. further developed their mESC differentiation method to produce Nkx2-1/Pax8 double positive thyroid progenitors (Kurmann et al., 2015). BMP4 and FGF2, but not Wnt3a, produced the highest percentage of Nkx2-1/Pax8 positive thyroid progenitors that expressed thyroglobulin (*Tg*), thyroid peroxidase (*Tpo*), thyroid stimulating hormone receptor (*Tsh-r*), and sodium (Na)-iodine symporter (*Nis*) transcripts. Perturbation of BMP4 and FGF2 signaling in mouse and *Xenopus* embryos demonstrated the necessity of these signaling molecules during *in vivo* development, thereby verifying their importance for *in vitro* differentiation toward the thyroid lineage. Further manipulation of culture conditions to induce thyroid progenitor maturation lead to increased expression of *Tg*, *Tpo*, *Tsh-r*, and *Nis*, protein expression of Tg and Pax8, and the formation of thyroid follicular cells *in vitro* (Kurmann et al., 2015).

Accordingly, BMP4 and FGF2-driven differentiation of both hESCs and hiPSCs successfully induced expression of functional thyroid genes including thyroglobulin and thyroid stimulating hormone receptor (TSH-R). The authors noted that due to low differentiation efficiency from hPSCs, the expression level of key thyroid markers was significantly reduced compared to adult human thyroid tissue (Kurmann et al., 2015). In a more transcription-centric approach, Ma et al. reasoned that upregulation of transcriptional co-factors commonly associated with NKX2-1 and PAX8 would improve differentiation of hPSCs toward thyroid cells (Ma et al., 2017). Indeed, small molecule activation of TAZ increased expression of NKX2-1, PAX8, and functional genes. Additionally, these cells effectively formed thyroid follicles and expressed abundant levels of thyroglobulin protein (Ma et al., 2017). Although lacking data to show *in vivo* functionality of these thyroid epithelial-like cells, the cells were responsive *in vitro* to TSH treatment as indicated by uptake of radioiodine and subsequent incorporation into protein production (Ma et al., 2017).

Though progress has been made toward deriving thyroid lineage from hPSCs, improving on the efficiency of differentiation, either through the selection of the desired cell type using reporter lines and/or cell surface marker expression, or by rationally modifying the differentiation protocol *via* small molecule screens could prove beneficial. Additionally, UBB-derived thyroid C cells comprise an important component of thyroid biology. In order to fully restore thyroid functionality *via* cell replacement therapy, UBB differentiation protocols must be developed in conjunction with thyroid differentiation protocols.

4.3 Thymus

The thymus establishes the self-tolerant adaptive immune system through positive and negative selection, processes whereby TECs select for functional yet non-autoreactive developing thymocytes, respectively. Not surprisingly, autoimmune diseases have been linked to deficient negative selection in the thymus (Cheng & Anderson, 2018). For example, insufficient expression of insulin self-antigens in the thymus allows insulin-reactive T cells to escape into the periphery, which is one proposed mechanism whereby pancreatic beta cells are destroyed in type 1 diabetes (Fan et al., 2009; Pugliese et al., 1997; Vafiadis et al., 1997). Similarly, mutations affecting the function of autoimmune regulator (AIRE), the transcription factor that controls expression and presentation of tissue-restricted self-antigens in the thymus, cause defective negative selection of autoreactive T cells and subsequent autoimmune disease (Bruserud, Oftedal, Wolff, & Husebye, 2016; Chan & Anderson, 2015; Pitkanen & Peterson, 2003; Takaba & Takayanagi, 2017). Conversely, immunodeficiency syndromes arise from congenital disorders affecting thymus development. In particular, individuals suffering from DiGeorge syndrome, resulting from deletion of the 22q11.2 locus, present with thymic hypo- or aplasia and immunodeficiency (Davies, 2013).

Mouse models designed for studying autoimmune disease, and type 1 diabetes in particular, have facilitated the discovery of many disease mechanisms. However, given the inherent complexity of type 1 diabetes and other autoimmune diseases, these mouse models often fail to fully recapitulate the disease state (Chen, Mathews, & Driver, 2018; King, 2012; Thomas, Brodnicki, & Kay, 2016). Due to limited feasibility of studying disease mechanisms in humans, there is a great need to develop other sources of human-specific model systems. In addition to the need for better model systems to study human disease mechanisms related to thymus dysfunction, there is also a need for improved treatment options for patients with immunodeficiency syndromes. Current methods for treating immunodeficiency syndromes including complete DiGeorge syndrome (cDGS) are based on allogeneic transplantation of thymus tissue. In a clinical trial of allogeneic thymus transplant treatment for cDGS, more than half of the patients developed autoimmune complications (Davies et al., 2017). Similarly, patients in earlier trials developed autoimmune responses in the thyroid, gut, and blood system (Markert et al., 2007; Markert, Devlin, & McCarthy, 2010). The use of autologous-derived thymus tissue could circumvent these complications. Therefore, developing protocols for the generation of TECs from hPSCs remains a high priority for the purpose of improved disease modeling and cell replacement therapies.

To date, several publications have reported the generation of TECs or thymic epithelial progenitors (TEPs) from hPSCs. In 2013, two groups published protocols for differentiating hESCs into TEPs. Interestingly, these protocols do not pass through the same intermediate cell states. While the

Sun et al. protocol derives third pharyngeal pouch endoderm (PPE)-like cells directly from DE, the Parent et al. protocol differentiates DE into anterior foregut endoderm (AFE) before producing ventral pharyngeal endoderm (VPE) (Parent et al., 2013; Sun et al., 2013). Sun et al. noted elevated expression of HOXA3, TBX1, PAX9, EYA1, SIX1, and PBX1 in third PPE cells, albeit at lower levels than human fetal thymus. Treatment of third PPE with BMP4 and WNT3a induced expression of the key thymic transcription factor FOXN1, as well as surface markers Keratin 5 and Keratin 8, which correspond to distinct thymic epithelial cell subtypes yet are not specific to thymic epithelial cells (Chu & Weiss, 2002). Despite expression of FOXN1, the Sun et al. protocol did not induce expression of functionally relevant genes such as PSMB11 or AIRE. Thus, Sun et al. termed these cells thymic epithelial progenitor-like cells (TEPLCs). Upon transplantation into immunodeficient mice, the TEPLCs further matured. After 16–24 weeks, immunofluorescent analysis of the grafts revealed that the hESC-derived cells acquired expression of the mature, functional TEC markers MHCII and AIRE. Importantly, the authors reported that transplanted TEPLCs could support the development of several mouse immune cell populations including CD4 and CD8 T cells. However, the nude mouse phenotype was not fully rescued and thymopoiesis was more efficient in recipients of human fetal thymus transplants. Lastly, after co-transplantation with human hematopoietic stem cells (hHSCs), the authors observed formation of human CD4 and CD8 T cells, indicating the TEPLCs could promote human T cell development *in vivo*.

Parent et al. describe hPSC-derived cells that express *FOXN1*, *HOXA3*, *EYA1*, and *EPCAM*. Immunostaining confirmed protein expression of HOXA3 and EPCAM, but not *FOXN1* or *EYA1*. Since neither HOXA3 nor EPCAM are specific to thymic identity, this result raises questions about the lineage commitment of these cells. When these hPSC-derived cells were transplanted into nude mice, the cells further matured and gained expression of *HLA-DRA*, *DLL4*, *CCL25*, and *CXCL12*, yet the expression levels remained below that of transplanted human fetal thymus. After transplantation, the cells expressed Keratin 5 and Keratin 8 proteins. Transplanted TEPs were found to support development of the CD4 and CD8 mouse T cells that express TCRβ, although not to the same degree as wild-type CD4 and CD8 T cells. *In vitro* TCR stimulation invoked proliferation in T cells derived from mice transplanted with TEPs. However, allogeneic stimulation *in vitro* did not induce the same proliferative response.

Finally, TEP-transplanted mice demonstrated rejection of allogeneic skin grafts, although graft rejection took about 7 days longer than wild-type mice.

In 2014, Soh et al. published an alternate approach for generating FOXN1-expressing cells from hPSCs. The group developed a FOXN1-GFP reporter line to optimize an embryoid body differentiation protocol toward the thymic lineage and identify novel surface markers of TEPs (Soh et al., 2014). Using this reporter line, they identified Integrin-B4, HLA-DR, and EPCAM as markers of hPSC-derived TEPs. Of note, Soh et al. found that high levels of KGF significantly improved the differentiation efficiency and maturation of TEPs as measured by the proportion of FOXN1-GFP+ cells. Sorted GFP+ and EPCAM+ cells expressed elevated levels of *KRT5*, *KRT14*, *BCL11B*, *TP63*, and *NOTCH1* among other genes as compared to EPCAM+ cells and double negative cells. This study did not test the capacity of these hPSC-derived thymic progenitor cells to restore immune cell populations in immunodeficient mice.

Most recently, Su et al. generated an additional protocol for generating TEPs from hESCs (Su et al., 2015). After differentiation to DE, their method involves overexpression of recombinant FOXN1 and HOXA3 protein in the presence of BMP4, FGF7, FGF10, EGF, and RA which produces *FOXN1*, *HOXA3*, *TBX1*, *PAX9*, *EYA1*, and *PAX1* expressing cells. In addition, these cells express EPCAM, Keratin 5, and Keratin 8 proteins. After transplantation into nude mice, the authors observed the development of CD4 and CD8 mouse T cells which could proliferate in response to anti-CD3 and anti-CD28 antibodies. Finally, the hESC-derived TEP-like cells were also found to support development of human CD4 and CD8 T cells *in vivo*.

While these studies represent tremendous progress toward the derivation of hPSC-derived TECs, the maturity and functionality of these cells remain limited, especially *in vitro*. With the exception of the Sun et al. method, existing hPSC thymus differentiation protocols produce early progenitors that do not show protein expression of key genes like FOXN1, the hallmark transcription factor of TEC identity. Furthermore, *in vitro* differentiation of hPSCs does not to induce expression of functionally relevant genes. Given the interdependence between TEP and thymocyte maturation *in vivo*, complete differentiation of hPSCs to TECs could require co-culture with thymocytes. Perhaps such an approach could also produce the functionally distinct TEC subtypes, medullary TECs (mTECs) and cortical TECs (cTECs). Until research uncovers methods to further mature and impart functionality on TEPs *in vitro*, these cells have limited disease modeling and therapeutic applications.

5. Applications of hPSCs for studying pharyngeal endoderm development and disease

hPSCs have revolutionized the study of human biology. In lieu of developing human tissues, hPSC differentiation platforms function as model systems to study transcription factors, signaling pathways, and epigenetic mechanisms that control development and lineage specification (DeLaForest et al., 2011; McCauley et al., 2017; Teo et al., 2015; Xie, Carrano, & Sander, 2015; Zhu et al., 2016). Additionally, patient-specific iPSCs enable studies concerning human disease mechanisms as well as drug screens to predict new therapies. For complete reviews covering the use of hPSCs as model systems for development and disease, we refer you to Zhu and Huangfu (2013) and Rowe and Daley (2019), respectively.

Several aspects of pharyngeal endoderm development remain obscure despite numerous studies attempting to reveal the underlying molecular mechanisms. More specifically, hPSC-derived pharyngeal endoderm and lineage-specific cells have the potential to elucidate transcription factor functions in the context of development. For instance, the transcription factors required for specification of the thymus, parathyroid, and thyroid lineages remain undefined (Fernandez et al., 2015; Gordon & Manley, 2011). In the pancreatic and hepatic endoderm fields, hPSCs have been used to study the function of transcription factors in terms of lineage specification and differentiation (DeLaForest et al., 2011; Teo et al., 2015). Likewise, hPSC-derived cells also have the potential to reveal the function of various signaling pathways in development of the pharyngeal endoderm derivatives. This was demonstrated for the thyroid lineage, and the same approach could be applied for other pharyngeal endoderm lineages (Kurmann et al., 2015; Serra et al., 2017).

Additionally, many aspects of pharyngeal endoderm-associated diseases remain poorly understood. To investigate the etiology behind T cell immunodeficiency associated with *EXTL3* mutations, Volpi et al. derived patient-specific iPSC lines that were subsequently differentiated toward thymic epithelial cells using the Parent et al. protocol (Volpi et al., 2017). This approach uncovered a defect in TEP differentiation potential of *EXTL3*-deficient cells. Similarly, Kurmann et al. demonstrated the potential to differentiate iPSCs derived from hypothyroid patients diagnosed with brain-lung-thyroid syndrome (Kurmann et al., 2015). While the study did not go on to dissect disease pathogenesis using the iPSC-derived thyroid cells, the success of this experiment strongly supports the plausibility of using iPSC-derived pharyngeal cell types to generate patient-specific models for pharyngeal cell type-associated diseases. The same approach could be applied to study the involvement of thymic epithelial cells in type 1 diabetes or other autoimmune and immunodeficiency syndromes.

Currently, early stage clinical trials using hPSCs are ongoing for a number of diseases including diabetes mellitus, spinal cord injury, macular degeneration, muscular dystrophy, and ischemic heart disease among others (Guhr et al., 2018; Trounson & McDonald, 2015). The lack of hPSC-treatment based clinical trials for the treatment of pharyngeal cell type-associated diseases stems from the limitations in the current state of differentiation efforts toward these cell types, as described above. The approach of using hPSC-derived cells as cell replacement therapies could prove particularly useful for diseases associated with pharyngeal cell types, since allogeneic transplant tissue can be scarce. Furthermore, using autologous tissue for cell replacement therapy could avoid the need for damaging immunosuppression prior to transplant.

6. Future directions for hPSC differentiation approaches toward pharyngeal derivatives

Over the past two decades, tremendous advancements have catalyzed the differentiation of hPSCs toward lineage- and organ-specific cells. In this review, we have discussed the advances made toward generating hPSC-derived pharyngeal derivatives, including the parathyroid, thyroid, and thymus progenitors. While many groups have produced promising evidence to verify the cellular identity, further work will be necessary to improve the molecular accuracy and functionality of these pharyngeal derivatives. Finally, we will suggest alternate methods and improved assessments for differentiating hPSCs toward pharyngeal endoderm cell types.

6.1 Reprogramming hPSCs toward pharyngeal derivatives

Reprogramming represents an alternate approach to directed differentiation of hPSCs for generating human cells of interest. Although directed differentiation has proven widely successful as an approach for producing biologically relevant cells, such protocols rely on expensive small molecules and growth factors. Additionally, directed differentiation protocols often lack robustness due to compound variations. In recent years, direct reprogramming has gained interest as an alternate differentiation approach. Direct reprogramming reduces the period of differentiation by eliminating the need to pass through a pluripotent state. Bypassing the pluripotent state in some instances could produce more clinically-relevant cells as this approach reduces the concern for teratoma formation upon transplantation (Xu, Du, & Deng, 2015).

Reprogramming approaches traditionally depend on overexpression of tissue-specific transcription factor in somatic cells such as fibroblasts, but PSCs can also act as the starting material. The first demonstration of reprogramming concept of cell fate changes in somatic cells through transcription factor overexpression was initially demonstrated by generating muscle cells from fibroblasts (Davis, Weintraub, & Lassar, 1987). Since then, the concept has also been demonstrated to generate neuronal cell types (Caiazzo et al., 2011; Pfisterer et al., 2011). Similarly, transdifferentiation relies on transcription factor overexpression in a developmentally relevant and abundant somatic cell type, rather than utilizing fibroblasts or PSCs as the starting material. For example, overexpression of pancreas-specific transcription factors in intestinal cells or pancreatic exocrine cells causes them to shift into an insulin-producing state (Ariyachet et al., 2016; Zhou, Brown, Kanarek, Rajagopal, & Melton, 2008). With respect to pharyngeal endoderm cell types, several groups have demonstrated reprogramming of mouse and human fibroblasts and PSCs to pharyngeal cell types including the thymus and thyroid (Antonica et al., 2012; Bredenkamp et al., 2014; Dame et al., 2017; Ma, Latif, & Davies, 2013, 2015). However, questions still remain regarding the stability and fidelity of cells undergoing reprogramming, in addition to the persistence of epigenetic memory. Importantly, gene regulatory network (GRN) analysis revealed that cells produced *via* directed differentiation can more accurately reflect the molecular state of the desired *in vivo* cell type (Cahan et al., 2014). Moreover, since somatic cells do not possess the capacity for unlimited self-renewal, starting cell numbers can be a limiting factor for reprogramming approaches.

6.2 Single cell-omics for informing and assessing hPSC differentiation

Both directed differentiation and reprogramming approaches can benefit from genomics advances. Single cell-omics methods can produce insight into the pathways that could be missing from (or erroneously activated/inhibited) in current protocols (Fig. 3) (Camp, Wollny, & Treutlein, 2018). More specifically, single cell RNA datasets of tissues of interest and developing cells can be used as frameworks for designing and assessing various differentiation approaches (Ramond et al., 2018). Currently several

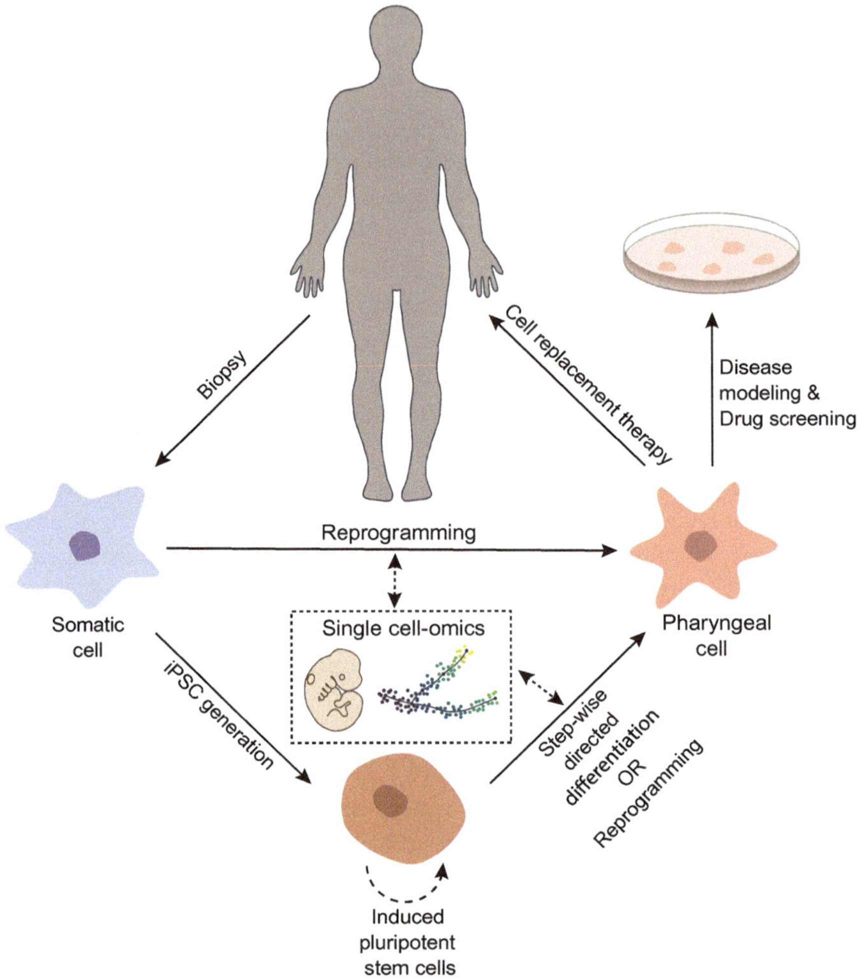

Fig. 3 Integration of technologies to improve the state of the art. Modern single cell-omic approaches increase our knowledge and classification of developing cell types and can be used to inform stem cell differentiation approaches. Additionally, they can yield insight into diverse techniques such as direct reprogramming, disease modeling, and cell replacement therapy. Ultimately all these approaches are focused on improving the outcome of diseased patients.

groups are working to generate complete reference datasets for developing endoderm tissues (Nowotschin et al., 2019; Pijuan-Sala et al., 2019). Bioinformatic methods for reconstructing *in vivo* development from single cell data and for comparing *in vitro* derived cells to relevant primary cell counterparts will be useful tools for designing and improving hPSC differentiation approaches. For instance, pseudotime methods can extrapolate developmental trajectories and cell state transitions from single cell datasets of developing tissues, and machine learning-based cell classifiers can determine the similarities between cells across species and differentiation platforms (Camp et al., 2018). Additionally, in surveying the stem cell field as a whole, the drive toward the translation of traditional 2D differentiation protocols to 3D organoid-based platforms becomes readily apparent (Eiraku et al., 2011; Lancaster et al., 2013; McCracken et al., 2014; Spence et al., 2011; Takasato et al., 2015; Takebe et al., 2014). These organoids facilitate studies concerning multicellular interactions and more accurately recapitulate disease states which often involve several cell types. A recent study using single cell RNA sequencing to compare 2D and 3D brain organ culture systems with primary fetal brain tissue determined that 3D culture systems produced more biologically relevant cells (Camp et al., 2015). Incorporating these new perspectives into differentiation or direct reprogramming platforms will likely allow more physiologically relevant cell types to be derived *in vitro*.

7. Concluding remarks

Ultimately, improving the functional maturation of pharyngeal-derived cells will unlock many opportunities to model diseases, study human development, and provide new drug screening platforms and cell replacement therapies to directly benefit patients. As protocols become more sophisticated and we gain the ability to produce the various cell types that often comprise a single organ, our understanding of human development and disease development will continue to grow, ultimately leading to disease interventions.

Acknowledgments

We are grateful to Maehr lab members for discussions. This work was supported by NIH grants (DP3DK111898, R01AI132963, U01DK104218), and funds from The Glass Charitable Foundation to R.M. In addition, this work was supported by a research grant from the University of Pennsylvania Orphan Disease Center in partnership with the Hypopara Research Foundation to R.M.

References

Abate, E. G., & Clarke, B. L. (2016). Review of hypoparathyroidism. *Frontiers in Endocrinology*, 7, 172.

Abramson, J., & Anderson, G. (2017). Thymic epithelial cells. *Annual Review of Immunology, 35*, 85–118.

Antonica, F., Kasprzyk, D. F., Opitz, R., Iacovino, M., Liao, X. H., Dumitrescu, A. M., et al. (2012). Generation of functional thyroid from embryonic stem cells. *Nature, 491*, 66–71.

Ariyachet, C., Tovaglieri, A., Xiang, G., Lu, J., Shah, M. S., Richmond, C. A., et al. (2016). Reprogrammed stomach tissue as a renewable source of functional beta cells for blood glucose regulation. *Cell Stem Cell, 18*, 410–421.

Arnold, J. S., Werling, U., Braunstein, E. M., Liao, J., Nowotschin, S., Edelmann, W., et al. (2006). Inactivation of Tbx1 in the pharyngeal endoderm results in 22q11DS malformations. *Development, 133*, 977–987.

Arufe, M. C., Lu, M., Kubo, A., Keller, G., Davies, T. F., & Lin, R. Y. (2006). Directed differentiation of mouse embryonic stem cells into thyroid follicular cells. *Endocrinology, 147*, 3007–3015.

Arufe, M. C., Lu, M., & Lin, R. Y. (2009). Differentiation of murine embryonic stem cells to thyrocytes requires insulin and insulin-like growth factor-1. *Biochemical and Biophysical Research Communications, 381*, 264–270.

Bain, V. E., Gordon, J., O'Neil, J. D., Ramos, I., Richie, E. R., & Manley, N. R. (2016). Tissue-specific roles for sonic hedgehog signaling in establishing thymus and parathyroid organ fate. *Development (Cambridge, England), 143*, 4027–4037.

Balciunaite, G., Keller, M. P., Balciunaite, E., Piali, L., Zuklys, S., Mathieu, Y. D., et al. (2002). Wnt glycoproteins regulate the expression of FoxN1, the gene defective in nude mice. *Nature Immunology, 3*, 1102–1108.

Begbie, J., Brunet, J. F., Rubenstein, J. L., & Graham, A. (1999). Induction of the epibranchial placodes. *Development, 126*, 895–902.

Biben, C., Hatzistavrou, T., & Harvey, R. P. (1998). Expression of NK-2 class homeobox gene Nkx2-6 in foregut endoderm and heart. *Mechanisms of Development, 73*, 125–127.

Bilezikian, J. P., Khan, A., Potts, J. T., Jr., Brandi, M. L., Clarke, B. L., Shoback, D., et al. (2011). Hypoparathyroidism in the adult: Epidemiology, diagnosis, pathophysiology, target-organ involvement, treatment, and challenges for future research. *Journal of Bone and Mineral Research: The Official Journal of the American Society for Bone and Mineral Research, 26*, 2317–2337.

Bingham, E. L., Cheng, S. P., Woods Ignatoski, K. M., & Doherty, G. M. (2009). Differentiation of human embryonic stem cells to a parathyroid-like phenotype. *Stem Cells and Development, 18*, 1071–1080.

Blackburn, C. C., Augustine, C. L., Li, R., Harvey, R. P., Malin, M. A., Boyd, R. L., et al. (1996). The nu gene acts cell-autonomously and is required for differentiation of thymic epithelial progenitors. *Proceedings of the National Academy of Sciences of the United States of America, 93*, 5742–5746.

Boyd, J. D. (1950). Development of the thyroid and parathyroid glands and the thymus. *Annals of the Royal College of Surgeons of England, 7*, 455–471.

Bredenkamp, N., Ulyanchenko, S., O'Neill, K. E., Manley, N. R., Vaidya, H. J., & Blackburn, C. C. (2014). An organized and functional thymus generated from FOXN1-reprogrammed fibroblasts. *Nature Cell Biology, 16*, 902–908.

Brolen, G., Sivertsson, L., Bjorquist, P., Eriksson, G., Ek, M., Semb, H., et al. (2010). Hepatocyte-like cells derived from human embryonic stem cells specifically via definitive endoderm and a progenitor stage. *Journal of Biotechnology, 145*, 284–294.

Bruserud, O., Oftedal, B. E., Wolff, A. B., & Husebye, E. S. (2016). AIRE-mutations and autoimmune disease. *Current Opinion in Immunology, 43*, 8–15.

Burridge, P. W., Keller, G., Gold, J. D., & Wu, J. C. (2012). Production of de novo cardiomyocytes: Human pluripotent stem cell differentiation and direct reprogramming. *Cell Stem Cell, 10*, 16–28.

Cahan, P., Li, H., Morris, S. A., Lummertz da Rocha, E., Daley, G. Q., & Collins, J. J. (2014). CellNet: Network biology applied to stem cell engineering. *Cell, 158*, 903–915.

Caiazzo, M., Dell'Anno, M. T., Dvoretskova, E., Lazarevic, D., Taverna, S., Leo, D., et al. (2011). Direct generation of functional dopaminergic neurons from mouse and human fibroblasts. *Nature, 476*, 224–227.

Camp, J. G., Badsha, F., Florio, M., Kanton, S., Gerber, T., Wilsch-Brauninger, M., et al. (2015). Human cerebral organoids recapitulate gene expression programs of fetal neocortex development. *Proceedings of the National Academy of Sciences of the United States of America, 112*, 15672–15677.

Camp, J. G., Wollny, D., & Treutlein, B. (2018). Single-cell genomics to guide human stem cell and tissue engineering. *Nature Methods, 15*, 661–667.

Chan, A. Y., & Anderson, M. S. (2015). Central tolerance to self revealed by the autoimmune regulator. *Annals of the New York Academy of Sciences, 1356*, 80–89.

Chen, Y. G., Mathews, C. E., & Driver, J. P. (2018). The role of NOD mice in type 1 diabetes research: Lessons from the past and recommendations for the future. *Frontiers in Endocrinology, 9*, 51.

Cheng, M., & Anderson, M. S. (2018). Thymic tolerance as a key brake on autoimmunity. *Nature Immunology, 19*, 659–664.

Choe, C. P., Collazo, A., Trinh le, A., Pan, L., Moens, C. B., & Crump, J. G. (2013). Wnt-dependent epithelial transitions drive pharyngeal pouch formation. *Developmental Cell, 24*, 296–309.

Chojnowski, J. L., Masuda, K., Trau, H. A., Thomas, K., Capecchi, M., & Manley, N. R. (2014). Multiple roles for HOXA3 in regulating thymus and parathyroid differentiation and morphogenesis in mouse. *Development (Cambridge, England), 141*, 3697–3708.

Chu, P. G., & Weiss, L. M. (2002). Keratin expression in human tissues and neoplasms. *Histopathology, 40*, 403–439.

Clarke, B. L., Brown, E. M., Collins, M. T., Juppner, H., Lakatos, P., Levine, M. A., et al. (2016). Epidemiology and diagnosis of hypoparathyroidism. *The Journal of Clinical Endocrinology and Metabolism, 101*, 2284–2299.

Clifton-Bligh, R. J., Wentworth, J. M., Heinz, P., Crisp, M. S., John, R., Lazarus, J. H., et al. (1998). Mutation of the gene encoding human TTF-2 associated with thyroid agenesis, cleft palate and choanal atresia. *Nature Genetics, 19*, 399–401.

Crump, J. G., Maves, L., Lawson, N. D., Weinstein, B. M., & Kimmel, C. B. (2004). An essential role for Fgfs in endodermal pouch formation influences later craniofacial skeletal patterning. *Development, 131*, 5703–5716.

Dame, K., Cincotta, S., Lang, A. H., Sanghrajka, R. M., Zhang, L., Choi, J., et al. (2017). Thyroid progenitors are robustly derived from embryonic stem cells through transient, developmental stage-specific overexpression of Nkx2-1. *Stem Cell Reports, 8*, 216–225.

D'Amour, K. A., Agulnick, A. D., Eliazer, S., Kelly, O. G., Kroon, E., & Baetge, E. E. (2005). Efficient differentiation of human embryonic stem cells to definitive endoderm. *Nature Biotechnology, 23*, 1534–1541.

Davenport, C., Diekmann, U., Budde, I., Detering, N., & Naujok, O. (2016). Anterior-posterior patterning of definitive endoderm generated from human embryonic stem cells depends on the differential signaling of retinoic acid, Wnt-, and BMP-signaling. *Stem cells (Dayton, Ohio), 34*, 2635–2647.

Davies, E. G. (2013). Immunodeficiency in DiGeorge syndrome and options for treating cases with complete athymia. *Frontiers in Immunology, 4*, 322.

Davies, E. G., Cheung, M., Gilmour, K., Maimaris, J., Curry, J., Furmanski, A., et al. (2017). Thymus transplantation for complete DiGeorge syndrome: European experience. *The Journal of Allergy and Clinical Immunology, 140*, 1660–1670.e1616.

Davis, R. L., Weintraub, H., & Lassar, A. B. (1987). Expression of a single transfected cDNA converts fibroblasts to myoblasts. *Cell, 51*, 987–1000.

De Felice, M., Ovitt, C., Biffali, E., Rodriguez-Mallon, A., Arra, C., Anastassiadis, K., et al. (1998). A mouse model for hereditary thyroid dysgenesis and cleft palate. *Nature Genetics*, *19*, 395–398.

DeLaForest, A., Nagaoka, M., Si-Tayeb, K., Noto, F. K., Konopka, G., Battle, M. A., et al. (2011). HNF4A is essential for specification of hepatic progenitors from human pluripotent stem cells. *Development (Cambridge, England)*, *138*, 4143–4153.

Doyle, D., Kirwin, S. M., Sol-Church, K., & Levine, M. A. (2012). A novel mutation in the GCM2 gene causing severe congenital isolated hypoparathyroidism. *Journal of Pediatric Endocrinology & Metabolism*, *25*, 741–746.

Du, Q., Huynh, L. K., Coskun, F., Molina, E., King, M. A., Raj, P., et al. (2019). FOXN1 compound heterozygous mutations cause selective thymic hypoplasia in humans. *The Journal of Clinical Investigation*, *129*, 4724–4738.

Eiraku, M., Takata, N., Ishibashi, H., Kawada, M., Sakakura, E., Okuda, S., et al. (2011). Self-organizing optic-cup morphogenesis in three-dimensional culture. *Nature*, *472*, 51–56.

Fan, Y., Rudert, W. A., Grupillo, M., He, J., Sisino, G., & Trucco, M. (2009). Thymus-specific deletion of insulin induces autoimmune diabetes. *The EMBO Journal*, *28*, 2812–2824.

Fernandez, L. P., Lopez-Marquez, A., & Santisteban, P. (2015). Thyroid transcription factors in development, differentiation and disease. *Nature Reviews. Endocrinology*, *11*, 29–42.

Figueiredo, M., Silva, J. C., Santos, A. S., Proa, V., Alcobia, I., Zilhao, R., et al. (2016). Notch and Hedgehog in the thymus/parathyroid common primordium: Crosstalk in organ formation. *Developmental Biology*, *418*, 268–282.

Fisher, D. A., & Klein, A. H. (1981). Thyroid development and disorders of thyroid function in the newborn. *The New England Journal of Medicine*, *304*, 702–712.

Frank, D. U., Fotheringham, L. K., Brewer, J. A., Muglia, L. J., Tristani-Firouzi, M., Capecchi, M. R., et al. (2002). An Fgf8 mouse mutant phenocopies human 22q11 deletion syndrome. *Development*, *129*, 4591–4603.

Frisdal, A., & Trainor, P. A. (2014). Development and evolution of the pharyngeal apparatus. *Wiley Interdisciplinary Reviews: Developmental Biology*, *3*, 403–418.

Gong, W., Gottlieb, S., Collins, J., Blescia, A., Dietz, H., Goldmuntz, E., et al. (2001). Mutation analysis of TBX1 in non-deleted patients with features of DGS/VCFS or isolated cardiovascular defects. *Journal of Medical Genetics*, *38*, E45.

Gordon, J. (2018). Hox genes in the pharyngeal region: How Hoxa3 controls early embryonic development of the pharyngeal organs. *The International Journal of Developmental Biology*, *62*, 775–783.

Gordon, J., Bennett, A. R., Blackburn, C. C., & Manley, N. R. (2001). Gcm2 and Foxn1 mark early parathyroid- and thymus-specific domains in the developing third pharyngeal pouch. *Mechanisms of Development*, *103*, 141–143.

Gordon, J., & Manley, N. R. (2011). Mechanisms of thymus organogenesis and morphogenesis. *Development (Cambridge, England)*, *138*, 3865–3878.

Gordon, J., Patel, S. R., Mishina, Y., & Manley, N. R. (2010). Evidence for an early role for BMP4 signaling in thymus and parathyroid morphogenesis. *Developmental Biology*, *339*, 141–154.

Gordon, J., Wilson, V. A., Blair, N. F., Sheridan, J., Farley, A., Wilson, L., et al. (2004). Functional evidence for a single endodermal origin for the thymic epithelium. *Nature Immunology*, *5*, 546–553.

Graham, A., Okabe, M., & Quinlan, R. (2005). The role of the endoderm in the development and evolution of the pharyngeal arches. *Journal of Anatomy*, *207*, 479–487.

Grati, F. R., Molina Gomes, D., Ferreira, J. C., Dupont, C., Alesi, V., Gouas, L., et al. (2015). Prevalence of recurrent pathogenic microdeletions and microduplications in over 9500 pregnancies. *Prenatal Diagnosis*, *35*, 801–809.

Green, M. D., Chen, A., Nostro, M. C., d'Souza, S. L., Schaniel, C., Lemischka, I. R., et al. (2011). Generation of anterior foregut endoderm from human embryonic and induced pluripotent stem cells. *Nature Biotechnology*, *29*, 267–272.

Grevellec, A., Graham, A., & Tucker, A. S. (2011). Shh signalling restricts the expression of Gcm2 and controls the position of the developing parathyroids. *Developmental Biology*, *353*, 194–205.

Guhr, A., Kobold, S., Seltmann, S., Seiler Wulczyn, A. E. M., Kurtz, A., & Loser, P. (2018). Recent trends in research with human pluripotent stem cells: Impact of research and use of cell lines in experimental research and clinical trials. *Stem Cell Reports*, *11*, 485–496.

Hidaka, K., Nitta, T., Sugawa, R., Shirai, M., Schwartz, R. J., Amagai, T., et al. (2010). Differentiation of pharyngeal endoderm from mouse embryonic stem cell. *Stem Cells and Development*, *19*, 1735–1743.

Hollenberg, A. N., Choi, J., Serra, M., & Kotton, D. N. (2017). Regenerative therapy for hypothyroidism: Mechanisms and possibilities. *Molecular and Cellular Endocrinology*, *445*, 35–41.

Ikonomou, L., & Kotton, D. N. (2015). Derivation of endodermal progenitors from pluripotent stem cells. *Journal of Cellular Physiology*, *230*, 246–258.

Jacob, A., Morley, M., Hawkins, F., McCauley, K. B., Jean, J. C., Heins, H., et al. (2017). Differentiation of human pluripotent stem cells into functional lung alveolar epithelial cells. *Cell Stem Cell*, *21*, 472–488.e410.

Jiang, N., Hu, Y., Liu, X., Wu, Y., Zhang, H., Chen, G., et al. (2010). Differentiation of E14 mouse embryonic stem cells into thyrocytes in vitro. *Thyroid: Official Journal of the American Thyroid Association*, *20*, 77–84.

Johansson, E., Andersson, L., Ornros, J., Carlsson, T., Ingeson-Carlsson, C., Liang, S., et al. (2015). Revising the embryonic origin of thyroid C cells in mice and humans. *Development*, *142*, 3519–3528.

Jones, N. C., & Trainor, P. A. (2004). The therapeutic potential of stem cells in the treatment of craniofacial abnormalities. *Expert Opinion on Biological Therapy*, *4*, 645–657.

Kearns, N. A., Genga, R. M., Ziller, M., Kapinas, K., Peters, H., Brehm, M. A., et al. (2013). Generation of organized anterior foregut epithelia from pluripotent stem cells using small molecules. *Stem Cell Research*, *11*, 1003–1012.

Kimura, S., Hara, Y., Pineau, T., Fernandez-Salguero, P., Fox, C. H., Ward, J. M., et al. (1996). The T/ebp null mouse: Thyroid-specific enhancer-binding protein is essential for the organogenesis of the thyroid, lung, ventral forebrain, and pituitary. *Genes & Development*, *10*, 60–69.

Kimura, S., Ward, J. M., & Minoo, P. (1999). Thyroid-specific enhancer-binding protein/thyroid transcription factor 1 is not required for the initial specification of the thyroid and lung primordia. *Biochimie*, *81*, 321–327.

King, A. J. (2012). The use of animal models in diabetes research. *British Journal of Pharmacology*, *166*, 877–894.

Knoepfler, P. S., Lu, Q., & Kamps, M. P. (1996). Pbx-1 Hox heterodimers bind DNA on inseparable half-sites that permit intrinsic DNA binding specificity of the Hox partner at nucleotides 3′ to a TAAT motif. *Nucleic Acids Research*, *24*, 2288–2294.

Kopinke, D., Sasine, J., Swift, J., Stephens, W. Z., & Piotrowski, T. (2006). Retinoic acid is required for endodermal pouch morphogenesis and not for pharyngeal endoderm specification. *Developmental Dynamics: An Official Publication of the American Association of the Anatomists*, *235*, 2695–2709.

Krude, H., Schutz, B., Biebermann, H., von Moers, A., Schnabel, D., Neitzel, H., et al. (2002). Choreoathetosis, hypothyroidism, and pulmonary alterations due to human NKX2-1 haploinsufficiency. *The Journal of Clinical Investigation*, *109*, 475–480.

Kubo, A., Shinozaki, K., Shannon, J. M., Kouskoff, V., Kennedy, M., Woo, S., et al. (2004). Development of definitive endoderm from embryonic stem cells in culture. *Development (Cambridge, England)*, *131*, 1651–1662.

Kurmann, A. A., Serra, M., Hawkins, F., Rankin, S. A., Mori, M., Astapova, I., et al. (2015). Regeneration of thyroid function by transplantation of differentiated pluripotent stem cells. *Cell Stem Cell*, *17*, 527–542.

Lancaster, M. A., Renner, M., Martin, C. A., Wenzel, D., Bicknell, L. S., Hurles, M. E., et al. (2013). Cerebral organoids model human brain development and microcephaly. *Nature*, *501*, 373–379.

Le Douarin, N. M., & Jotereau, F. V. (1975). Tracing of cells of the avian thymus through embryonic life in interspecific chimeras. *The Journal of Experimental Medicine*, *142*, 17–40.

Liu, Z., Yu, S., & Manley, N. R. (2007). Gcm2 is required for the differentiation and survival of parathyroid precursor cells in the parathyroid/thymus primordia. *Developmental Biology*, *305*, 333–346.

Loh, K. M., Ang, L. T., Zhang, J., Kumar, V., Ang, J., Auyeong, J. Q., et al. (2014). Efficient endoderm induction from human pluripotent stem cells by logically directing signals controlling lineage bifurcations. *Cell Stem Cell*, *14*, 237–252.

Longmire, T. A., Ikonomou, L., Hawkins, F., Christodoulou, C., Cao, Y., Jean, J. C., et al. (2012). Efficient derivation of purified lung and thyroid progenitors from embryonic stem cells. *Cell Stem Cell*, *10*, 398–411.

Ma, R., Latif, R., & Davies, T. F. (2009). Thyrotropin-independent induction of thyroid endoderm from embryonic stem cells by activin A. *Endocrinology*, *150*, 1970–1975.

Ma, R., Latif, R., & Davies, T. F. (2013). Thyroid follicle formation and thyroglobulin expression in multipotent endodermal stem cells. *Thyroid: Official Journal of the American Thyroid Association*, *23*, 385–391.

Ma, R., Latif, R., & Davies, T. F. (2015). Human embryonic stem cells form functional thyroid follicles. *Thyroid: Official Journal of the American Thyroid Association*, *25*, 455–461.

Ma, R., Morshed, S. A., Latif, R., & Davies, T. F. (2017). TAZ induction directs differentiation of thyroid follicular cells from human embryonic stem cells. *Thyroid: Official Journal of the American Thyroid Association*, *27*, 292–299.

Macchia, P. E., Lapi, P., Krude, H., Pirro, M. T., Missero, C., Chiovato, L., et al. (1998). PAX8 mutations associated with congenital hypothyroidism caused by thyroid dysgenesis. *Nature Genetics*, *19*, 83–86.

Manley, N. R., & Condie, B. G. (2010). Transcriptional regulation of thymus organogenesis and thymic epithelial cell differentiation. *Progress in Molecular Biology and Translational Science*, *92*, 103–120.

Manley, N. R., Selleri, L., Brendolan, A., Gordon, J., & Cleary, M. L. (2004). Abnormalities of caudal pharyngeal pouch development in Pbx1 knockout mice mimic loss of Hox3 paralogs. *Developmental Biology*, *276*, 301–312.

Mannstadt, M., Bilezikian, J. P., Thakker, R. V., Hannan, F. M., Clarke, B. L., Rejnmark, L., et al. (2017). Hypoparathyroidism. *Nature Reviews. Disease Primers*, *3*, 17055.

Mansouri, A., Chowdhury, K., & Gruss, P. (1998). Follicular cells of the thyroid gland require Pax8 gene function. *Nature Genetics*, *19*, 87–90.

Markert, M. L., Devlin, B. H., Alexieff, M. J., Li, J., McCarthy, E. A., Gupton, S. E., et al. (2007). Review of 54 patients with complete DiGeorge anomaly enrolled in protocols for thymus transplantation: Outcome of 44 consecutive transplants. *Blood*, *109*, 4539–4547.

Markert, M. L., Devlin, B. H., & McCarthy, E. A. (2010). Thymus transplantation. *Clinical Immunology (Orlando, FL)*, *135*, 236–246.

Martinez Barbera, J. P., Clements, M., Thomas, P., Rodriguez, T., Meloy, D., Kioussis, D., et al. (2000). The homeobox gene Hex is required in definitive endodermal tissues for

normal forebrain, liver and thyroid formation. *Development (Cambridge, England), 127*, 2433–2445.
McCauley, K. B., Hawkins, F., Serra, M., Thomas, D. C., Jacob, A., & Kotton, D. N. (2017). Efficient derivation of functional human airway epithelium from pluripotent stem cells via temporal regulation of Wnt signaling. *Cell Stem Cell, 20*, 844–857, e846.
McCracken, K. W., Cata, E. M., Crawford, C. M., Sinagoga, K. L., Schumacher, M., Rockich, B. E., et al. (2014). Modelling human development and disease in pluripotent stem-cell-derived gastric organoids. *Nature, 516*, 400–404.
McDonald-McGinn, D. M., Sullivan, K. E., Marino, B., Philip, N., Swillen, A., Vorstman, J. A., et al. (2015). 22q11.2 deletion syndrome. *Nature Reviews. Disease Primers, 1*, 15071.
McLean, A. B., D'Amour, K. A., Jones, K. L., Krishnamoorthy, M., Kulik, M. J., Reynolds, D. M., et al. (2007). Activin a efficiently specifies definitive endoderm from human embryonic stem cells only when phosphatidylinositol 3-kinase signaling is suppressed. *Stem Cells (Dayton, Ohio), 25*, 29–38.
Moore-Scott, B. A., & Manley, N. R. (2005). Differential expression of Sonic hedgehog along the anterior-posterior axis regulates patterning of pharyngeal pouch endoderm and pharyngeal endoderm-derived organs. *Developmental Biology, 278*, 323–335.
Morizane, R., Miyoshi, T., & Bonventre, J. V. (2017). Concise review: Kidney generation with human pluripotent stem cells. *Stem Cells (Dayton, Ohio), 35*, 2209–2217.
Mun, S. J., Ryu, J. S., Lee, M. O., Son, Y. S., Oh, S. J., Cho, H. S., et al. (2019). Generation of expandable human pluripotent stem cell-derived hepatocyte-like liver organoids. *Journal of Hepatology, 71*, 970–985.
Munera, J. O., Sundaram, N., Rankin, S. A., Hill, D., Watson, C., Mahe, M., et al. (2017). Differentiation of human pluripotent stem cells into colonic organoids via transient activation of BMP signaling. *Cell Stem Cell, 21*, 51–64.e56.
Nair, G. G., Liu, J. S., Russ, H. A., Tran, S., Saxton, M. S., Chen, R., et al. (2019). Recapitulating endocrine cell clustering in culture promotes maturation of human stem-cell-derived beta cells. *Nature Cell Biology, 21*, 263–274.
Nilsson, M., & Fagman, H. (2017). Development of the thyroid gland. *Development (Cambridge, England), 144*, 2123–2140.
Nowotschin, S., Setty, M., Kuo, Y. Y., Liu, V., Garg, V., Sharma, R., et al. (2019). The emergent landscape of the mouse gut endoderm at single-cell resolution. *Nature, 569*, 361–367.
Ogata, T., Niihori, T., Tanaka, N., Kawai, M., Nagashima, T., Funayama, R., et al. (2014). TBX1 mutation identified by exome sequencing in a Japanese family with 22q11.2 deletion syndrome-like craniofacial features and hypocalcemia. *PLoS One, 9*, e91598.
Ogawa, S., Surapisitchat, J., Virtanen, C., Ogawa, M., Niapour, M., Sugamori, K. S., et al. (2013). Three-dimensional culture and cAMP signaling promote the maturation of human pluripotent stem cell-derived hepatocytes. *Development (Cambridge, England), 140*, 3285–3296.
Paganini, I., Sestini, R., Capone, G. L., Putignano, A. L., Contini, E., Giotti, I., et al. (2017). A novel PAX1 null homozygous mutation in autosomal recessive otofaciocervical syndrome associated with severe combined immunodeficiency. *Clinical Genetics, 92*, 664–668.
Parent, A. V., Russ, H. A., Khan, I. S., LaFlam, T. N., Metzger, T. C., Anderson, M. S., et al. (2013). Generation of functional thymic epithelium from human embryonic stem cells that supports host T cell development. *Cell Stem Cell, 13*, 219–229.
Parlato, R., Rosica, A., Rodriguez-Mallon, A., Affuso, A., Postiglione, M. P., Arra, C., et al. (2004). An integrated regulatory network controlling survival and migration in thyroid organogenesis. *Developmental Biology, 276*, 464–475.

Patel, S. R., Gordon, J., Mahbub, F., Blackburn, C. C., & Manley, N. R. (2006). Bmp4 and Noggin expression during early thymus and parathyroid organogenesis. *Gene Expression Patterns, 6*, 794–799.

Peters, H., Neubuser, A., Kratochwil, K., & Balling, R. (1998). Pax9-deficient mice lack pharyngeal pouch derivatives and teeth and exhibit craniofacial and limb abnormalities. *Genes & Development, 12*, 2735–2747.

Pfisterer, U., Kirkeby, A., Torper, O., Wood, J., Nelander, J., Dufour, A., et al. (2011). Direct conversion of human fibroblasts to dopaminergic neurons. *Proceedings of the National Academy of Sciences of the United States of America, 108*, 10343–10348.

Pijuan-Sala, B., Griffiths, J. A., Guibentif, C., Hiscock, T. W., Jawaid, W., Calero-Nieto, F. J., et al. (2019). A single-cell molecular map of mouse gastrulation and early organogenesis. *Nature, 566*, 490–495.

Piotrowski, T., Ahn, D. G., Schilling, T. F., Nair, S., Ruvinsky, I., Geisler, R., et al. (2003). The zebrafish van gogh mutation disrupts tbx1, which is involved in the DiGeorge deletion syndrome in humans. *Development, 130*, 5043–5052.

Piotrowski, T., & Nusslein-Volhard, C. (2000). The endoderm plays an important role in patterning the segmented pharyngeal region in zebrafish (Danio rerio). *Developmental Biology, 225*, 339–356.

Pitkanen, J., & Peterson, P. (2003). Autoimmune regulator: From loss of function to autoimmunity. *Genes and Immunity, 4*, 12–21.

Pugliese, A., Zeller, M., Fernandez, A., Jr., Zalcberg, L. J., Bartlett, R. J., Ricordi, C., et al. (1997). The insulin gene is transcribed in the human thymus and transcription levels correlated with allelic variation at the INS VNTR-IDDM2 susceptibility locus for type 1 diabetes. *Nature Genetics, 15*, 293–297.

Pyzik, A., Grywalska, E., Matyjaszek-Matuszek, B., & Rolinski, J. (2015). Immune disorders in Hashimoto's thyroiditis: What do we know so far? *Journal of Immunology Research, 2015*, 979167.

Ramond, C., Beydag-Tasoz, B. S., Azad, A., van de Bunt, M., Petersen, M. B. K., Beer, N. L., et al. (2018). Understanding human fetal pancreas development using subpopulation sorting, RNA sequencing and single-cell profiling. *Development (Cambridge, England), 145* dev165480.

Revest, J. M., Suniara, R. K., Kerr, K., Owen, J. J., & Dickson, C. (2001). Development of the thymus requires signaling through the fibroblast growth factor receptor R2-IIIb. *Journal of Immunology (Baltimore, Md.: 1950), 167*, 1954–1961.

Rota, I. A., & Dhalla, F. (2017). FOXN1 deficient nude severe combined immunodeficiency. *Orphanet Journal of Rare Diseases, 12,* 6.

Rowe, R. G., & Daley, G. Q. (2019). Induced pluripotent stem cells in disease modelling and drug discovery. *Nature Reviews. Genetics, 20*, 377–388.

Sandell, L. L., Sanderson, B. W., Moiseyev, G., Johnson, T., Mushegian, A., Young, K., et al. (2007). RDH10 is essential for synthesis of embryonic retinoic acid and is required for limb, craniofacial, and organ development. *Genes & Development, 21*, 1113–1124.

Serra, M., Alysandratos, K. D., Hawkins, F., McCauley, K. B., Jacob, A., Choi, J., et al. (2017). Pluripotent stem cell differentiation reveals distinct developmental pathways regulating lung- versus thyroid-lineage specification. *Development (Cambridge, England), 144*, 3879–3893.

Shoback, D. M., Bilezikian, J. P., Costa, A. G., Dempster, D., Dralle, H., Khan, A. A., et al. (2016). Presentation of hypoparathyroidism: Etiologies and clinical features. *The Journal of Clinical Endocrinology and Metabolism, 101*, 2300–2312.

Sikjaer, T., Rejnmark, L., Rolighed, L., Heickendorff, L., & Mosekilde, L. (2011). The effect of adding PTH(1-84) to conventional treatment of hypoparathyroidism: A randomized, placebo-controlled study. *Journal of Bone and Mineral Research: The Official Journal of the American Society for Bone and Mineral Research, 26*, 2358–2370.

Soh, C. L., Giudice, A., Jenny, R. A., Elliott, D. A., Hatzistavrou, T., Micallef, S. J., et al. (2014). FOXN1 (GFP/w) reporter hESCs enable identification of integrin-beta4, HLA-DR, and EpCAM as markers of human PSC-derived FOXN1(+) thymic epithelial progenitors. *Stem Cell Reports*, *2*, 925–937.

Spence, J. R., Mayhew, C. N., Rankin, S. A., Kuhar, M. F., Vallance, J. E., Tolle, K., et al. (2011). Directed differentiation of human pluripotent stem cells into intestinal tissue in vitro. *Nature*, *470*, 105–109.

Su, M., Hu, R., Jin, J., Yan, Y., Song, Y., Sullivan, R., et al. (2015). Efficient in vitro generation of functional thymic epithelial progenitors from human embryonic stem cells. *Scientific Reports*, *5*, 9882.

Sun, L., Luo, H., Li, H., & Zhao, Y. (2013). Thymic epithelial cell development and differentiation: Cellular and molecular regulation. *Protein & Cell*, *4*, 342–355.

Swann, J. B., Happe, C., & Boehm, T. (2017). Elevated levels of Wnt signaling disrupt thymus morphogenesis and function. *Scientific Reports*, *7*, 785.

Swann, J. B., Krauth, B., Happe, C., & Boehm, T. (2017). Cooperative interaction of BMP signalling and Foxn1 gene dosage determines the size of the functionally active thymic epithelial compartment. *Scientific Reports*, *7*, 8492.

Takaba, H., & Takayanagi, H. (2017). The mechanisms of T cell selection in the thymus. *Trends in Immunology*, *38*, 805–816.

Takahashi, K., Tanabe, K., Ohnuki, M., Narita, M., Ichisaka, T., Tomoda, K., et al. (2007). Induction of pluripotent stem cells from adult human fibroblasts by defined factors. *Cell*, *131*, 861–872.

Takasato, M., Er, P. X., Chiu, H. S., Maier, B., Baillie, G. J., Ferguson, C., et al. (2015). Kidney organoids from human iPS cells contain multiple lineages and model human nephrogenesis. *Nature*, *526*, 564–568.

Takebe, T., Zhang, R. R., Koike, H., Kimura, M., Yoshizawa, E., Enomura, M., et al. (2014). Generation of a vascularized and functional human liver from an iPSC-derived organ bud transplant. *Nature Protocols*, *9*, 396–409.

Tanaka, M., Schinke, M., Liao, H. S., Yamasaki, N., & Izumo, S. (2001). Nkx2.5 and Nkx2.6, homologs of Drosophila tinman, are required for development of the pharynx. *Molecular and Cellular Biology*, *21*, 4391–4398.

Tao, Y., & Zhang, S. C. (2016). Neural subtype specification from human pluripotent stem cells. *Cell Stem Cell*, *19*, 573–586.

Teo, A. K., Tsuneyoshi, N., Hoon, S., Tan, E. K., Stanton, L. W., Wright, C. V., et al. (2015). PDX1 binds and represses hepatic genes to ensure robust pancreatic commitment in differentiating human embryonic stem cells. *Stem Cell Reports*, *4*, 578–590.

Teo, A. K., Valdez, I. A., Dirice, E., & Kulkarni, R. N. (2014). Comparable generation of activin-induced definitive endoderm via additive Wnt or BMP signaling in absence of serum. *Stem Cell Reports*, *3*, 5–14.

Thomas, H. E., Brodnicki, T. C., & Kay, T. W. (2016). How to make mice tell the truth. *Diabetes*, *65*, 1161–1163.

Thomson, J. A., Itskovitz-Eldor, J., Shapiro, S. S., Waknitz, M. A., Swiergiel, J. J., Marshall, V. S., et al. (1998). Embryonic stem cell lines derived from human blastocysts. *Science*, *282*, 1145–1147.

Toivonen, S., Lundin, K., Balboa, D., Ustinov, J., Tamminen, K., Palgi, J., et al. (2013). Activin A and Wnt-dependent specification of human definitive endoderm cells. *Experimental Cell Research*, *319*, 2535–2544.

Trisno, S. L., Philo, K. E. D., McCracken, K. W., Cata, E. M., Ruiz-Torres, S., Rankin, S. A., et al. (2018). Esophageal organoids from human pluripotent stem cells delineate Sox2 functions during esophageal specification. *Cell Stem Cell*, *23*, 501–515.e507.

Trokovic, N., Trokovic, R., Mai, P., & Partanen, J. (2003). Fgfr1 regulates patterning of the pharyngeal region. *Genes & Development*, *17*, 141–153.

Trounson, A., & McDonald, C. (2015). Stem cell therapies in clinical trials: Progress and challenges. *Cell Stem Cell*, *17*, 11–22.
Vafiadis, P., Bennett, S. T., Todd, J. A., Nadeau, J., Grabs, R., Goodyer, C. G., et al. (1997). Insulin expression in human thymus is modulated by INS VNTR alleles at the IDDM2 locus. *Nature Genetics*, *15*, 289–292.
Van Esch, H., Groenen, P., Nesbit, M. A., Schuffenhauer, S., Lichtner, P., Vanderlinden, G., et al. (2000). GATA3 haplo-insufficiency causes human HDR syndrome. *Nature*, *406*, 419–422.
Vermot, J., Niederreither, K., Garnier, J. M., Chambon, P., & Dolle, P. (2003). Decreased embryonic retinoic acid synthesis results in a DiGeorge syndrome phenotype in newborn mice. *Proceedings of the National Academy of Sciences of the United States of America*, *100*, 1763–1768.
Volpi, S., Yamazaki, Y., Brauer, P. M., van Rooijen, E., Hayashida, A., Slavotinek, A., et al. (2017). EXTL3 mutations cause skeletal dysplasia, immune deficiency, and developmental delay. *The Journal of Experimental Medicine*, *214*, 623–637.
Wallin, J., Eibel, H., Neubuser, A., Wilting, J., Koseki, H., & Balling, R. (1996). Pax1 is expressed during development of the thymus epithelium and is required for normal T-cell maturation. *Development (Cambridge, England)*, *122*, 23–30.
Winer, K. K., Sinaii, N., Peterson, D., Sainz, B., Jr., & Cutler, G. B., Jr. (2008). Effects of once versus twice-daily parathyroid hormone 1–34 therapy in children with hypoparathyroidism. *The Journal of Clinical Endocrinology and Metabolism*, *93*, 3389–3395.
Wong, A. P., Chin, S., Xia, S., Garner, J., Bear, C. E., & Rossant, J. (2015). Efficient generation of functional CFTR-expressing airway epithelial cells from human pluripotent stem cells. *Nature Protocols*, *10*, 363–381.
Woods Ignatoski, K. M., Bingham, E. L., Frome, L. K., & Doherty, G. M. (2010). Differentiation of precursors into parathyroid-like cells for treatment of hypoparathyroidism. *Surgery*, *148*, 1186–1189. discussion 1189–1190.
Xie, R., Carrano, A. C., & Sander, M. (2015). A systems view of epigenetic networks regulating pancreas development and beta-cell function. *Wiley Interdisciplinary Reviews. Systems Biology and Medicine*, 7, 1–11.
Xu, J., Du, Y., & Deng, H. (2015). Direct lineage reprogramming: Strategies, mechanisms, and applications. *Cell Stem Cell*, *16*, 119–134.
Yagi, H., Furutani, Y., Hamada, H., Sasaki, T., Asakawa, S., Minoshima, S., et al. (2003). Role of TBX1 in human y22q11.2 syndrome. *Lancet*, *362*, 1366–1373.
Yiangou, L., Ross, A. D. B., Goh, K. J., & Vallier, L. (2018). Human pluripotent stem cell-derived endoderm for modeling development and clinical applications. *Cell Stem Cell*, *22*, 485–499.
Yoshida, K., Nakahata, A., Treen, N., Sakuma, T., Yamamoto, T., & Sasakura, Y. (2017). Hox-mediated endodermal identity patterns pharyngeal muscle formation in the chordate pharynx. *Development*, *144*, 1629–1634.
Zhou, Q., Brown, J., Kanarek, A., Rajagopal, J., & Melton, D. A. (2008). In vivo reprogramming of adult pancreatic exocrine cells to beta-cells. *Nature*, *455*, 627–632.
Zhu, Z., & Huangfu, D. (2013). Human pluripotent stem cells: An emerging model in developmental biology. *Development (Cambridge, England)*, *140*, 705–717.
Zhu, Z., Li, Q. V., Lee, K., Rosen, B. P., Gonzalez, F., Soh, C. L., et al. (2016). Genome editing of lineage determinants in human pluripotent stem cells reveals mechanisms of pancreatic development and diabetes. *Cell Stem Cell*, *18*, 755–768.
Zorn, A. M., & Wells, J. M. (2009). Vertebrate endoderm development and organ formation. *Annual Review of Cell and Developmental Biology*, *25*, 221–251.
Zweier, C., Sticht, H., Aydin-Yaylagul, I., Campbell, C. E., & Rauch, A. (2007). Human TBX1 missense mutations cause gain of function resulting in the same phenotype as 22q11.2 deletions. *American Journal of Human Genetics*, *80*, 510–517.

CHAPTER SIX

Epigenetic metabolites license stem cell states

Logeshwaran Somasundaram[a,b,†], Shiri Levy[a,b,†], Abdiasis M. Hussein[a,b], Devon D. Ehnes[a,b], Julie Mathieu[b,c], Hannele Ruohola-Baker[a,b,*]

[a]Department of Biochemistry, University of Washington, Seattle, WA, United States
[b]Institute for Stem Cell and Regenerative Medicine, University of Washington, Seattle, WA, United States
[c]Department of Comparative Medicine, University of Washington, Seattle, WA, United States
*Corresponding author: e-mail address: hannele@uw.edu

Contents

1. Introduction 210
2. Stem cell energetics 210
3. Metabolism of quiescent stem cells 212
   3.1 Adult stem cells 212
   3.2 Pluripotent stem cell quiescence, diapause 216
4. Metabolism of active stem cells 217
   4.1 Metabolism after fertilization 217
   4.2 Metabolism of pre-implantation and post-implantation pluripotent stem cells 218
   4.3 Metabolism of actively cycling adult stem cells: MSC as case-study 220
5. HIF, the master regulator of metabolism 222
6. Epigenetic signatures and epigenetic metabolites 224
   6.1 Epigenetic signatures of naïve and primed pluripotent stem cells 224
   6.2 Epigenetic signatures of adult stem cells 227
   6.3 Epigenetic metabolites 228
7. Conclusion 229
Acknowledgments 230
References 230
Further reading 240

Abstract

It has become clear during recent years that stem cells undergo metabolic remodeling during their activation process. While these metabolic switches take place in pluripotency as well as adult stem cell populations, the rules that govern the switch are not clear.

† Equal contribution.

Current Topics in Developmental Biology, Volume 138
ISSN 0070-2153
https://doi.org/10.1016/bs.ctdb.2020.02.003

In this review, we summarize some of the transitions in adult and pluripotent cell types and will propose that the key function in this process is the generation of epigenetic metabolites that govern critical epigenetic modifications, and therefore stem cell states.

1. Introduction

Stem cells have dynamic metabolic programs that support their constant capacity to regenerate. These metabolic signatures are proposed to propagate cell fate changes (Bracha, Ramanathan, Huang, Ingber, & Schreiber, 2010; Folmes et al., 2011; Greer, Metcalf, Wang, & Ohh, 2012; Panopoulos et al., 2011; Rafalski, Mancini, & Brunet, 2012; Yanes et al., 2010), which alter as the cells become increasingly specified. Recent studies are revealing that these metabolic shifts go beyond direct metabolic needs for energy production and expand into the production of specific metabolites that have epigenetic implications for cell fate (Buck et al., 2016; Gascón et al., 2015; Mathieu & Ruohola-Baker, 2017; Zhang, Mei, et al., 2016; Zhang, Ryu, et al., 2016; Zhang, Termanis, et al., 2016; Zheng et al., 2016). This review aims to understand the potential rules for this interdependence.

Many adult stem cells stay in a quiescent state, until the signals from the environment demand their contribution to regenerate the tissue. As it turns out, the early embryo can also stop the developmental trajectory temporarily in a quiescent state. This mechanism, known as diapause, is used by over a hundred mammals including roe deer, kangaroos and koala bears in order to temporarily halt their pregnancy when conditions make it unlikely that the progeny will survive. We will compare and contrast the metabolic differences in quiescent and activated adult and pluripotent stem cells, and their epigenetic patterns.

2. Stem cell energetics

Cellular metabolism is the set of all biochemical reactions that produce the energy, and irrespective of cell type or differentiation state, it is required to support the intricate molecular machinery that keeps the cell alive. Metabolic processes can be broken down into either synthesis of new biomolecules (anabolism) or breaking down of molecules and existing biomolecules (catabolism) to generate energy. Several pathways are involved in the building up and breaking down of biomolecules and cellular components.

In this review, we consider glycolysis, pentose phosphate pathway, tricarboxylic acid (TCA) cycle, fatty acid β-oxidation and oxidative phosphorylation (Fig. 1). Glycolysis is a metabolic pathway that converts glucose into pyruvate while generating ATP and NADPH. Biosynthetic intermediates from glycolysis can be directed into the pentose phosphate pathway (PPP) for cell growth and proliferation. PPP is shown to be an essential metabolic pathway for pluripotent stem cells (Filosa et al., 2003; Varum et al., 2011) because it generates metabolites that are needed for lipid and nucleotide biosynthesis. Some adult stem cells also require this pathway since the PPP enzyme hexose-6-phosphate dehydrogenase (H6PD) was found to be required for the self-renewal of myoblast *in vitro* during muscle

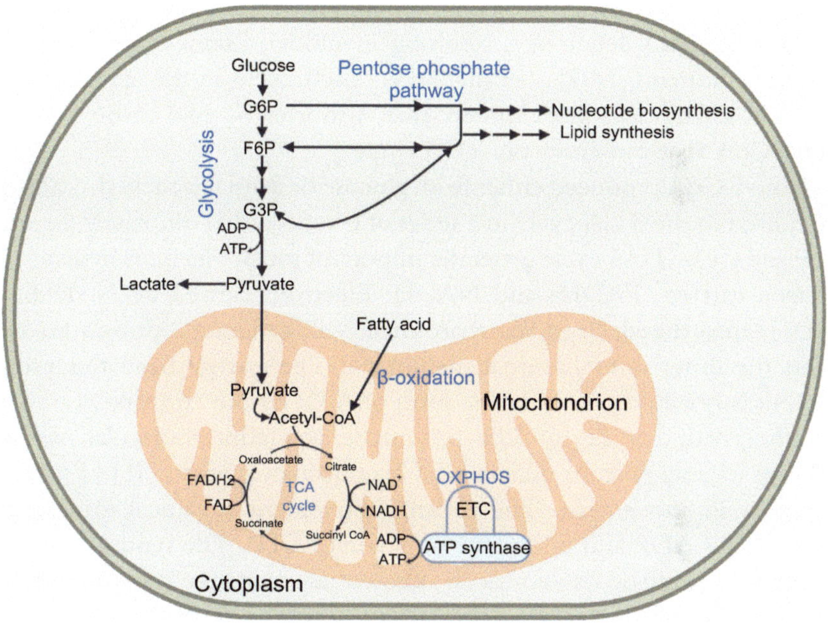

Fig. 1 Overview of the major cellular metabolic pathways (indicated in blue): Glycolysis, pentose phosphate pathway (PPP), TCA (Krebs) cycle, β-oxidation and oxidative phosphorylation (OXPHOS). Metabolites from glycolysis can move to PPP to generate NADPH and precursors for lipid and nucleotide synthesis. In addition to generating ATP, glycolysis generates pyruvate which is oxidized in the TCA cycle. OXPHOS generates the most ATP in the electron transport chain (ETC). Adult and pluripotent stem cells both utilize these metabolic pathways (see text for details). ADP, adenosine diphosphate; ATP, adenosine triphosphate; ATPase, ATP synthase; ETC, electron transport chain; F6P, fructose 6 phosphate; G6P, glucose 6 phosphate; G3P, glyceraldehyde-3-phosphate; FAD, Flavin adenine dinucleotide; NAD, Nicotinamide adenine dinucleotide; TCA, tricarboxylic acid. *The figure was created with Biorender.com.*

regeneration (Bracha et al., 2010). In the presence of oxygen, pyruvate generated from glycolysis can be transported into mitochondria and converted into acetyl-CoA. However, in low oxygen conditions (hypoxia) pyruvate will be reduced to lactate and a free energy carrier, NAD^+ is generated.

In addition to glucose, lipids are a major source of energy in cells. Energy is generated from lipid breakdown through fatty acid β-oxidation, producing acetyl-CoA. Long chain fatty acids are transported to mitochondria by the carnitine acyl transferase, CAT, system, also known as the carnitine shuttle. The rate limiting enzyme in this step is CAT1. Fatty Acyl-CoA is then dehydrogenated to acetyl-CoA utilizing the mitochondrial trifunctional protein (TFP) consisting of enoyl-CoA hydratase, hydroxyacyl-CoA dehydrogenase A and B (HADHA and HADHB) and ketoacyl-CoA thiolase. Defects in components of this pathway are causative for LCHAD deficiency, resulting in sudden infant death syndrome (SIDS) in humans (Miklas et al., 2019). Each cycle in the TFP complex results in a fatty acyl-CoA moiety that is shorter by two carbons and an acetyl-CoA that can enter the TCA cycle.

Acetyl-CoA produced either from glucose or from fatty acid β-oxidation is oxidized in the TCA cycle in a series of reactions that ultimately generate ATP and CO_2. TCA cycle generates important metabolic intermediates and electron carriers, $FADH_2$ and NADH. Electrons carried by NADH and $FADH_2$ into the electron transport chain will generate a proton gradient across the inner mitochondrial membrane. The energy from this proton gradient is finally captured in the form of ATP by the conversion of ADP and phosphate to ATP by ATP-β-synthase, a rotating molecular machine that taps its energy from utilizing the chemiosmotic proton [H+] gradient to power its movement. The rotating movement is critical for catalytic site to access ADP and phosphate to generate ATP. The synthesis of ATP by ATP-β synthase is a process known as oxidative phosphorylation (OXPHOS) (Fig. 1).

3. Metabolism of quiescent stem cells

3.1 Adult stem cells

In general, most tissues have resident stem cell populations, most of which exist in a relatively quiescent state and serve to repopulate old or injured cells within that organ. Some examples of these adult stem cell types can be identified in brain, teeth, gut, bone marrow (HSC and MSC), skin, hair follicle, testicles, and skeletal muscle (satellite cells). Here, we will discuss three of these adult stem cell types: satellite cells, hematopoietic stem cells (HSC) and hair follicle stem cells. These types of stem cells reside in a quiescent stage until they receive an activating signal for regeneration. The quiescent stage may have cellular advantages, protecting the cell from damage, allowing repair, or allowing an existence of specific metabolic state (Fig. 2).

3.1.1 Satellite cell metabolic switch during activation

Satellite cells are small, mononucleated cells found on a surface of skeletal muscles that can partake in the regeneration of injured muscle by self-renewing, undergoing myogenic differentiation and finally fusing to a damaged muscle fiber to induce repair. These muscle stem cells remain quiescent during healthy resting periods (Cheung & Rando, 2013) by communications and factors by surrounding environment (Baghdadi et al., 2018; Sampath et al., 2018; Scott, Arostegui, Schweitzer, Rossi, & Underhill, 2019; Verma et al., 2018; Wosczyna & Rando, 2018). In this G_0 state, quiescent satellite cells have a low metabolism and are more resistant to DNA damage. The quiescent state is required for the long-term maintenance of muscle stem cells, since loss of the capacity to remain quiescent can lead to precocious differentiation and loss of satellite cells over time. These quiescent satellite stem cells can enter a non-cycling "alert" state that, though still technically quiescent, has a particular metabolic signature with active mitochondrial respiration. They utilize fatty acids as an energy source through mitochondrial β-oxidation/FAO activity. During commitment toward active, regenerating myoblasts, satellite cells undergo a dramatic metabolic switch from mitochondrial FAO to glycolysis. This process has been coined metabolic reprogramming (Almada & Wagers, 2016; García-Prat, Sousa-Victor, & Muñoz-Cánoves, 2017; Mathieu & Ruohola-Baker, 2017; Ryall et al., 2015; Wang, Dumont, & Rudnicki, 2014). While metabolic reprogramming in satellite cell activation from G_0 to G-alert, and further to regenerating states is well documented to take place, the exact metabolic changes in each stage is still under debate due to potential alterations based on how these adult stem cells were isolated or analyzed (Almada & Wagers, 2016; Forcina, Miano, Pelosi, & Musarò, 2019; Pala et al., 2018; Rodgers et al., 2014; Yucel et al., 2019).

3.1.2 Hematopoietic stem cell metabolism

Quiescent hematopoietic stem cells (HSC) reside mainly in bone marrow in adults. Once HSCs are activated, they will generate all cell lineages in blood. Niche signals and nutrient-sensing pathways regulate HSC quiescent state

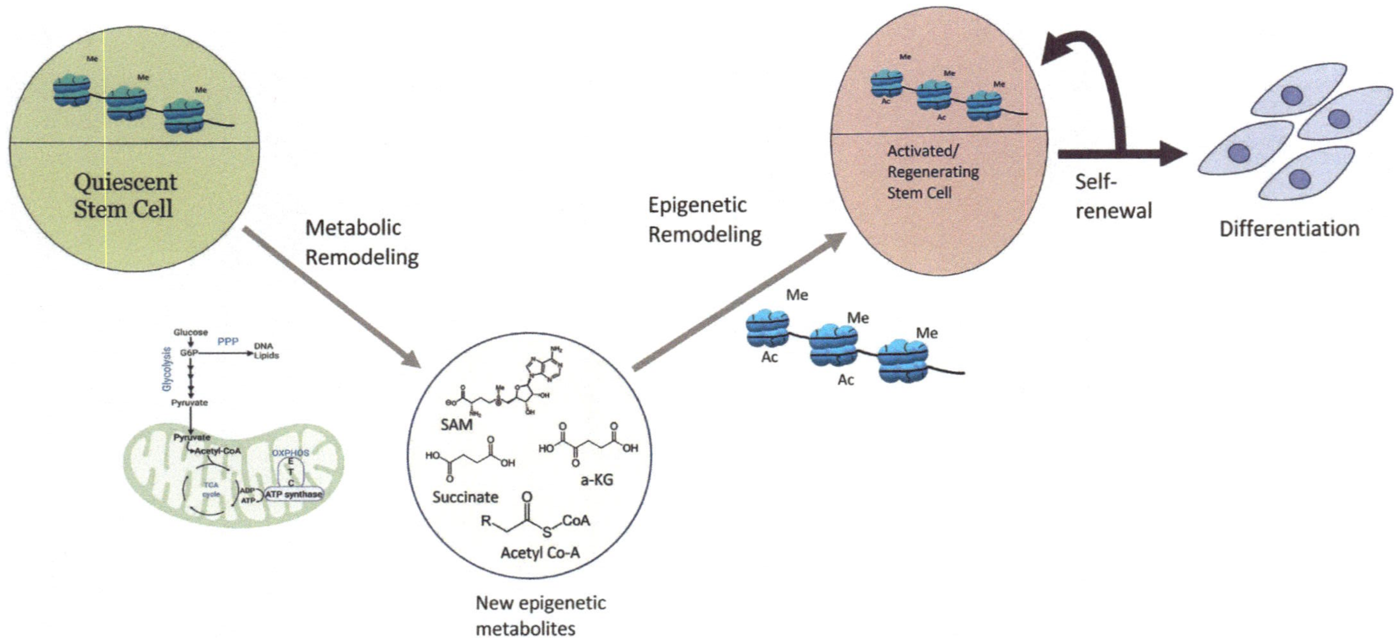

Fig. 2 Hypothesis: Remodeling of metabolism generates epigenetic metabolites that changes the epigenetic makeup and gene expression, leading to transition in stem cell state. Quiescent stem cells have an epigenetic signature. Upon metabolic remodeling the epigenetic landscape changes due to new epigenetic metabolite make-up, resulting in an activated-regenerative stem cell state. This late stem cell state has the ability to both self-renew and differentiate into committed cell fate. *The figure was created with Biorender.com.*

(Comazzetto et al., 2019; Laplante & Sabatini, 2012; Ochocki & Simon, 2013). Interestingly the physiological state of dormancy that maintains the self-renewal of adult blood-forming HSCs is governed primarily by metabolism (García-Prat et al., 2017; Liu et al., 2015; Warr et al., 2013). Self-renewing HSCs rely mainly on HIF1-regulated anaerobic glycolysis for energy production (Takubo et al., 2013). As a matter of fact, oxidative phosphorylation by the mitochondria has to be actively prevented to maintain HSC quiescence. Control by nutrient-sensing pathways, including mTOR, AMPK, FoxO, and SIRT (sirtuins) (García-Prat et al., 2017; Laplante & Sabatini, 2012; Ochocki & Simon, 2013) play an important role in restricting mitochondrial respiration in quiescent HSC stage.

3.1.3 Hair follicle stem cell

The mammalian skin is constructed into two distinct layers, the epidermis and the underlying dermis (Watt & Fujiwara, 2011). The epidermis, the skin's outer layer, serves a barrier function that protects underlying skin from external or environmental stressors, including chemical, biochemical or thermal stress (Fuchs, 2009; Proksch, Brandner, & Jensen, 2008; Wickett & Visscher, 2006). As a multilayered epithelium, the epidermis consists of interfollicular epidermis (IFE), hair follicles (HFs), sebaceous glands (SGs), and eccrine sweat glands (Blanpain & Fuchs, 2006). Hair follicles go through a periodic phases rest and growth that correlate with the beginning of the hair cycle (Fuchs, Merrill, Jamora, & DasGupta, 2001; van der Veen et al., 1999), and degeneration. The process of getting out of the rest (telogen) or quiescent state is dependent on hair follicle stem cells (HFSCs), located in a microenvironment called the bulge (Blanpain & Fuchs, 2006). The molecular mechanism that controls how HFSCs exit quiescence to proliferate and then return to quiescent state is poorly understood.

While many studies showed unique gene expression profiles in HFSCs compared to other cells in the interfollicular epidermis (Blanpain, Lowry, Geoghegan, Polak, & Fuchs, 2004; Morris et al., 2003, 2004; Tumbar et al., 2004), a recent study sought to characterize the metabolism of HFSCs and performed metabolomics by using sorted populations from mouse skin (Flores et al., 2017). They found that several glycolytic metabolites including lactate were higher in HFSCs relative to the total epidermis. Interestingly, no TCA cycle metabolite differences were detected between HFSCs and the epidermis (Flores et al., 2017), suggesting HFSCs have an increased lactate dehydrogenase (Ldh) activity and glycolytic metabolism.

Deletion of LDHA reduced lactate and other glycolytic metabolite levels, thereby demonstrating that the abrogation of the glycolytic metabolism signature is associated with activated HFSCs. Importantly, Ldha deletion also blocked HFSC activation, indicating that Ldha, and therefore, glycolysis and lactate production, are critical for HFSC activation. In accordance, deletion or pharmacological inhibition of mitochondrial pyruvate carrier 1 (MPC1), which transports pyruvate into mitochondria, promoted lactate production and increased activation of HFSCs and thus initiated the hair cycle. Although LDHA activity and glycolytic signature are required for HFSC activation, future studies should explore whether activated HFSCs are metabolically bivalent, perhaps also utilizing some level of oxidative phosphorylation.

3.2 Pluripotent stem cell quiescence, diapause

Pluripotent stem cells are often thought of as relatively proliferative; however, they are capable of entering quiescence. Embryonic diapause, or delayed implantation, is a reversible quiescent state of pluripotent stem cells. Over 130 mammalian species have been reported to undergo diapause (Fenelon, Banerjee, & Murphy, 2014), which can be induced experimentally in mice through ovariectomy (Yoshinaga & Ada, 1966) or inhibition of estrogen (Hunter & Evans, 1999; Paria, Huet-Hudson, & Dey, 1993) on day 3.5 (E3.5) after fertilization. Diapause is associated with proliferation arrest as protein and DNA synthesis are reduced (Blerkom, Chavez, & Bell, 1978; Fenelon et al., 2014; Fu et al., 2014; Hamatani et al., 2004; Liu, Mao, & Chen, 2011; Martin & Sutherland, 2001; Menke & McLaren, 1970; Pike, 1981). Inhibition of the major nutrient sensors and transducers, mTOR and Myc have shown to result in a diapause-like state (Bulut-Karslioglu et al., 2016; Scognamiglio et al., 2016). More recently metabolic and epigenetic remodeling has shown to play an important role in the regulation of entry and exit of the embryonic diapause state (Hussein et al., 2020; Fig. 2). In diapause, both lipolysis and glycolysis are upregulated and mTOR-dependent H4K16ac is inhibited by nutritional starvation that is dependent on LKB1/AMPK activation and glutamine transporter (Slc38A1/2) activity (Hussein et al., 2020). Glutamine transporters were shown to be essential for maintenance of embryonic diapause, because their inhibition led to exit from diapause state (Hussein et al., 2020). Future studies should explore the role of glutamine in embryonic development and especially in diapause, and whether glutamine is required to inhibit mTOR as an additional layer of inhibition after LKB/AMPK. Autophagy, a pathway that generates nutrients for cell survival during periods of starvation and regulated by mTOR (Kim & Guan, 2015; Nicklin et al., 2009), is activated during diapause (Hussein et al., 2020; Lee et al., 2011). Since diapause is a reversible state, it is critical that the pluripotent stem cells remain protected while they are in diapause. Lipolysis generates free fatty acids, phosphatidylcholines and phosphatidylethanolamines, which are all important for activation of the NFkB pathway, a protective pathway (Hussein et al., 2020). Interestingly, diapause was enriched with metabolites with antioxidant activities to prevent these quiescent cells from oxidation (Hussein et al., 2020). These studies reveal the importance of glycolytic metabolism and glutamine transporter function in the pluripotent quiescent state of diapause.

4. Metabolism of active stem cells

4.1 Metabolism after fertilization

Proliferation requires substantial energy, but it also requires significant amounts of macromolecules, including nucleotides, amino acids and lipids, to assemble the daughter cells during cell division (Zhu & Thompson, 2019). Though oxidative phosphorylation produces substantially more ATP than glycolysis, because of the need to ramp up the macromolecules, using all the glucose for ATP production would actually limit cell proliferation. Therefore, to facilitate repeated cell division, actively cycling stem cells divert some glucose for the generation of glycolytic precursors such as acetyl-CoA and other glycolytic intermediates. For this reason, many stem cell populations primarily use aerobic glycolysis, resulting in lactate production, rather than pyruvate oxidation in a mitochondrion, even in the presence of oxygen (Mathieu & Ruohola-Baker, 2017). However, active stem cells can require alternative modes of metabolism. One major example of this is pyruvate metabolism at the 2-cell stage of the developing embryo. Studies at this stage have revealed that early stage embryos exhibit limited oxidative phosphorylation due to a low NAD+/NADH ratio (Brown & Whittingham, 1991), and that these embryos cannot metabolize supplemental glucose (Brinster, 1969). Rather, early developmental stages including the 2-cell stage require pyruvate for metabolism, but since they don't undergo oxidative phosphorylation and have relatively low bioenergetic activity, the reason behind this has been unclear. One recent study (Nagaraj et al., 2017) sought to evaluate whether this might be related with pyruvate's role in the TCA cycle. In their investigation of this complicated interaction in early embryos, they found that multiple TCA metabolites including pyruvate carboxylase, pyruvate dehydrogenase, pyruvate dehydrogenase phosphatase, citrate synthase, aconitase-2, and isocitrate dehydrogenase 3A were all transiently seen in the nucleus. Following pyruvate starvation, they found that citrate, aconitate, and α-ketoglutarate (α-KG) levels are significantly reduced upon starvation and quickly restored upon re-supplementation (Fig. 2). Metabolites that were not localized to the nucleus were not diminished following starvation and their levels were nonresponsive to pyruvate supplementation following starvation. Interestingly, they observed that enzymes localized to the mitochondria were unaffected. In correlation with previous studies which demonstrated a strong interaction with metabolites and epigenetic regulation (Mathieu & Ruohola-Baker, 2017; Tzika, Dreker, & Imhof, 2018), these findings suggest that early stage embryos require pyruvate to generate such metabolites in order to activate the zygotic genome and promote the transition to subsequent embryonic developmental stages.

4.2 Metabolism of pre-implantation and post-implantation pluripotent stem cells

Multiple recent studies have revealed that pluripotency does not represent a single defined state. Instead, pluripotent cells can be stabilized in at least two distinct stages known as naïve, which represent pre-implantation pluripotent cells, and primed, which represent the post-implantation epiblast stage with somewhat limited stemness (Gafni et al., 2013; Nakamura et al., 2016; Theunissen et al., 2016; Ware et al., 2014), along with a heterogeneic spectrum of intermediates. Naïve and primed stem cells and their intermediates exhibit differential epigenetic and metabolic signatures: naïve cells can use both glycolysis and oxidative phosphorylation to generate energy, but primed cells rely almost exclusively on glycolysis (Mathieu et al., 2019; Mathieu & Ruohola-Baker, 2017; Moody et al., 2017; Sperber et al., 2015; Takashima et al., 2014; Zhou et al., 2012). These metabolic differences coincide with distinct epigenetic profiles: H3K27me3 marks are significantly lower in naïve compared to primed hESCs (Gafni et al., 2013; Moody et al., 2017; Sperber et al., 2015; Theunissen et al., 2014; Ware et al., 2014; Fig. 2). Metabolites play a major role in resulting the epigenetic changes that drive the shift between states of pluripotent cells. For example, modulation of the methylation substrate SAM affects primed stage H3K4me3 levels, and SAM modification in naive hESC by knockout of the metabolic enzyme NNMT alters PRC2-dependent H3K27me3 levels in naïve-to-primed transition (Shiraki et al., 2014; Sperber et al., 2015). This will be discussed in further detail later.

The transition from naïve to primed cells is marked by a shift from bivalent metabolism to glycolytic metabolism. Accordingly, the opposite has been observed during somatic cell reprogramming. Though the mechanism by which cells are reprogrammed is still not completely understood, time course studies have found that reprogramming induces a metabolic shift from an oxidative metabolism reliant upon mitochondria for energy production, to highly glycolytic metabolism (Folmes et al., 2011; Mathieu et al., 2013, 2014; Panopoulos et al., 2011; Varum et al., 2011), showing that proliferation is maintained by HIF-guided switch in metabolism rapidly generating ATP through glycolysis, while mitochondria, decoupled from the respiration, instead promote anabolic processes to produce cellular building blocks.

The common thread between the switch from oxidative and glycolytic metabolism, and therefore the transitions from naïve pluripotency to primed pluripotency to lineage specification, is the change in mitochondrial morphology and activity. Studies have found that mitochondrial morphology is highly dynamic at different developmental stages (Bavister & Squirrell, 2000; Collins et al., 2012) and depending upon metabolic requirements of the cell (Buck et al., 2016; Khacho et al., 2016; Lee, Kang, et al., 2016; Lee, Lee, et al., 2016; Zhang, Mei, et al., 2016; Zhang, Ryu, et al., 2016; Zhang, Termanis, et al., 2016). At the early 2-cell stage, spherical mitochondria cluster around the two nuclei (Motta, Nottola, Makabe, & Heyn, 2000; Squirrell, Schramm, Paprocki, Wokosin, & Bavister, 2003), and they continue to elongate and develop transverse cristae as they move into the blastocyst stage (Sathananthan & Trounson, 2000). In contrast, somatic cells contain mitochondria with well-defined transverse cristae that are capable of supporting a higher level of oxygen consumption and oxidative metabolism (Varum et al., 2011). Both mouse epiblast stem cells and human embryonic stem cells have more morphologically mature (elongated) mitochondria compared to mESC and naïve hESC (Zhou et al., 2012) suggesting mitochondrial dynamics may contribute to the metabolic differences observed between naïve and primed stages. In fact, a recent study (Bahat et al., 2018) showed a regulator of mitochondrial apoptosis and metabolism called that mitochondrial carrier homolog 2 (MTCH2) controls mitochondrial elongation. They found that loss of MTCH2 both prevented mitochondrial fusion and perturbed the naïve-primed transition, highlighting the link between cell fate changes and mitochondrial maturation. Moreover, MTCH2-null stem cells have a lower rate of fusion and have different levels of glutamine and histone acetylation during naïve-to-primed transition compared to wildtype ESCs, and forced elongation was sufficient to cause naïve cells to at least partially exit pluripotency, even while growing in conditions that promote the naïve phenotype. Though the underlying mechanism of how MTCH2 regulates the naïve to primed transition is unclear, this study links the process of mitochondrial elongation with epigenetic regulation.

Conversely, studies have found that in cell reprogramming, which, as previously mentioned, exhibits a metabolic reversal to a primed pluripotent state, is driven by mitophagy (Naik, Birbrair, & Bhutia, 2018), or the selective degradation of mitochondria, typically in response to damage or stress. It remains unclear as to why mitophagy is so critical in this process, but studies have identified some possibilities. For example, one noted that since cell reprogramming prompts mitochondrial fission (splitting), mitophagy might be necessary to selectively remove faulty mitochondria (Prieto et al., 2016). Other studies have noted that cells that inducing mitochondrial fusion, which has been shown to inhibit mitophagy, during cell reprogramming reduced reprogramming success (Son et al., 2015). Still others have shown that depletion of proteins that promote mitochondrial fusion result in failure of somatic cells to reprogram and disrupt pluripotency maintenance in stem cells (Vazquez-Martin et al., 2012). More intriguingly, there has recently been proposed a "mitophagy-mediated metabolic reshuffling" (Naik et al., 2018) that mediates cell fate transitions. Interestingly, mitophagy-induced cell fate changes in different types of cells results in different outcomes. Specifically, mitophagy-mediated metabolic reprogramming toward glycolysis in dedifferentiated cells induces differentiation, while the same shift in terminally differentiated cells and cancer cells stimulates dedifferentiation and connotes a level of stemness. However, similar to the mitophagy-mediated shift toward glycolysis, mitophagy-mediated metabolic reprogramming toward OXPHOS in dedifferentiated cells also induces differentiation (Naik et al., 2018). Though many of the hows and whys of mitophagy remain unclear, it is becoming apparent that mitophagy-regulated metabolic shifts are another key factor in regulating the shift between and toward pluripotent states.

4.3 Metabolism of actively cycling adult stem cells: MSC as case-study

In vivo, MSCs reside within the hypoxic environment of the bone marrow niche and have been found to preferentially use glycolysis relative to

MSC-derived terminal cells like osteoblasts (Chen, Shih, Kuo, Lee, & Wei, 2008). Under normoxic conditions *in vitro*, MSC proliferation is significantly increased (Pattappa et al., 2012), but the switch to oxidative phosphorylation leads to significant MSC senescence, a major paradox and limitation for using such a promising stem cell type. This is especially apparent in tissue engineering grafts, which, though promising in clinical trials, are plagued by reduced cell survival (García-Sánchez, Fernández, Rodríguez-Rey, & Pérez-Campo, 2019). Expanding upon other studies (Deschepper et al., 2013) that showed that MSCs could survive better in anoxia so long as they had sufficient glucose, one recent study (Bernardo et al., 2009; Moya et al., 2018) offered a possible explanation for this. They found that in near-anoxia conditions, MSCs produce almost all their ATP almost exclusively through anaerobic glycolysis and that they are therefore unable to use other exogenous molecules as energy substrates. Most notably, they found that MSCs are unable to adapt their metabolism to the lack of glucose, and since they maintain almost no internal reserve of glucose or ATP, it is likely that this contributes to their poor survival rate following transplants. They suggest that the solution to this problem is the development of a transplantable glucose-releasing scaffold to improve cell survival.

As with other types of stem cells, several recent studies have spotlighted the role of epigenetics in regulating MSC senescence. For example, the inhibition of HDAC resulted in apoptosis and senescence in human MSCs by upregulating cyclin-dependent kinase inhibitors (Bernardo et al., 2009) and another (So, Jung, Lee, Kim, & Kang, 2011) observed that the inhibition of DNMTs with induced senescence. Though there has been an established relationship between the role of epigenetics and differentiation of MSCs (Mortada & Mortada, 2018), and the previously established relationship between metabolites and their regulation of the shift from glycolytic metabolism and oxidative phosphorylation in other cells will almost certainly be found to apply to MSCs. Though there don't currently exist many studies that have investigated this relationship, one very recent study (Jeong et al., 2019) has made the first connection. Rather than adjusting oxygen tension within the culture as other studies have, they sought to understand the signaling that would allow for expansion in physiological (hypoxic) conditions. They found that several metabolites including fructose-1, 6-bisphosphate, phosphoenolpyruvic acid, and sodium oxalate, all of which have an established epigenetic role, were able to stimulate MSC expansion in hypoxic conditions by activating the AKT/STAT signaling pathway.

Though preliminary, this study establishes links between the many studies that have established an independent role for metabolism and epigenetics in mesenchymal stem cell proliferation and survival.

5. HIF, the master regulator of metabolism

Oxygen plays a critical role in cellular energy production and is used as a cofactor or substrate by many enzymes. Cells exposed to low levels of oxygen, or hypoxia, must adapt their metabolism to switch from OXPHOS to glycolysis. The sensing and response to changes in oxygen concentrations are mainly mediated by a heterodimeric transcription factor, Hypoxia Inducible Factor (HIF) (Wang, Jiang, Rue, & Semenza, 1995). HIF is composed of two subunits, an inducible subunit (HIF1α or HIF2α) and a constitutively expressed subunit (HIFβ). Under normoxic conditions HIFα is proline-hydroxylated, leading to its ubiquitylation by von Hippel–Lindau protein (VHL) and its subsequent degradation by the proteasome. The hydroxylation of HIF is dependent on α-ketoglutarate (α-KG)-dependent dioxygenases called prolyl hydroxylases (PHDs). Under hypoxia PHDs are inactive, HIFα is stabilized, binds to HIFβ and the HIF dimer translocates into the nucleus where it activates the transcription of genes involved in adaptation of low oxygen conditions (Kaelin & Ratcliffe, 2008). HIF target genes include genes involved in glucose and energy homeostasis, angiogenesis, cell survival and some stem cell factors. HIF activates the transcription of glucose transporters such as GLUT1, promoting glucose uptake. It also upregulates glycolytic enzymes such as HK, PGK1, PKM2 or ENO1 (Semenza, 2012). In addition, HIF is also involved in the regulation of lipid metabolism by increasing FA synthesis (FASN, SREB1), FA uptake (PPARg) and inhibiting FAO (Mylonis, Simos, & Paraskeva, 2019).

Many stem cells reside in a specialized microenvironment, or niche that are often hypoxic, inducing a unique metabolic state that allows them to maintain their self-renewal and multipotency capacities. One of the best studied examples is the hematopoietic stem cell that resides in the bone marrow in low oxygen tension locations. The exposure to chronic hypoxia allows the cells to maintain their quiescence as well as low level of ROS production to keep their genomic integrity. HIF has been shown to be important for the maintenance of HSC by regulating their metabolic state. Indeed, inactivation of HIF leads to the loss of HSC quiescence (Takubo et al., 2010) while over-activation of HIF by inhibition of VHL results in an upregulation of PDK1 and an increase of glycolysis (Takubo et al., 2013).

Hypoxia and HIF have also been shown to be important for the establishment and maintenance of pluripotent stem cells. Embryonic stem cells are evolving in a hypoxic environment in the uterus and rely mostly on glycolysis as a source of energy (Fischer & Bavister, 1993; Mathieu & Ruohola-Baker, 2017; Sperber et al., 2015; Zhou et al., 2012). Culture of hESC under low levels of oxygen prevents their spontaneous differentiation (Ezashi, Das, & Roberts, 2005) and hypoxia induces re-entry of committed cells into pluripotency (Mathieu et al., 2011). Hypoxia can also enhance the generation of induced pluripotent stem cells from fibroblasts (Mathieu et al., 2013; Yoshida, Takahashi, Okita, Ichisaka, & Yamanaka, 2009). Interestingly, HIF has been shown to regulate the transcription of Oct4, one of the reprogramming factors (Covello et al., 2006). We and others have shown that both HIF1 and HIF2 are responsible for the early metabolic switch required for a successful reprogramming, shutting down mitochondrial genes and increasing transcription of glycolytic genes such as pyruvate dehydrogenase kinase PDK1 (Mathieu et al., 2014; Prigione et al., 2014).

HIF1 is stabilized in the post-implantation primed PSC state compared to the pre-implantation naïve PSC (Sperber et al., 2015; Zhou et al., 2012). HIF is important for the glycolytic metabolism of primed PSC in mice and humans. Indeed, CRISPR knockout experiments revealed that HIF1 is required for the naïve to primed hESC transition while ectopic overexpression of HIF1 in naïve ESC pushes then toward the primed state by increasing glycolysis (Sperber et al., 2015; Zhou et al., 2012). In addition, primed PSCs accumulate FA due to increase of FA synthesis and decrease of FAO compared to naïve PSC (Sperber et al., 2015). It would be interesting to investigate whether HIF is responsible for lipogenesis in PSC and whether FA accumulation and usage plays a role in their cell state and survival, perhaps comparable to diapause state.

In addition to directly targeting genes involved in metabolic switches, hypoxia and HIF can also modify the epigenetic landscape of the cells by regulating DNA and histone methylation (Choudhry & Harris, 2018). For example, one of the HIF targets is G9a, an H3K9 methytransferase that has been shown to play a crucial role in the control of cell metabolism (Buck et al., 2016; Gascón et al., 2015; Mathieu & Ruohola-Baker, 2017; Zhang, Ryu, et al., 2016; Zheng et al., 2016), as a modulator of oxidative stress response (Riahi et al., 2019), and to protect imprinted DNA methylation in ESCs. Furthermore, HIF1 stability has shown to be regulated by lysine methylation (Nam & Baek, 2019). Two recent studies demonstrate that certain histone demethylases, such as KDM5A and KDM6A, can directly sense oxygen to reprogram chromatin and control cell fate (Batie & Rocha, 2019; Chakraborty et al., 2019).

6. Epigenetic signatures and epigenetic metabolites

6.1 Epigenetic signatures of naïve and primed pluripotent stem cells

Naive (pre-implantation) and primed (post-implantation) pluripotent stem cells have been extensively studied for their differences in culture conditions, morphology, gene expression profile, functional abilities, epigenetic landscape state (see Section 3.2) (Nichols & Smith, 2009; Weinberger, Ayyash, Novershtern, & Hanna, 2016).

Epigenetic regulation is a key factor in cell fate determination and developmental transitions hence forming the basis for cell type specific gene expression and specification. ESCs are thought to maintain their pluripotency by regulation of open chromatin (Denholtz et al., 2013; Ji et al., 2015). Multiple pluripotent states have been stabilized from early mouse and human embryos, resulting in recent studies that have analyzed if changes in open chromatin regulate the transitions between these states. In particular, PRC2 complex has reached the center stage.

The polycomb repressive complex 2 (PRC2) histone methyltransferase plays a central role in epigenetic regulation at many developmental stages and in cancer (Vizán, Beringer, Ballaré, & Croce, 2014). This epigenetic silencing requires polycomb repressive complexes PRC1 and PRC2 and results in histone repressive marks, followed by methylation of the DNA (Bernstein et al., 2006; Richly et al., 2010; Schwartz & Pirrotta, 2008). H3K27 methylation is catalyzed by a methyltransferase, EZH2 that is in a complex with other PRC2 components, including EED (embryonic ectoderm development), SUZ12 (Suppressor of Zeste 12) and RbAp46/48 (Margueron & Reinberg, 2011). During mouse blastocyst formation, PRC2 complex dependent repression of CDX2 and GATA3 is essential for ICM lineage (Saha et al., 2013), and reprogramming assays have revealed an essential function for PRC2 in acquisition of pluripotency (Pereira et al., 2010). However, mouse ground state ESCs maintain pluripotency without PRC2 (Chamberlain, Yee, & Magnuson, 2008; Galonska, Ziller, Karnik, & Meissner, 2015). While PRC2 knockout mouse ground state ESC cannot properly differentiate, they still express the key pluripotency markers. Previous studies showed that pluripotency depends on the chromatin-based silencing of developmental gene expression (Boyer et al., 2006), raising the possibility that different stages of pluripotency have different requirements for PRC2-dependent repressive histone marks. Accordingly, dramatic differences have been observed in H3K27me3 modifications between naïve and primed pluripotent stem cells. Naïve cells maintain low H3K27me3, due to Nicotinamide *N*-methyltransferase (NNMT) activity. NNMT acts as methyl sink for S-adenosyl methionine (SAM) resulting in restricted availability of SAM for H3K27-trimethylation (Sperber et al., 2015). Low H3K27me3 in naïve state upregulates Wnt pathway and reduces HIF stabilization, which contributes to self-renewal capacity. However, in naive to primed transition NNMT is reduced, making SAM available for the Polycomb Repressive Complex 2's (PRC2) writer, Enhancer of Zeste homolog 2 (EZH2). EZH2 generates H3K27me3 marks in the primed stage to silence many developmental genes (Ferreccio et al., 2018; Sperber et al., 2015).

To pinpoint the requirement of PRC2 in different developmental stages of pluripotency, Moody et al. created a computational design of proteins that bind to the EZH2 interaction site on EED with subnanomolar affinity *in vitro* and form tight and specific complexes with EED in living cells. Induction of the EED binding protein abolishes H3K27 methylation, degrades PRC2 members EZH2 and JARID2 and abolishes pluripotent colony morphology and gene expression patterns in primed stages, but not at ground naïve WIBR 5iLA cell stage (Moody et al., 2017). These data suggest that while PRC2 is required at primed stages, it is dispensable in naïve state, both in mouse and human pluripotency.

Another histone repressive mark that resembles H3K27me3 behavior is the heterochromatin associated mark H3K9me3. H3K9me3 marks are depleted at naïve stages (Hawkins et al., 2010) but found sparse in prime cells marking heterochromatin regions (Battle et al., 2019). Depletion of the Jmjd1a and Jmjd2c demethylases for H3K9me3 results in stem cell differentiation (Loh, Zhang, Chen, George, & Ng, 2007). Unlike somatic cells, stem cells' epigenome is highly flexible and practices bivalency as the H3K27me3/H3K4me3 duo (Meshorer & Misteli, 2006). Many developmental promoter regions are poised to be either repressed or activated by containing both repressive and activating marks, H3K27me3 or H3K4me3, respectively (Azuara et al., 2006; Bernstein et al., 2006; Pan et al., 2007).

H3K4me3 mark is mostly associated with Pol II bound promoters and found to be equally stable between naïve and primed stem cell state.

Two other marks that unify naïve stem cells are H3K4me1 and H3K27ac. H3K4me1 provides an enhanced open chromatin state on 9% of the genome at 50 kb breadth and enriched 3 times more in naïve than in primed cells (Battle et al., 2019). Another aspect of H3K4me1 is that 77% of naïve enhancers are discharged in stepwise manner as cells become more primed (Battle et al., 2019). Finally, naïve stem cells have more open chromatin structure due to greater deposits of H3K4me1 and H3K27ac compared to primed stem cells as well as larger association of topology associated domains expansion impacting on 3-dimensional genome architecture (Battle et al., 2019). Overall, naïve and primed stem cells have distinct epigenetic landscapes that shape their gene expression and functionality abilities.

DNA methylation is not essential in mouse naïve BUT is essential in human and mouse primed ESC. Work with DNA methylation marks (Liao et al., 2015) revealed further epigenetic differences between ESC. While DNMT1 knockout is lethal in all studied somatic cells, mouse ground stage ESCs are viable despite global loss of DNA methylation (Smith & Meissner, 2013). However, work in human primed ESC revealed an essential function for DNMT1 in hESC (Liao et al., 2015). The authors proposed that in both mouse and human, primed ESC may represent the first developmental period in which maintenance becomes essential. DNMT1 has also recently been shown to be required for mouse primed ESC (Geula et al., 2015). These data further support the idea that both human and mouse primed ESC have reached a pluripotent stage in which repressive DNA methylation marks are essential. PRC2-dependent histone methylation and DNA methylation are tightly connected processes, since DNA methyltransferases can bind EZH2/1, the PRC2 methyltransferases (Neri et al., 2013). It is therefore interesting to note that ESCs have a developmental state in which repressive PRC2-dependent histone marks and DNA methylation marks are dispensable, yet shortly after, in a later pluripotency stage, the marks are essential.

Embryonic diapause, the quiescent stage between pre-implantation and post-implantation has also an epigenetic signature (Hussein et al., 2020). While histone H4 in both naive and primed pluripotent cells are well decorated by positive H4K16Ac marks, the halted intermediate, diapause stage is surprisingly void of the mark. Furthermore, this epigenetic signature is dramatically regulated by mTOR activity and by Glutamine transporter Slc38A1/2. mTOR activity increases H4K16Ac marks, while glutamine transporter activity represses the mark (Hussein et al., 2020).

6.2 Epigenetic signatures of adult stem cells

Adult stem cells (ASC) also use epigenetics to control and maintain stemness while holding the potential to differentiate. As discussed, quiescent muscle stem cells (satellite cells) are activated in response to signals from injured muscles for the repair of damaged myofibers (Robinson & Dilworth, 2018). The polycomb group of proteins (PcG) is a main epigenetic repressor of satellite cells, responsible to promote stemness and self-renewal by expressing the H3K27me3 marks (Bracken, Dietrich, Pasini, Hansen, & Helin, 2006; Caretti, Padova, Micales, Lyons, & Sartorelli, 2004; Margueron & Reinberg, 2011; Fig. 2). Conversely, downregulation of PcG in satellite stem cells results in de-repression of many genes, including p16INK4a, which lead to loss of cell identity, senescence and exhaustion of the quiescent satellite stem cell pool (Bianchi et al., 2020). Another histone modifier that is associated with satellite cell quiescence is H4K20me2 (Jørgensen, Schotta, & Sørensen, 2013). H4K20me2 and its methyltransferase Suv4-20h1 promote heterochromatin formation to repress MyoD expression as the ablation of Suv4-20h1 and loss of H4K20me2 led to defective differentiation (Boonsanay et al., 2015). To activate quiescent satellite cells the transcription factor Pax7 recognizes Myf5 promoter region mediating the methyltransferase Ash2L/MLL2 to methylate H3K4 in genes that are involved in muscle cell fate (Diao et al., 2012; Kawabe, Wang, McKinnell, Bedford, & Rudnicki, 2012; McKinnell et al., 2007). To control myogenin expression while keeping myoblast in proliferative state, phosphorylated MyoD protein recruits the H3K9 methyltransferase Suv39h1/KMT1A which marks the regional chromatin with repressive H3K9me2 and H3K9me3 methylation marks (Esteve, Chin, & Pradhan, 2005; Fritsch et al., 2010).

As discussed above, hematopoietic stem and progenitor cells (HSPCs) possess the capacity for self-renewal and differentiation to all cell lineages in blood. The polycomb repressive complex 1 (PRC1) includes among other proteins, Ring1, which ubiquitinates H2AK119ub to promote gene repression by PRC2/H3K27me3 mechanism and to allow HPSCs self-renewal (Eskeland et al., 2010). Ring1B compacts chromatin structure and represses gene expression independent of histone ubiquitination. Additionally, the histone demethylase JARID1d was shown to be a positive regulator of HSC as its deletion compromises HSC self-renewal. Hematopoiesis requires histone acetyl transferase function such as NuA4/P300/CBP/HBO1 (Sun, Man, Tan, Nimer, & Wang, 2015). The DNA methyltransferase DNMT3a/3b are also associated with cell fate determination their methylation patterns responsible for silencing of self-renewal genes in HSC (Trowbridge & Orkin, 2011). Moreover, deletion of DNMT3a results in the expansion of HSC population by obstructing differentiation and upregulation of self-renewal genes such as Runx1 and Gata3 (Challen et al., 2014).

Unlike satellite cells and HSPCs that introduce a similar model as ESC by heavily repressing differentiation genes, hair follicle SCs (HFSCs) display reduced histone H3K4me3, H3K9me3, and H3K27me3 methylation levels (hypomethylation) preceding hair growth (Kang, Long, Wang, Sada, & Tumbar, 2019). This is hypothesized to keep HFSC in a highly plastic state of low epigenetic identity, such that HFSCs can easily adopt more differentiated cell fates in the subsequent stages of hair cycle (Buck et al., 2016; Khacho et al., 2016; Lee, Lee, et al., 2016; Zhang, Ryu, et al., 2016). On the other hand, it was shown that PRC1 functions to mediate PRC2 H3K27me3 on repressed genes, which is essential for skin development and stem cell (SC) specification; however, PRC1 H2AK119ub catalytic activity is dispensable (Cohen et al., 2018). Another epigenetic regulator is 5-hydroxymethycytosine, 5-hmC, a marker that is linked to aberrate distribution in the HFSCs niche and found as a requirement for mediating cell growth and differentiation upon activation (Leavitt, Wells, Abarzua, Murphy, & Lian, 2019).

6.3 Epigenetic metabolites

Traditionally, cellular metabolism has been studied for its role in providing energy to the cell. More recently, however, metabolism has been implicated in a new context: cell-fate determination (Buck et al., 2016; Gascón et al., 2015; Mathieu & Ruohola-Baker, 2017; Zhang, Mei, et al., 2016; Zheng et al., 2016). Recent data suggest that key epigenetic marks are regulated by the levels of specific metabolites, coined epigenetic metabolites (for example, α-KG, fumarate, succinate, SAM, acetyl-CoA, NAD). The epigenetic metabolites often act as substrates or activators for epigenetic writer or eraser enzymes, such as HMT, DNMT, JMDH, TET, SIRT, HAT and AMPK (Mathieu & Ruohola-Baker, 2017).

Epigenetic metabolites are a link between metabolic state and epigenetic control of gene activity in T cells (Peng et al., 2016). Glycolysis supports T helper cell differentiation by controlling the levels of acetyl-CoA. Acetyl-CoA levels are critical since acetyl-CoA serves as a substrate for Histone Acetyl Transferases, HATs. Thereby, increase in glycolytic metabolism will reduce activating epigenetic marks and therefore have a specific effect in cell fate. Similarly, epigenetic metabolites have shown to be critical for epigenetic marks governing transcription factor networks in development (Etchegaray et al., 2015). An additional field in which dramatic changes in cellular fates have been observed based on metabolic remodeling is cancer biology. Alterations in epigenetic metabolites have been shown to change molecular rewiring of cancer cells through epigenetic alterations. These can affect cancer cell differentiation, proliferation, apoptosis and therapeutic responses (Kinnaird, Zhao, Wellen, & Michelakis, 2016).

7. Conclusion

In this review, we have discussed many stem cell "state changes" and their accompanied metabolic switches and epigenetic modifications. While metabolic and epigenetic changes are well confirmed, it isn't always easy to understand why the stem cell in question undergoes the observed dramatic metabolic switch.

Muscle stem cells, satellite cells, switch to mitochondrial respiration at the priming, quiescent state, while HSCs activate mitochondria in regenerative, active state. Contrary to both aforementioned choices, hair follicle stem cells turn to highly glycolytic metabolism during regenerative, active state. To make matters more confusing, the metabolic choice of quiescent pluripotent cells in embryonic diapause, as well as actively dividing MSC is glycolysis. Naïve to primed transition shows a strong characteristic of down-regulation of mitochondrial activity, seemingly dangerous move during the beginning of embryonic development. These discussed examples drive the point home that while the metabolic remodeling is invariable, the direction of the change most often is not easy to guess. Sometimes the observed switch is from glycolysis to bivalency, while other times it is the opposite (bivalency to glycolysis).

We have also discussed the epigenetic changes coexisting with the metabolic remodeling. We now propose that one of the critical functions of the metabolic switch is to generate correct changes in the epigenetic metabolite make-up to accommodate the situation. This will then regulate the key epigenetic reader, writer and eraser enzymes, resulting in the correct epigenetic alterations. Epigenetic alterations are critical for gene expression changes. We propose these gene expression changes controlled by epigenetic metabolites govern the switch in stem cell state.

Acknowledgments

This work is supported by the Washington Research Foundation Fellowship for S.L., ISCRM Fellows Program Award for A.M.H., ISCRM Innovation Pilot Award for J.M. and grants from the National Institutes of Health R01GM097372, R01HL135143 and R01GM083867 and 1P01GM081619 for H.R.-B.

References

Almada, A. E., & Wagers, A. J. (2016). Molecular circuitry of stem cell fate in skeletal muscle regeneration, ageing and disease. *Nature Reviews. Molecular Cell Biology*, *17*, 267–279. https://doi.org/10.1038/nrm.2016.7.

Azuara, V., Perry, P., Sauer, S., Spivakov, M., Jørgensen, H. F., John, R. M., et al. (2006). Chromatin signatures of pluripotent cell lines. *Nature Cell Biology*, *8*, 532–538. https://doi.org/10.1038/ncb1403.

Baghdadi, M. B., Firmino, J., Soni, K., Evano, B., Girolamo, D., Mourikis, P., et al. (2018). Notch-induced miR-708 antagonizes satellite cell migration and maintains quiescence. *Cell Stem Cell, 23.* https://doi.org/10.1016/j.stem.2018.09.017. 859-868.e5.

Bahat, A., Goldman, A., Zaltsman, Y., Khan, D. H., Halperin, C., Amzallag, E., et al. (2018). MTCH2-mediated mitochondrial fusion drives exit from naïve pluripotency in embryonic stem cells. *Nature Communications*, *9*, 5132. https://doi.org/10.1038/s41467-018-07519-w.

Batie, M., & Rocha, S. (2019). JmjC histone demethylases act as chromatin oxygen sensors. *Molecular & Cellular Oncology*, *6*, 1608501. https://doi.org/10.1080/23723556.2019.1608501.

Battle, S. L., Jayavelu, N., Azad, R. N., Hesson, J., Ahmed, F. N., Overbey, E. G., et al. (2019). Enhancer chromatin and 3D genome architecture changes from naive to primed human embryonic stem cell states. *Stem Cell Reports*, *12*, 1129–1144. https://doi.org/10.1016/j.stemcr.2019.04.004.

Bavister, B., & Squirrell, J. (2000). Mitochondrial distribution and function in oocytes and early embryos. *Human Reproduction*, *15*, 189–198. https://doi.org/10.1093/humrep/15.suppl_2.189.

Bernardo, G., Squillaro, T., Dell'Aversana, C., Miceli, M., Cipollaro, M., Cascino, A., et al. (2009). Histone deacetylase inhibitors promote apoptosis and senescence in human mesenchymal stem cells. *Stem Cells and Development*, *18*, 573–582. https://doi.org/10.1089/scd.2008.0172.

Bernstein, B. E., Mikkelsen, T. S., Xie, X., Kamal, M., Huebert, D. J., Cuff, J., et al. (2006). A bivalent chromatin structure marks key developmental genes in embryonic stem cells. *Cell*, *125*, 315–326. https://doi.org/10.1016/j.cell.2006.02.041.

Bianchi, A., Mozzetta, C., Pegoli, G., Lucini, F., Valsoni, S., Rosti, V., et al. (2020). Dysfunctional polycomb transcriptional repression contributes to Lamin A/C dependent muscular dystrophy. *Journal of Clinical Investigation*. https://doi.org/10.1172/jci128161.

Blanpain, C., & Fuchs, E. (2006). Epidermal stem cells of the skin. *Annual Review of Cell and Developmental Biology*, *22*, 339–373. https://doi.org/10.1146/annurev.cellbio.22.010305.104357.

Blanpain, C., Lowry, W. E., Geoghegan, A., Polak, L., & Fuchs, E. (2004). Self-renewal, multipotency, and the existence of two cell populations within an epithelial stem cell niche. *Cell*, *118*, 635–648. https://doi.org/10.1016/j.cell.2004.08.012.

Blerkom, V. J., Chavez, D., & Bell, H. (1978). Molecular and cellular aspects of facultative delayed implantation in the mouse. *Ciba Foundation Symposium*, (64), 141–172.

Boonsanay, V., Zhang, T., Georgieva, A., Kostin, S., Qi, H., Yuan, X., et al. (2015). Regulation of skeletal muscle stem cell quiescence by Suv4-20h1-dependent facultative heterochromatin formation. *Cell Stem Cell*, *18*, 229–242. https://doi.org/10.1016/j.stem.2015.11.002.

Boyer, L. A., Plath, K., Zeitlinger, J., Brambrink, T., Medeiros, L. A., Lee, T., et al. (2006). Polycomb complexes repress developmental regulators in murine embryonic stem cells. *Nature, 441*, 349–353. https://doi.org/10.1038/nature04733.

Bracha, A. L., Ramanathan, A., Huang, S., Ingber, D. E., & Schreiber, S. L. (2010). Carbon metabolism-mediated myogenic differentiation. *Nature Chemical Biology, 6*, 202–204. https://doi.org/10.1038/nchembio.301.

Bracken, A. P., Dietrich, N., Pasini, D., Hansen, K. H., & Helin, K. (2006). Genome-wide mapping of Polycomb target genes unravels their roles in cell fate transitions. *Genes & Development, 20*, 1123–1136. https://doi.org/10.1101/gad.381706.

Brinster, R. (1969). Incorporation of carbon from glucose and pyruvate into the preimplantation mouse embryo. *Experimental Cell Research, 58*, 153–158. https://doi.org/10.1016/0014-4827(69)90125-6.

Brown, J., & Whittingham, D. (1991). The roles of pyruvate, lactate and glucose during preimplantation development of embryos from F1 hybrid mice in vitro. *Development (Cambridge, England), 112*, 99–105.

Buck, M. D., O'Sullivan, D., Geltink, R. I., Curtis, J. D., Chang, C.-H., Sanin, D. E., et al. (2016). Mitochondrial dynamics controls T cell fate through metabolic programming. *Cell, 166*, 63–76. https://doi.org/10.1016/j.cell.2016.05.035.

Bulut-Karslioglu, A., Biechele, S., Jin, H., Macrae, T. A., Hejna, M., Gertsenstein, M., et al. (2016). Inhibition of mTOR induces a paused pluripotent state. *Nature, 540*, 119–123. https://doi.org/10.1038/nature20578.

Caretti, G., Padova, M., Micales, B., Lyons, G. E., & Sartorelli, V. (2004). The Polycomb Ezh2 methyltransferase regulates muscle gene expression and skeletal muscle differentiation. *Genes & Development, 18*, 2627–2638. https://doi.org/10.1101/gad.1241904.

Chakraborty, A. A., Laukka, T., Myllykoski, M., Ringel, A. E., Booker, M. A., Tolstorukov, M. Y., et al. (2019). Histone demethylase KDM6A directly senses oxygen to control chromatin and cell fate. *Science (New York, N.Y.), 363*, 1217–1222. https://doi.org/10.1126/science.aaw1026.

Challen, G. A., Sun, D., Mayle, A., Jeong, M., Luo, M., Rodriguez, B., et al. (2014). Dnmt3a and Dnmt3b have overlapping and distinct functions in hematopoietic stem cells. *Cell Stem Cell, 15*, 350–364. https://doi.org/10.1016/j.stem.2014.06.018.

Chamberlain, S. J., Yee, D., & Magnuson, T. (2008). Polycomb repressive complex 2 is dispensable for maintenance of embryonic stem cell pluripotency. *Stem cells (Dayton, Ohio), 26*, 1496–1505. https://doi.org/10.1634/stemcells.2008-0102.

Chen, C.-T., Shih, Y.-R. V., Kuo, T. K., Lee, O. K., & Wei, Y.-H. (2008). Coordinated changes of mitochondrial biogenesis and antioxidant enzymes during osteogenic differentiation of human mesenchymal stem cells. *Stem Cells (Dayton, Ohio), 26*, 960–968. https://doi.org/10.1634/stemcells.2007-0509.

Cheung, T. H., & Rando, T. A. (2013). Molecular regulation of stem cell quiescence. *Nature Reviews. Molecular Cell Biology, 14*, 329–340. https://doi.org/10.1038/nrm3591.

Choudhry, H., & Harris, A. L. (2018). Advances in hypoxia-inducible factor biology. *Cell Metabolism, 27*, 281–298. https://doi.org/10.1016/j.cmet.2017.10.005.

Cohen, I., Zhao, D., Bar, C., Valdes, V. J., Dauber-Decker, K. L., Nguyen, M., et al. (2018). PRC1 fine-tunes gene repression and activation to safeguard skin development and stem cell specification. *Cell Stem Cell, 22*. https://doi.org/10.1016/j.stem.2018.04.005. 726-739.e7.

Collins, Y., Chouchani, E. T., James, A. M., Menger, K. E., Cochemé, H. M., & Murphy, M. P. (2012). Mitochondrial redox signalling at a glance. *Journal of Cell Science, 125*, 801–806. https://doi.org/10.1242/jcs.098475.

Comazzetto, S., Murphy, M. M., Berto, S., Jeffery, E., Zhao, Z., & Morrison, S. J. (2019). Restricted hematopoietic progenitors and erythropoiesis require SCF from leptin receptor+ niche cells in the bone marrow. *Cell Stem Cell, 24*. https://doi.org/10.1016/j.stem.2018.11.022. 477-486.e6.

Covello, K. L., Kehler, J., Yu, H., Gordan, J. D., Arsham, A. M., Hu, C.-J., et al. (2006). HIF-2 regulates Oct-4: Effects of hypoxia on stem cell function, embryonic development, and tumor growth. *Genes & Development, 20*, 557–570. https://doi.org/10.1101/gad.1399906.

Denholtz, M., Bonora, G., Chronis, C., Splinter, E., de Laat, W., Ernst, J., et al. (2013). Long-range chromatin contacts in embryonic stem cells reveal a role for pluripotency factors and polycomb proteins in genome organization. *Cell Stem Cell, 13*, 602–616. https://doi.org/10.1016/j.stem.2013.08.013.

Deschepper, M., Manassero, M., Oudina, K., Paquet, J., Monfoulet, L.-E., Bensidhoum, M., et al. (2013). Proangiogenic and prosurvival functions of glucose in human mesenchymal stem cells upon transplantation. *Stem Cells (Dayton, Ohio), 31*, 526–535. https://doi.org/10.1002/stem.1299.

Diao, Y., Guo, X., Li, Y., Sun, K., Lu, L., Jiang, L., et al. (2012). Pax3/7BP is a Pax7- and Pax3-binding protein that regulates the proliferation of muscle precursor cells by an epigenetic mechanism. *Cell Stem Cell, 11*, 231–241. https://doi.org/10.1016/j.stem.2012.05.022.

Eskeland, R., Leeb, M., Grimes, G. R., Kress, C., Boyle, S., Sproul, D., et al. (2010). Ring1B compacts chromatin structure and represses gene expression independent of histone ubiquitination. *Molecular Cell, 38*, 452–464. https://doi.org/10.1016/j.molcel.2010.02.032.

Esteve, O. P., Chin, H., & Pradhan, S. (2005). Human maintenance DNA (cytosine-5)-methyltransferase and p53 modulate expression of p53-repressed promoters. *Proceedings of the National Academy of Sciences of the United States of America, 102*, 1000–1005. https://doi.org/10.1073/pnas.0407729102.

Etchegaray, J.-P., Chavez, L., Huang, Y., Ross, K. N., Choi, J., Martinez-Pastor, B., et al. (2015). The histone deacetylase SIRT6 controls embryonic stem cell fate via TET-mediated production of 5-hydroxymethylcytosine. *Nature Cell Biology, 17*, 545–557. https://doi.org/10.1038/ncb3147.

Ezashi, T., Das, P., & Roberts, R. (2005). Low O2 tensions and the prevention of differentiation of hES cells. *Proceedings of the National Academy of Sciences of the United States of America, 102*, 4783–4788. https://doi.org/10.1073/pnas.0501283102.

Fenelon, J. C., Banerjee, A., & Murphy, B. D. (2014). Embryonic diapause: Development on hold. *The International Journal of Developmental Biology, 58*, 163–174. https://doi.org/10.1387/ijdb.140074bm.

Ferreccio, A., Mathieu, J., Detraux, D., Somasundaram, L., Cavanaugh, C., Sopher, B., et al. (2018). Inducible CRISPR genome editing platform in naive human embryonic stem cells reveals JARID2 function in self-renewal. *Cell Cycle (Georgetown, Tex.), 17*, 535–549. https://doi.org/10.1080/15384101.2018.1442621.

Filosa, S., Fico, A., Paglialunga, F., Lestrieri, M., Crooke, A., Verde, P., et al. (2003). Failure to increase glucose consumption through the pentose-phosphate pathway results in the death of glucose-6-phosphate dehydrogenase gene-deleted mouse embryonic stem cells subjected to oxidative stress. *The Biochemical Journal, 370*, 935–943. https://doi.org/10.1042/bj20021614.

Fischer, B., & Bavister, B. (1993). Oxygen tension in the oviduct and uterus of rhesus monkeys, hamsters and rabbits. *Reproduction, 99*, 673–679. https://doi.org/10.1530/jrf.0.0990673.

Flores, A., Schell, J., Krall, A. S., Jelinek, D., Miranda, M., Grigorian, M., et al. (2017). Lactate dehydrogenase activity drives hair follicle stem cell activation. *Nature Cell Biology, 19*, 1017–1026. https://doi.org/10.1038/ncb3575.

Folmes, C. D., Nelson, T. J., Martinez-Fernandez, A., Arrell, K. D., Lindor, J., Dzeja, P. P., et al. (2011). Somatic oxidative bioenergetics transitions into pluripotency-dependent glycolysis to facilitate nuclear reprogramming. *Cell Metabolism, 14*, 264–271. https://doi.org/10.1016/j.cmet.2011.06.011.

Forcina, L., Miano, C., Pelosi, L., & Musarò, A. (2019). An overview about the biology of skeletal muscle satellite cells. *Current Genomics, 20*, 24–37. https://doi.org/10.2174/1389202920666190116094736.

Fritsch, L., Robin, P., Mathieu, J. R., Souidi, M., Hinaux, H., Rougeulle, C., et al. (2010). A subset of the histone H3 lysine 9 methyltransferases Suv39h1, G9a, GLP, and SETDB1 participate in a multimeric complex. *Molecular Cell, 37*, 46–56. https://doi.org/10.1016/j.molcel.2009.12.017.

Fu, Z., Wang, B., Wang, S., Wu, W., Wang, Q., Chen, Y., et al. (2014). Integral proteomic analysis of blastocysts reveals key molecular machinery governing embryonic diapause and reactivation for implantation in mice. *Biology of Reproduction, 90*, 52. https://doi.org/10.1095/biolreprod.113.115337.

Fuchs, E. (2009). Finding one's niche in the skin. *Cell Stem Cell, 4*, 499–502. https://doi.org/10.1016/j.stem.2009.05.001.

Fuchs, E., Merrill, B. J., Jamora, C., & DasGupta, R. (2001). At the roots of a never-ending cycle. *Developmental Cell, 1*, 13–25. https://doi.org/10.1016/s1534-5807(01)00022-3.

Gafni, O., Weinberger, L., Mansour, A., Manor, Y. S., Chomsky, E., Ben-Yosef, D., et al. (2013). Derivation of novel human ground state naive pluripotent stem cells. *Nature, 504*, 282–286. https://doi.org/10.1038/nature12745.

Galonska, C., Ziller, M. J., Karnik, R., & Meissner, A. (2015). Ground state conditions induce rapid reorganization of core pluripotency factor binding before global epigenetic reprogramming. *Cell Stem Cell, 17*, 462–470. https://doi.org/10.1016/j.stem.2015.07.005.

García-Prat, L., Sousa-Victor, P., & Muñoz-Cánoves, P. (2017). Proteostatic and metabolic control of stemness. *Cell Stem Cell, 20*, 593–608. https://doi.org/10.1016/j.stem.2017.04.011.

García-Sánchez, D., Fernández, D., Rodríguez-Rey, J. C., & Pérez-Campo, F. M. (2019). Enhancing survival, engraftment, and osteogenic potential of mesenchymal stem cells. *World Journal of Stem Cells, 11*, 748–763. https://doi.org/10.4252/wjsc.v11.i10.748.

Gascón, S., Murenu, E., Masserdotti, G., Ortega, F., Russo, G. L., Petrik, D., et al. (2015). Identification and successful negotiation of a metabolic checkpoint in direct neuronal reprogramming. *Cell Stem Cell, 18*, 396–409. https://doi.org/10.1016/j.stem.2015.12.003.

Geula, S., Moshitch-Moshkovitz, S., Dominissini, D., Mansour, A., Kol, N., Salmon-Divon, M., et al. (2015). Stem cells. m6A mRNA methylation facilitates resolution of naïve pluripotency toward differentiation. *Science (New York, N.Y.), 347*, 1002–1006. https://doi.org/10.1126/science.1261417.

Greer, S. N., Metcalf, J. L., Wang, Y., & Ohh, M. (2012). The updated biology of hypoxia-inducible factor. *The EMBO Journal, 31*, 2448–2460. https://doi.org/10.1038/emboj.2012.125.

Hamatani, T., Daikoku, T., Wang, H., Matsumoto, H., Carter, M., Ko, H., et al. (2004). Global gene expression analysis identifies molecular pathways distinguishing blastocyst dormancy and activation. *Proceedings of the National Academy of Sciences of the United States of America, 101*, 10326–10331. https://doi.org/10.1073/pnas.0402597101.

Hawkins, D. R., Hon, G. C., Lee, L. K., Ngo, Q., Lister, R., Pelizzola, M., et al. (2010). Distinct epigenomic landscapes of pluripotent and lineage-committed human cells. *Cell Stem Cell, 6*, 479–491. https://doi.org/10.1016/j.stem.2010.03.018.

Hunter, S., & Evans, M. (1999). Non-surgical method for the induction of delayed implantation and recovery of viable blastocysts in rats and mice by the use of tamoxifen and Depo-Provera. *Molecular Reproduction and Development, 52*, 29–32. https://doi.org/10.1002/(sici)1098-2795(199901)52:1<29::aid-mrd4>3.0.co;2-2.

Hussein, A. M., Wang, Y., Mathieu, J., Margaretha, L., Song, C., Jones, D. C., et al. (2020). Metabolic control over mTOR-dependent diapause-like state. *Developmental Cell, 52*. https://doi.org/10.1016/j.devcel.2019.12.018. 236-250.e7.

Jeong, G.-J., Kang, D., Kim, A.-K., Han, K.-H., Jeon, H., & Kim, D.-I. (2019). Metabolites can regulate stem cell behavior through the STAT3/AKT pathway in a similar trend to that under hypoxic conditions. *Scientific Reports*, *9*, 6112. https://doi.org/10.1038/s41598-019-42669-x.

Ji, X., Dadon, D. B., Powell, B. E., Fan, Z., Borges-Rivera, D., Shachar, S., et al. (2015). 3D chromosome regulatory landscape of human pluripotent cells. *Cell Stem Cell*, *18*, 262–275. https://doi.org/10.1016/j.stem.2015.11.007.

Jørgensen, S., Schotta, G., & Sørensen, C. (2013). Histone H4 lysine 20 methylation: Key player in epigenetic regulation of genomic integrity. *Nucleic Acids Research*, *41*, 2797–2806. https://doi.org/10.1093/nar/gkt012.

Kaelin, W. G., & Ratcliffe, P. J. (2008). Oxygen sensing by metazoans: The central role of the HIF hydroxylase pathway. *Molecular Cell*, *30*, 393–402. https://doi.org/10.1016/j.molcel.2008.04.009.

Kang, S., Long, K., Wang, S., Sada, A., & Tumbar, T. (2019). Histone H3 K4/9/27 trimethylation levels affect wound healing and stem cell dynamics in adult skin. *Stem Cell Reports*, *14*, 34–48. https://doi.org/10.1016/j.stemcr.2019.11.007.

Kawabe, Y.-I., Wang, Y., McKinnell, I. W., Bedford, M. T., & Rudnicki, M. A. (2012). Carm1 regulates Pax7 transcriptional activity through MLL1/2 recruitment during asymmetric satellite stem cell divisions. *Cell Stem Cell*, *11*, 333–345. https://doi.org/10.1016/j.stem.2012.07.001.

Khacho, M., Clark, A., Svoboda, D. S., Azzi, J., MacLaurin, J. G., Meghaizel, C., et al. (2016). Mitochondrial dynamics impacts stem cell identity and fate decisions by regulating a nuclear transcriptional program. *Cell Stem Cell*, *19*, 232–247. https://doi.org/10.1016/j.stem.2016.04.015.

Kim, Y., & Guan, K.-L. (2015). mTOR: A pharmacologic target for autophagy regulation. *Journal of Clinical Investigation*, *125*, 25–32. https://doi.org/10.1172/jci73939.

Kinnaird, A., Zhao, S., Wellen, K. E., & Michelakis, E. D. (2016). Metabolic control of epigenetics in cancer. *Nature Reviews. Cancer*, *16*, 694–707. https://doi.org/10.1038/nrc.2016.82.

Laplante, M., & Sabatini, D. (2012). mTOR signaling in growth control and disease. *Cell*, *149*, 274–293. https://doi.org/10.1016/j.cell.2012.03.017.

Leavitt, D., Wells, M., Abarzua, P., Murphy, G. F., & Lian, C. G. (2019). Differential distribution of the epigenetic marker 5-hydroxymethylcytosine occurs in hair follicle stem cells during bulge activation. *Journal of Cutaneous Pathology*, *46*, 327–334. https://doi.org/10.1111/cup.13434.

Lee, J., Kang, S., Lilja, K. C., Colletier, K. J., Scheitz, C., Zhang, Y. V., et al. (2016). Signalling couples hair follicle stem cell quiescence with reduced histone H3 K4/K9/K27me3 for proper tissue homeostasis. *Nature Communications*, 7, 11278. https://doi.org/10.1038/ncomms11278.

Lee, S., Lee, K.-S., Huh, S., Liu, S., Lee, D.-Y., Hong, S., et al. (2016). Polo kinase phosphorylates Miro to control ER-mitochondria contact sites and mitochondrial Ca(2+) homeostasis in neural stem cell development. *Developmental Cell*, *37*, 174–189. https://doi.org/10.1016/j.devcel.2016.03.023.

Lee, J.-E., Oh, H.-A., Song, H., Jun, J., Roh, C.-R., Xie, H., et al. (2011). Autophagy regulates embryonic survival during delayed implantation. *Endocrinology*, *152*, 2067–2075. https://doi.org/10.1210/en.2010-1456.

Liao, J., Karnik, R., Gu, H., Ziller, M. J., Clement, K., Tsankov, A. M., et al. (2015). Targeted disruption of DNMT1, DNMT3A and DNMT3B in human embryonic stem cells. *Nature Genetics*, *47*, 469–478. https://doi.org/10.1038/ng.3258.

Liu, J., Mao, J. J., & Chen, L. (2011). Epithelial-mesenchymal interactions as a working concept for oral mucosa regeneration. *Tissue Engineering. Part B, Reviews*, *17*, 25–31. https://doi.org/10.1089/ten.teb.2010.0489.

Liu, X., Zheng, H., Yu, W.-M., Cooper, T. M., Bunting, K. D., & Qu, C.-K. (2015). Maintenance of mouse hematopoietic stem cells ex vivo by reprogramming cellular metabolism. *Blood*, *125*, 1562–1565. https://doi.org/10.1182/blood-2014-04-568949.

Loh, H. Y., Zhang, W., Chen, X., George, J., & Ng, H. H. (2007). Jmjd1a and Jmjd2c histone H3 Lys 9 demethylases regulate self-renewal in embryonic stem cells. *Genes & Development*, *21*, 2545–2557. https://doi.org/10.1101/gad.1588207.

Margueron, R., & Reinberg, D. (2011). The Polycomb complex PRC2 and its mark in life. *Nature*, *469*, 343–349. https://doi.org/10.1038/nature09784.

Martin, P. M., & Sutherland, A. E. (2001). Exogenous amino acids regulate trophectoderm differentiation in the mouse blastocyst through an mTOR-dependent pathway. *Developmental Biology*, *240*, 182–193. https://doi.org/10.1006/dbio.2001.0461.

Mathieu, J., Detraux, D., Kuppers, D., Wang, Y., Cavanaugh, C., Sidhu, S., et al. (2019). Folliculin regulates mTORC1/2 and WNT pathways in early human pluripotency. *Nature Communications*, *10*, 632. https://doi.org/10.1038/s41467-018-08020-0.

Mathieu, J., & Ruohola-Baker, H. (2017). Metabolic remodeling during the loss and acquisition of pluripotency. *Development (Cambridge, England)*, *144*, 541–551. https://doi.org/10.1242/dev.128389.

Mathieu, J., Zhang, Z., Nelson, A., Lamba, D. A., Reh, T. A., Ware, C., et al. (2013). Hypoxia induces re-entry of committed cells into pluripotency. *Stem Cells (Dayton, Ohio)*, *31*, 1737–1748. https://doi.org/10.1002/stem.1446.

Mathieu, J., Zhang, Z., Zhou, W., Wang, A. J., Heddleston, J. M., Pinna, C. M., et al. (2011). HIF induces human embryonic stem cell markers in cancer cells. *Cancer Research*, *71*, 4640–4652. https://doi.org/10.1158/0008-5472.can-10-3320.

Mathieu, J., Zhou, W., Xing, Y., Sperber, H., Ferreccio, A., Agoston, Z., et al. (2014). Hypoxia-inducible factors have distinct and stage-specific roles during reprogramming of human cells to pluripotency. *Cell Stem Cell*, *14*, 592–605. https://doi.org/10.1016/j.stem.2014.02.012.

McKinnell, I. W., Ishibashi, J., Grand, F., Punch, V. G., Addicks, G. C., Greenblatt, J. F., et al. (2007). Pax7 activates myogenic genes by recruitment of a histone methyltransferase complex. *Nature Cell Biology*, *10*, 77–84. https://doi.org/10.1038/ncb1671.

Menke, T. M., & McLaren, A. (1970). Carbon dioxide production by mouse blastocysts during lactational delay of implantation or after ovariectomy. *The Journal of Endocrinology*, *47*, 287–294. https://doi.org/10.1677/Joe.0.0470287.

Meshorer, E., & Misteli, T. (2006). Chromatin in pluripotent embryonic stem cells and differentiation. *Nature Reviews. Molecular Cell Biology*, 7, 540–546. https://doi.org/10.1038/nrm1938.

Miklas, J. W., Clark, E., Levy, S., Detraux, D., Leonard, A., Beussman, K., et al. (2019). TFPa/HADHA is required for fatty acid beta-oxidation and cardiolipin re-modeling in human cardiomyocytes. *Nature Communications*, *10*, 4671. https://doi.org/10.1038/s41467-019-12482-1.

Moody, J. D., Levy, S., Mathieu, J., Xing, Y., Kim, W., Dong, C., et al. (2017). First critical repressive H3K27me3 marks in embryonic stem cells identified using designed protein inhibitor. *Proceedings of the National Academy of Sciences of the United States of America*, *114*, 10125–10130. https://doi.org/10.1073/pnas.1706907114.

Morris, R. J., Bortner, C. D., Cotsarelis, G., Reece, J. M., Trempus, C. S., Faircloth, R. S., et al. (2003). Enrichment for living murine keratinocytes from the hair follicle bulge with the cell surface marker CD34. *The Journal of Investigative Dermatology*, *120*, 501–511. https://doi.org/10.1046/j.1523-1747.2003.12088.x.

Morris, R. J., Liu, Y., Marles, L., Yang, Z., Trempus, C., Li, S., et al. (2004). Capturing and profiling adult hair follicle stem cells. *Nature Biotechnology*, *22*, 411–417. https://doi.org/10.1038/nbt950.

Mortada, I., & Mortada, R. (2018). Epigenetic changes in mesenchymal stem cells differentiation. *European Journal of Medical Genetics, 61*, 114–118. https://doi.org/10.1016/j.ejmg.2017.10.015.

Motta, P., Nottola, S., Makabe, S., & Heyn, R. (2000). Mitochondrial morphology in human fetal and adult female germ cells. *Human Reproduction, 15*, 129–147. https://doi.org/10.1093/humrep/15.suppl_2.129.

Moya, A., Paquet, J., Deschepper, M., Larochette, N., Oudina, K., Denoeud, C., et al. (2018). Human mesenchymal stem cell failure to adapt to glucose shortage and rapidly use intracellular energy reserves through glycolysis explains poor cell survival after implantation. *Stem Cells, 36*, 363–376. https://doi.org/10.1002/stem.2763.

Mylonis, I., Simos, G., & Paraskeva, E. (2019). Hypoxia-inducible factors and the regulation of lipid metabolism. *Cell, 8*, 214. https://doi.org/10.3390/cells8030214.

Nagaraj, R., Sharpley, M. S., Chi, F., Braas, D., Zhou, Y., Kim, R., et al. (2017). Nuclear localization of mitochondrial TCA cycle enzymes as a critical step in mammalian zygotic genome activation. *Cell, 168*. https://doi.org/10.1016/j.cell.2016.12.026. 210-223.e11.

Naik, P., Birbrair, A., & Bhutia, S. (2018). Mitophagy-driven metabolic switch reprograms stem cell fate. *Cellular and Molecular Life Sciences: CMLS, 76*, 27–43. https://doi.org/10.1007/s00018-018-2922-9.

Nakamura, T., Okamoto, I., Sasaki, K., Yabuta, Y., Iwatani, C., Tsuchiya, H., et al. (2016). A developmental coordinate of pluripotency among mice, monkeys and humans. *Nature, 537*, 57–62. https://doi.org/10.1038/nature19096.

Nam, H. J., & Baek, S. H. (2019). Epigenetic regulation of hypoxic response. *Current Opinion in Physiology, 7*, 1–8. https://doi.org/10.1016/j.cophys.2018.11.007.

Neri, F., Krepelova, A., Incarnato, D., Maldotti, M., Parlato, C., Galvagni, F., et al. (2013). Dnmt3L antagonizes DNA methylation at bivalent promoters and favors DNA methylation at gene bodies in ESCs. *Cell, 155*, 121–134. https://doi.org/10.1016/j.cell.2013.08.056.

Nichols, J., & Smith, A. (2009). Naive and primed pluripotent states. *Cell Stem Cell, 4*, 487–492. https://doi.org/10.1016/j.stem.2009.05.015.

Nicklin, P., Bergman, P., Zhang, B., Triantafellow, E., Wang, H., Nyfeler, B., et al. (2009). Bidirectional transport of amino acids regulates mTOR and autophagy. *Cell, 136*, 521–534. https://doi.org/10.1016/j.cell.2008.11.044.

Ochocki, J. D., & Simon, C. M. (2013). Nutrient-sensing pathways and metabolic regulation in stem cells. *The Journal of Cell Biology, 203*, 23–33. https://doi.org/10.1083/jcb.201303110.

Pala, F., Girolamo, D., Mella, S., Yennek, S., Chatre, L., Ricchetti, M., et al. (2018). Distinct metabolic states govern skeletal muscle stem cell fates during prenatal and postnatal myogenesis. *Journal of Cell Science, 131*. pii: jcs212977. https://doi.org/10.1242/jcs.212977.

Pan, G., Tian, S., Nie, J., Yang, C., Ruotti, V., Wei, H., et al. (2007). Whole-genome analysis of histone H3 lysine 4 and lysine 27 methylation in human embryonic stem cells. *Cell Stem Cell, 1*, 299–312. https://doi.org/10.1016/j.stem.2007.08.003.

Panopoulos, A. D., Yanes, O., Ruiz, S., Kida, Y. S., Diep, D., Tautenhahn, R., et al. (2011). The metabolome of induced pluripotent stem cells reveals metabolic changes occurring in somatic cell reprogramming. *Cell Research, 22*, 168–177. https://doi.org/10.1038/cr.2011.177.

Paria, B., Huet-Hudson, Y., & Dey, S. (1993). Blastocyst's state of activity determines the "window" of implantation in the receptive mouse uterus. *Proceedings of the National Academy of Sciences of the United States of America, 90*, 10159–10162. https://doi.org/10.1073/pnas.90.21.10159.

Pattappa, G., Thorpe, S. D., Jegard, N. C., Heywood, H. K., de Bruijn, J. D., & Lee, D. A. (2012). Continuous and uninterrupted oxygen tension influences the colony formation and oxidative metabolism of human mesenchymal stem cells. *Tissue Engineering. Part C, Methods*, *19*, 68–79. https://doi.org/10.1089/ten.tec.2011.0734.

Peng, M., Yin, N., Chhangawala, S., Xu, K., Leslie, C., & Li, M. (2016). Aerobic glycolysis promotes T helper 1 cell differentiation through an epigenetic mechanism. *Science*, *354*, 481–484. https://doi.org/10.1126/science.aaf6284.

Pereira, C. F., Piccolo, F. M., Tsubouchi, T., Sauer, S., Ryan, N. K., Bruno, L., et al. (2010). ESCs require PRC2 to direct the successful reprogramming of differentiated cells toward pluripotency. *Cell Stem Cell*, *6*, 547–556. https://doi.org/10.1016/j.stem.2010.04.013.

Pike, I. (1981). Comparative studies of embryo metabolism in early pregnancy. *Journal of Reproduction and Fertility. Supplement*, *29*, 203–213.

Prieto, J., León, M., Ponsoda, X., García-García, F., Bort, R., Serna, E., et al. (2016). Dysfunctional mitochondrial fission impairs cell reprogramming. *Cell Cycle*, *15*, 3240–3250. https://doi.org/10.1080/15384101.2016.1241930.

Prigione, A., Rohwer, N., Hoffmann, S., Mlody, B., Drews, K., Bukowiecki, R., et al. (2014). HIF1α modulates cell fate reprogramming through early glycolytic shift and upregulation of PDK1-3 and PKM2. *Stem Cells*, *32*, 364–376. https://doi.org/10.1002/stem.1552.

Proksch, E., Brandner, J. M., & Jensen, J.-M. (2008). The skin: An indispensable barrier. *Experimental Dermatology*, *17*, 1063–1072. https://doi.org/10.1111/j.1600-0625.2008.00786.x.

Rafalski, V. A., Mancini, E., & Brunet, A. (2012). Energy metabolism and energy-sensing pathways in mammalian embryonic and adult stem cell fate. *Journal of Cell Science*, *125*, 5597–5608. https://doi.org/10.1242/jcs.114827.

Riahi, H., Brekelmans, C., Foriel, S., Merkling, S. H., Lyons, T. A., Itskov, P. M., et al. (2019). The histone methyltransferase G9a regulates tolerance to oxidative stress–induced energy consumption. *PLoS Biology*, *17*, e2006146. https://doi.org/10.1371/journal.pbio.2006146.

Richly, H., Rocha-Viegas, L., Ribeiro, J., Demajo, S., Gundem, G., Lopez-Bigas, N., et al. (2010). Transcriptional activation of polycomb-repressed genes by ZRF1. *Nature*, *468*, 1124–1128. https://doi.org/10.1038/nature09574.

Robinson, D. C., & Dilworth, F. J. (2018). Current topics in developmental biology. *Current Topics in Developmental Biology*, *126*, 235–284. https://doi.org/10.1016/bs.ctdb.2017.08.002.

Rodgers, J. T., King, K. Y., Brett, J. O., Cromie, M. J., Charville, G. W., Maguire, K. K., et al. (2014). mTORC1 controls the adaptive transition of quiescent stem cells from G0 to G(Alert). *Nature*, *510*, 393–396. https://doi.org/10.1038/nature13255.

Ryall, J. G., Dell'Orso, S., Derfoul, A., Juan, A., Zare, H., Feng, X., et al. (2015). The NAD (+)-dependent SIRT1 deacetylase translates a metabolic switch into regulatory epigenetics in skeletal muscle stem cells. *Cell Stem Cell*, *16*, 171–183. https://doi.org/10.1016/j.stem.2014.12.004.

Saha, B., Home, P., Ray, S., Larson, M., Paul, A., Rajendran, G., et al. (2013). EED and KDM6B coordinate the first mammalian cell lineage commitment to ensure embryo implantation. *Molecular and Cellular Biology*, *33*, 2691–2705. https://doi.org/10.1128/mcb.00069-13.

Sampath, S. C., Sampath, S. C., Ho, A. T., Corbel, S. Y., Millstone, J. D., & Lamb, J. (2018). Induction of muscle stem cell quiescence by the secreted niche factor Oncostatin M. *Nature Communications*, *9*, 1531. https://doi.org/10.1038/s41467-018-03876-8.

Sathananthan, A., & Trounson, A. (2000). Mitochondrial morphology during preimplantational human embryogenesis. *Human Reproduction*, *15*, 148–159. https://doi.org/10.1093/humrep/15.suppl_2.148.

Schwartz, Y. B., & Pirrotta, V. (2008). Polycomb complexes and epigenetic states. *Current Opinion in Cell Biology*, *20*, 266–273. https://doi.org/10.1016/j.ceb.2008.03.002.

Scognamiglio, R., Cabezas-Wallscheid, N., Thier, M., Altamura, S., Reyes, A., Prendergast, Á. M., et al. (2016). Myc depletion induces a pluripotent dormant state mimicking diapause. *Cell, 164*, 668–680. https://doi.org/10.1016/j.cell.2015.12.033.

Scott, W. R., Arostegui, M., Schweitzer, R., Rossi, F. M., & Underhill, M. T. (2019). Hic1 defines quiescent mesenchymal progenitor subpopulations with distinct functions and fates in skeletal muscle regeneration. *Cell Stem Cell, 25*. https://doi.org/10.1016/j.stem.2019.11.004. 797-813.e9.

Semenza, G. L. (2012). Hypoxia-inducible factors in physiology and medicine. *Cell, 148*, 399–408. https://doi.org/10.1016/j.cell.2012.01.021.

Shiraki, N., Shiraki, Y., Tsuyama, T., Obata, F., Miura, M., Nagae, G., et al. (2014). Methionine metabolism regulates maintenance and differentiation of human pluripotent stem cells. *Cell Metabolism, 19*, 780–794. https://doi.org/10.1016/j.cmet.2014.03.017.

Smith, Z. D., & Meissner, A. (2013). DNA methylation: Roles in mammalian development. *Nature Reviews. Genetics, 14*, 204–220. https://doi.org/10.1038/nrg3354.

So, A.-Y., Jung, J.-W., Lee, S., Kim, H.-S., & Kang, K.-S. (2011). DNA methyltransferase controls stem cell aging by regulating BMI1 and EZH2 through microRNAs. *PLoS One, 6*, e19503. https://doi.org/10.1371/journal.pone.0019503.

Son, M., Kwon, Y., Son, M.-Y., Seol, B., Choi, H.-S., Ryu, S.-W., et al. (2015). Mitofusins deficiency elicits mitochondrial metabolic reprogramming to pluripotency. *Cell Death and Differentiation, 22*, 1957–1969. https://doi.org/10.1038/cdd.2015.43.

Sperber, H., Mathieu, J., Wang, Y., Ferreccio, A., Hesson, J., Xu, Z., et al. (2015). The metabolome regulates the epigenetic landscape during naive-to-primed human embryonic stem cell transition. *Nature Cell Biology, 17*, 1523–1535. https://doi.org/10.1038/ncb3264.

Squirrell, J., Schramm, R., Paprocki, A., Wokosin, D., & Bavister, B. (2003). Imaging mitochondrial organization in living primate oocytes and embryos using multiphoton microscopy. *Microscopy and Microanalysis, 9*, 190–201. https://doi.org/10.1017/s1431927603030174.

Sun, X.-J., Man, N., Tan, Y., Nimer, S. D., & Wang, L. (2015). The role of histone acetyltransferases in normal and malignant hematopoiesis. *Frontiers in Oncology, 5*, 108. https://doi.org/10.3389/fonc.2015.00108.

Takashima, Y., Guo, G., Loos, R., Nichols, J., Ficz, G., Krueger, F., et al. (2014). Resetting transcription factor control circuitry toward ground-state pluripotency in human. *Cell, 158*, 1254–1269. https://doi.org/10.1016/j.cell.2014.08.029.

Takubo, K., Goda, N., Yamada, W., Iriuchishima, H., Ikeda, E., Kubota, Y., et al. (2010). Regulation of the HIF-1alpha level is essential for hematopoietic stem cells. *Cell Stem Cell, 7*, 391–402. https://doi.org/10.1016/j.stem.2010.06.020.

Takubo, K., Nagamatsu, G., Kobayashi, C. I., Nakamura-Ishizu, A., Kobayashi, H., Ikeda, E., et al. (2013). Regulation of glycolysis by Pdk functions as a metabolic checkpoint for cell cycle quiescence in hematopoietic stem cells. *Cell Stem Cell, 12*, 49–61. https://doi.org/10.1016/j.stem.2012.10.011.

Theunissen, T. W., Friedli, M., He, Y., Planet, E., O'Neil, R. C., Markoulaki, S., et al. (2016). Molecular criteria for defining the naive human pluripotent state. *Cell Stem Cell, 19*, 502–515. https://doi.org/10.1016/j.stem.2016.06.011.

Theunissen, T. W., Powell, B. E., Wang, H., Mitalipova, M., Faddah, D. A., Reddy, J., et al. (2014). Systematic identification of culture conditions for induction and maintenance of naive human pluripotency. *Cell Stem Cell, 15*, 471–487. https://doi.org/10.1016/j.stem.2014.07.002.

Trowbridge, J. J., & Orkin, S. H. (2011). Dnmt3a silences hematopoietic stem cell self-renewal. *Nature Genetics, 44*, 13–14. https://doi.org/10.1038/ng.1043.

Tumbar, T., Guasch, G., Greco, V., Blanpain, C., Lowry, W. E., Rendl, M., et al. (2004). Defining the epithelial stem cell niche in skin. *Science, 303*, 359–363. https://doi.org/10.1126/science.1092436.

Tzika, E., Dreker, T., & Imhof, A. (2018). Epigenetics and metabolism in health and disease. *Frontiers in Genetics*, *9*, 361. https://doi.org/10.3389/fgene.2018.00361.

van der Veen, C., Handjiski, B., Paus, R., Müller-Röver, S., Maurer, M., Eichmüller, S., et al. (1999). A comprehensive guide for the recognition and classification of distinct stages of hair follicle morphogenesis. *The Journal of Investigative Dermatology*, *113*, 523–532. https://doi.org/10.1046/j.1523-1747.1999.00740.x.

Varum, S., Rodrigues, A. S., Moura, M. B., Momcilovic, O., Easley, C. A., Ramalho-Santos, J., et al. (2011). Energy metabolism in human pluripotent stem cells and their differentiated counterparts. *PLoS One*, *6*, e20914. https://doi.org/10.1371/journal.pone.0020914.

Vazquez-Martin, A., Cufí, S., Corominas-Faja, B., Oliveras-Ferraros, C., Vellon, L., & Menendez, J. A. (2012). Mitochondrial fusion by pharmacological manipulation impedes somatic cell reprogramming to pluripotency: New insight into the role of mitophagy in cell stemness. *Aging*, *4*, 393–401. https://doi.org/10.18632/aging.100465.

Verma, M., Asakura, Y., Murakonda, B., Pengo, T., Latroche, C., Chazaud, B., et al. (2018). Muscle satellite cell cross-talk with a vascular niche maintains quiescence via VEGF and notch signaling. *Cell Stem Cell*, *23*. https://doi.org/10.1016/j.stem.2018.09.007. 530-543.e9.

Vizán, P., Beringer, M., Ballaré, C., & Croce, L. (2014). Role of PRC2-associated factors in stem cells and disease. *The FEBS Journal*, *282*, 1723–1735. https://doi.org/10.1111/febs.13083.

Wang, Y., Dumont, N. A., & Rudnicki, M. A. (2014). Muscle stem cells at a glance. *Journal of Cell Science*, *127*, 4543–4548. https://doi.org/10.1242/jcs.151209.

Wang, G., Jiang, B., Rue, E., & Semenza, G. (1995). Hypoxia-inducible factor 1 is a basic-helix-loop-helix-PAS heterodimer regulated by cellular O2 tension. *Proceedings of the National Academy of Sciences of the United States of America*, *92*, 5510–5514. https://doi.org/10.1073/pnas.92.12.5510.

Ware, C. B., Nelson, A. M., Mecham, B., Hesson, J., Zhou, W., Jonlin, E. C., et al. (2014). Derivation of naive human embryonic stem cells. *Proceedings of the National Academy of Sciences of the United States of America*, *111*, 4484–4489. https://doi.org/10.1073/pnas.1319738111.

Warr, M. R., Binnewies, M., Flach, J., Reynaud, D., Garg, T., Malhotra, R., et al. (2013). FOXO3A directs a protective autophagy program in haematopoietic stem cells. *Nature*, *494*, 323–327. https://doi.org/10.1038/nature11895.

Watt, F. M., & Fujiwara, H. (2011). Cell-extracellular matrix interactions in normal and diseased skin. *Cold Spring Harbor Perspectives in Biology*, *3*, a005124. https://doi.org/10.1101/cshperspect.a005124.

Weinberger, L., Ayyash, M., Novershtern, N., & Hanna, J. H. (2016). Dynamic stem cell states: Naive to primed pluripotency in rodents and humans. *Nature Reviews. Molecular Cell Biology*, *17*, 155–169. https://doi.org/10.1038/nrm.2015.28.

Wickett, R. R., & Visscher, M. O. (2006). Structure and function of the epidermal barrier. *American Journal of Infection Control*, *34*, S98–S110. https://doi.org/10.1016/j.ajic.2006.05.295.

Wosczyna, M. N., & Rando, T. A. (2018). A muscle stem cell support group: Coordinated cellular responses in muscle regeneration. *Developmental Cell*, *46*, 135–143. https://doi.org/10.1016/j.devcel.2018.06.018.

Yanes, O., Clark, J., Wong, D. M., Patti, G. J., Sánchez-Ruiz, A., Benton, P. H., et al. (2010). Metabolic oxidation regulates embryonic stem cell differentiation. *Nature Chemical Biology*, *6*, 411–417. https://doi.org/10.1038/nchembio.364.

Yoshida, Y., Takahashi, K., Okita, K., Ichisaka, T., & Yamanaka, S. (2009). Hypoxia enhances the generation of induced pluripotent stem cells. *Cell Stem Cell*, *5*, 237–241. https://doi.org/10.1016/j.stem.2009.08.001.

Yoshinaga, K., & Ada, C. (1966). Delayed implantation in the spayed, progesterone treated adult mouse. *Reproduction, 12*, 593–595. https://doi.org/10.1530/jrf.0.0120593.

Yucel, N., Wang, Y., Mai, T., Porpiglia, E., Lund, P. J., Markov, G., et al. (2019). Glucose metabolism drives histone acetylation landscape transitions that dictate muscle stem cell function. *Cell Reports, 27*. https://doi.org/10.1016/j.celrep.2019.05.092. 3939-3955.e6.

Zhang, Y., Mei, H., Chang, X., Chen, F., Zhu, Y., & Han, X. (2016). Adipocyte-derived microvesicles from obese mice induce M1 macrophage phenotype through secreted miR-155. *Journal of Molecular Cell Biology, 8*, 505–517. https://doi.org/10.1093/jmcb/mjw040.

Zhang, H., Ryu, D., Wu, Y., Gariani, K., Wang, X., Luan, P., et al. (2016). NAD^+ repletion improves mitochondrial and stem cell function and enhances life span in mice. *Science (New York, N.Y.), 352*, 1436–1443. https://doi.org/10.1126/science.aaf2693.

Zhang, T., Termanis, A., Özkan, B., Bao, X. X., Culley, J., de Alves, F., et al. (2016). G9a/GLP complex maintains imprinted DNA methylation in embryonic stem cells. *Cell Reports, 15*, 77–85. https://doi.org/10.1016/j.celrep.2016.03.007.

Zheng, X., Boyer, L., Jin, M., Mertens, J., Kim, Y., Ma, L., et al. (2016). Metabolic reprogramming during neuronal differentiation from aerobic glycolysis to neuronal oxidative phosphorylation. *eLife, 5*. e13374. https://doi.org/10.7554/elife.13374.

Zhou, W., Choi, M., Margineantu, D., Margaretha, L., Hesson, J., Cavanaugh, C., et al. (2012). HIF1α induced switch from bivalent to exclusively glycolytic metabolism during ESC-to-EpiSC/hESC transition. *The EMBO Journal, 31*, 2103–2116. https://doi.org/10.1038/emboj.2012.71.

Zhu, J., & Thompson, C. B. (2019). Metabolic regulation of cell growth and proliferation. *Nature Reviews. Molecular Cell Biology, 20*, 436–450. https://doi.org/10.1038/s41580-019-0123-5.

Further reading

Ding, J., Li, T., Wang, X., Zhao, E., Choi, J.-H., Yang, L., et al. (2013). The histone H3 methyltransferase G9A epigenetically activates the serine-glycine synthesis pathway to sustain cancer cell survival and proliferation. *Cell Metabolism, 18*, 896–907. https://doi.org/10.1016/j.cmet.2013.11.004.

FSC
www.fsc.org
MIX
Papier | Fördert
gute Waldnutzung
FSC® C083411

Kolibri 360 GmbH
Ferdinand-Jühlke-Straße 7
99095 Erfurt, Deutschland
produktsicherheit@kolibri360.de